The Measurement of Environmental and Resource Values

Theory and Methods

D0096727

The Measurement of Environmental and Resource Values

Theory and Methods

A. Myrick Freeman III

Resources for the Future
Washington, D.C.

Cloth: 0-915707-68-3/93 $1.00 + .10
Paper: 0-915707-69-1 $1.00 + .10

Printed in the United States of America

Published by Resources for the Future
1616 P Street, NW; Washington, DC 20036-1400

Library of Congress Cataloging-in-Publication Data

Freeman, A. Myrick
    The measurement of environmental and resource values : theory and methods / by A. Myrick Freeman III.
        p.   cm.
    "July, 1992."
    Includes bibliographical references and index.
    ISBN 0-915707-68-3 : ISBN 0-915707-69-1 (pbk.)
    1. Environmental policy—Cost effectiveness.   2. Environmental policy—Evaluation.   I. Title.
HC79.E5F713   1993
333.7—dc20   93-36099
                                             CIP

RESOURCES FOR THE FUTURE (RFF) is an independent nonprofit organization engaged in research and public education on natural resource and environmental issues. Its mission is to create and disseminate knowledge that helps people make better decisions about the conservation and use of their natural resources and the environment. RFF neither lobbies nor takes positions on current policy issues.

Because the work of RFF focuses on how people make use of scarce resources, its primary research discipline is economics. However, its staff also includes social scientists from other fields, ecologists, environmental health scientists, meteorologists, and engineers. Staff members pursue a wide variety of interests, including forest economics, recycling, multiple use of public lands, the costs and benefits of pollution control, endangered species, energy and national security, hazardous waste policy, climate resources, and quantitative risk assessment.

Acting on the conviction that good research and policy analysis must be put into service to be truly useful, RFF communicates its findings to government and industry officials, public interest advocacy groups, nonprofit organizations, academic researchers, and the press. It produces a range of publications and sponsors conferences, seminars, workshops, and briefings. Staff members write articles for journals, magazines, and newspapers, provide expert testimony, and serve on public and private advisory committees. The views they express are in all cases their own, and do not represent positions held by RFF, its officers, or trustees.

Established in 1952, RFF derives its operating budget in approximately equal amounts from three sources: investment income from a reserve fund, government grants, and contributions from corporations, foundations, and individuals. (Corporate support cannot be earmarked for specific research projects.) Some 45 percent of RFF's total funding is unrestricted, which provides crucial support for its foundational research and outreach and educational operations. RFF is a publicly funded organization under Section 501(c)(3) of the Internal Revenue Code, and all contributions to its work are tax deductible.

Table of Contents

List of Figures and Tables

FIGURES

TABLES

Foreword

Researchers at Resources for the Future (RFF) have long been examining the benefits of environmental and natural resource programs. Nearly 35 years ago, Marion Clawson published his path-breaking work on the so-called "travel cost" approach to estimating the recreation benefits associated with natural resource enhancement. In 1967, Ronald Ridker (with John Henning) was among the very first researchers to use changes in property values as a means of determining the value of improvements in environmental quality. More recently, Robert Mitchell and Richard Carson did much to call attention to and improve the practice of what has come to be known as the contingent valuation technique. (This latter approach, which involves the design, implementation, and interpretation of sample surveys asking respondents to value certain types of environmental programs, had been around for some time but had never before been pulled together systematically.)

Nevertheless, before 1979 and despite work by researchers at RFF and at universities around the world, no one book had ever attempted to integrate in a comprehensive, unified manner all that was known about the theory of environmental benefit estimation. Into the breach stepped Rick Freeman. During his time at RFF, and with support from RFF while he was teaching at Bowdoin College, Freeman produced *The Benefits of Environmental Improvement: Theory and Practice*.

The work quickly became the definitive reference on environmental benefit estimation. It has been used regularly by professors of environmental economics and their graduate and advanced undergraduate students, by government economists and policy analysts, by the growing number of economists at environmental advocacy groups, and by many others with a strong interest in the application of economics to environmental programs. Not long after its publication, most applied benefit-cost analyses of environmental regulatory actions began to make obligatory reference to the book.

But time marches on, and much progress has been made in the theory and practice of benefit estimation. This is true both for approaches covered in Freeman's earlier book and for newer approaches—contingent valuation being but one example. By the late 1980s, the author and RFF recognized that a wholesale revision of the 1979 book was called for. The present book is the result of that realization.

Once again, through time spent in residence at RFF and during summers at Bowdoin, Freeman has produced a book that is comprehensive, conceptually consistent, and up-to-date. The work not only reflects recent developments in those techniques discussed in his earlier book, but also contains altogether new material on the contingent valuation technique, the valuation of mortality risks, and intertemporal welfare measures.

The Measurement of Environmental and Resource Values: Theory and Methods represents a major commitment by both Freeman and RFF. We hope and expect that, in the manner of its distinguished predecessor, this book will set the standard for the next decade of research in environmental economics.

Paul R. Portney
Vice President
Resources for the Future
September 1993

Preface

The fourteen years since the publication of my earlier book, *The Benefits of Environmental Improvement: Theory and Practice,* by Resources for the Future (RFF) has seen a virtual explosion of new theoretical developments and empirical applications in the area of resource and environmental valuation. When I returned to RFF in the fall of 1988, my intention was to revise and update the earlier book to incorporate these more recent developments.

It soon became apparent to me that more than a mere revision was necessary. Except for some paragraphs of basic exposition and the materials on the concepts of welfare change (presented in Chapter 3 of this book), entirely new treatments of most topics would be required. This was especially true of such topics as contingent valuation, valuing health effects, property value models, and the travel cost model. In addition, many new topics called for attention. These included intertemporal welfare measures, the valuation of changes in risks, hedonic wage models, nonuse values, and the measurement of the costs of environmental policies. Thus, this is a new book.

Although the title and the content of the book are new, the objectives of this book are essentially the same as those of *Benefits*. These objectives are, first, to provide an introduction and overview of the principal methods and techniques of resource valuation to professional economists and graduate students who are not directly engaged in the field and, second, to give practitioners in the field an up-to-date reference on recent developments in the theory and methods underlying the practice of resource valuation.

While I have tried to be comprehensive in my coverage of topics, this book is not a "how-to" manual. That kind of book would have to deal with a host of econometric and related technical issues that are outside of my area of expertise. For details of the empirical implementation of the methods and techniques described in this book, readers will have to go to the original literature.

I want to thank RFF for providing financial and intellectual support for this project. Being at RFF gave me the opportunity to collaborate with Maureen Cropper and Winston Harrington on two projects, the results of which have found their way into Chapters 9 and 10 of this book. I also am indebted to my other colleagues at RFF, especially Adam Finkel, Carol Jones, Allen Kneese, Ray Kopp, Alan Krupnick, and Paul Portney, for support, assistance, and stimulation. In addition, the staff support for this project was outstanding. Betty Cawthorne typed the manuscript. To her, a special thanks for a job well done. And Sam Allen's careful editing has had a substantial, beneficial effect on the final product.

A number of people read parts of the manuscript and gave me helpful comments. They include Robert Deacon, John Fitzgerald, Haynes Goddard, Robert Mendelsohn, David Reaume, and William Vaughan. My thanks to them. Nancy Bockstael, Dallas Burtraw, and Kerry Smith read the whole manuscript with great care and provided numerous helpful suggestions. I am especially in their debt.

Finally, I want to thank the publishers of the *Southern Economic Journal* and the *Journal of Risk and Uncertainty* for permission to adapt materials which first appeared in their journals.[1]

A. Myrick Freeman III
Georgetown, Maine

---

[1]These publications include: "Measuring Welfare Values of Productivity Changes," with Winston Harrington, *Southern Economic Journal* vol. 56, no. 4, April 1990; "Welfare Measurement and the Benefit-Cost Analysis of Projects Affecting Risks," *Southern Economic Journal* vol. 58, no. 1, July, 1991; and "Indirect Methods for Valuing Changes in Environmental Risks with Non-Expected Utility Preferences, *Journal of Risk and Uncertainty* vol. 4, no. 2, April, 1991.

1

Resource Evaluation and Public Policy

The premise underlying this book is that estimates of the economic values of environmental and resource services can be a valuable part of the information base supporting resource and environmental management decisions. This premise is substantiated by a number of current environmental and resource policy issues, all of which involve in one way or another questions of economic values and trade-offs. Consider these issues:

• Achieving the air and water pollution control objectives established by Congress is requiring massive expenditures on the part of both the public and private sectors. Is this diversion of resources from the production of other goods and services making us better off?

• In an effort to minimize the total social cost of meeting the demand for the services that electricity provides, some states are now requiring that externality "adders" be incorporated in estimates of the costs of alternative supply options to reflect the impact of these options on the environment (Cohen and coauthors, 1990). How should these adders be calculated?

• The development of new reserves of petroleum and minerals is increasingly impinging on wild and natural areas that provide other environmental and resource services. Areas affected include the Arctic National Wildlife Refuge with its fragile habitat for caribou and other species, and the outer continental shelf, where commercial and recreational fisheries may be threatened by petroleum exploration and production. Are restrictions on development in ecologically sensitive areas worth the costs they impose on society in the form of reduced availability of and higher prices for energy and minerals?

• The development and management of large river systems such as the Columbia River basin involve choosing among alternative combinations of hydroelectric power, water supply, and commercial and recreational fishing. When is it worthwhile to curb water withdrawals for irrigation or reduce discharges for power production in order to protect populations of salmon?

• The commercial exploitation of some natural resource systems may be proceeding at unsustainable rates. Examples include some tropical forests and the demersal fisheries of the Gulf of Maine and the Canadian Maritimes. Shifting to sustainable rates of harvest may involve substantial short-term costs in the form of forgone incomes. Are the long-term gains from achieving sustainable rates of harvest greater or less than the short-term costs?

• The Comprehensive Environmental Response, Compensation, and Liability Act (CERCLA) of 1980 imposes financial liability for damages to natural resources on agents responsible for releases of hazardous materials into the environment. In the event of a release, how should the financial liability be calculated?

• Many people are now advocating that the U.S. system of national income accounts be substantially expanded to include measures of the values of nonmarket environmental services and deductions for the costs of environmental degradation and resource depletion. How are these values and costs to be measured?

This book is about how economics as a discipline can contribute to answering questions such as these, by providing measures of the economic values of the services of environmental and natural resource systems. I begin by introducing the idea of the natural environment as a system of assets.

THE ASSETS OF NATURE

Natural resources such as forests and commercially exploitable fisheries and environmental attributes such as air quality are valuable assets in that they yield flows of services to people. Public policies and the actions of individuals and firms can lead to changes in the flows of these services, thereby creating benefits and costs. Because of externalities and the common property and public good characteristics of at least some of these services, market forces can be relied on neither to guide them to their most highly valued uses nor to reveal prices that reflect their true social values.[1] It is the failure of the

---

[1]Externalities arise when a real variable (not a price) chosen by one economic agent enters the utility or production function of other economic

market system to allocate and price resource and environmental services correctly that creates the need for economic measures of values
to guide policymaking.

It has been natural to refer to the activity of measuring changes in
the economic values of these service flows as benefit-cost analysis.
But since the 1950s when the techniques of conventional benefit-cost
analysis were being developed and refined, there have been significant changes in the nature of the problems being dealt with and the
analytical tools that have become available. V. Kerry Smith, calling
attention to these changes in his keynote lecture at Resources for the
Future's thirty-fifth anniversary celebration in 1987, went on to say
that

> This expansion of applications has far-reaching implications
> for the techniques used and for the treatment of measures of
> the benefits and costs. Consequently, it has led me to argue
> for the use of a broader term, *resource evaluation*, to describe
> more adequately the amendments and expansions to
> benefit-cost methods in evaluating today's environmental
> and natural resource issues. (Smith, 1988, p. 2)

One of the changes noted by Smith is in the range of resource
and environmental management problems being subjected to economic analysis. As Smith pointed out, benefit-cost analysis was first
developed to assess the net economic value of public works projects,
especially water resource developments, that withdrew productive
factor inputs (land, labor, capital, and materials) from the economy to
produce tangible outputs (for example, water, hydroelectric power,
transportation). Many of the outputs had market counterparts, so
estimation of monetary values was relatively straightforward. For example, the savings in the monetary costs of repairing flood damages
was taken to be a measure of the benefits of controlling floods. In
contrast, today the effects of many public actions are much more
subtle and wide-ranging. And this is true for both the favorable effects (benefits) and unfavorable effects (costs and damages). What

---

agents and there is no requirement to or incentive for the first agent to take
the effect on others into account when making choices. A public good is
nonexcludable and nondepletable—that is, once the good has been provided
to one individual, others cannot be prevented from making use of the good,
and one person's use does not diminish the use that others can make of the
good. See any intermediate or advanced text in microeconomics for a discussion of how externalities and public goods affect the outcomes of market
processes.

were once considered unquantifiable and perhaps relatively unimportant intangibles such as improved recreation and visual amenities are now recognized as significant sources of value and are thought to be susceptible to economic measurement. And consequences that were once unrecognized (for example, small changes in the risk of cancer) or were thought to lie outside the realm of economic analysis (say, loss of biodiversity and the preservation of endangered species and unique ecological systems) are often central issues in the analysis of policy choices today.

Another change is that the distinction between natural resources and the environment that has prevailed in the economics discipline for so long is probably no longer meaningful. The objects of analysis for natural resource economists have typically been such resources as the forest, the ore body, and the fish species that produced a flow of commodities to the economy such as wood, metal, and fish sticks. The environment has been viewed as the medium through which the externalities associated with air, noise, and water pollution have flowed and, sometimes, as the source of amenities. Increasingly this distinction appears to be artificial as we recognize both the variety of service flows provided by natural resources and the importance of a variety of forms of externalities. This recognition is apparently what Smith had in mind when he suggested the need "to model both natural and environmental resources as assets" (1988, p. 3) that yield a variety of valuable services.[2]

If this change in perspective is adopted, it will be necessary to take a more expansive view of natural and environmental resources as complex systems with multiple outputs and joint products. The natural resource-environment complex can be viewed as producing four kinds of service flows to the economy. First, as in the conventional view of resource economics, the resource-environment system serves as a source of material inputs to the economy such as fossil fuels, wood products, minerals, water, and fish. Second, some components of the resource-environment system provide life-support services in the form of a breathable atmosphere and a livable climatic

---

[2]For an early recognition of this point see Freeman, Haveman, and Kneese (1973), who suggested that we "view the environment as an asset or a kind of nonreproducible capital good that produces a stream of various services for man. Services are tangible (such as flows of water or minerals), or functional (such as the removal, dispersion, storage, and degradation of wastes or residuals), or intangible (such as a scenic view)" (p. 20). As Mohring and Boyd (1971) pointed out, viewing the environment as an asset also provides useful insights into the analysis of congestion externalities.

regime. Third, the resource-environment system provides a wide variety of amenity services, including opportunities for recreation, wildlife observation, the pleasures of scenic views, and perhaps even services that are not related to any direct use of the environment (sometimes called nonuse or existence values). Finally, this system disperses, transforms, and stores the residuals that are generated as by-products of economic activity (Kneese, Ayres, and d'Arge, 1970; Freeman, Haveman, and Kneese, 1973).

A forest such as a unit in the U.S. National Forest system is a good example of a resource-environmental system that provides a wide range of services, from materials such as wood and fiber to amenities like scenic vistas, hiking, and wildlife observation, and from the regulation of streamflow and control of erosion to the absorption of atmospheric carbon dioxide. And since trees are known to emit nonmethane hydrocarbons, at least in some circumstances forests may contribute to the impairment of the life-support services (Chameides and coauthors, 1988). In the list of service flows there are examples of joint products, that is, pairs of services that can be increased or decreased together. But generally speaking increasing the flow of one type of service must be accompanied by a decrease in the flow of some other service, other things being equal.[3] In other words, the resource-environmental system is characterized by scarcity, trade-offs, and opportunity costs.

The economic value of a resource-environmental system as an asset can be defined as the sum of the discounted present values of the flows of all of the services.[4] The benefit of any public policy that increases the flow of one type of service is the increase in the present value of that service. But the policy may have costs in the form of decreases in the flows of other services. Similarly, what is termed the damage due to pollution or some other human intervention is the reduction in the value of the flow of services it causes. All of these changes in resource flows, whether benefits, costs, or damages, have

---

[3]See Bowes and Krutilla (1989) for an analysis of the problems of managing a forest to optimize the amenity and commodity services from this resource system.

[4]Since many of these service flows are not bought or sold in markets and therefore do not have market prices, the economic value of a natural asset may be quite different from its market value. For example, an acre of wetland might trade in the market for land on the basis of its value for commercial or residential development; but this value could be quite different from the value of its services as wildlife habitat and as means of controlling floods and recharging groundwater aquifers.

their counterparts in changes in the value of the resource-environmental system as an asset. Some attention must therefore be devoted to the theories of asset pricing and the role of time and discounting in calculating changes in environmental and resource values; these topics will be taken up in chapter 7.

Some of the service flows of resource-environmental systems are linked directly or indirectly to markets and hence are responsive to market forces. But many service flows are not properly regulated by markets because of externalities, their public goods characteristics of nonexcludability and nondepletability, and other factors. As is well known, this means that a decentralized market system is unlikely to lead to the optimal pattern of service flows. Hence there is a potential role for public policy in the management of resource-environmental systems and a need for information on the values of the service flows.

THE ECONOMIC CONCEPT OF VALUE

The economic concept of value employed here has its foundation in neoclassical welfare economics. The basic premises of welfare economics are that the purpose of economic activity is to increase the well-being of the individuals who make up the society, and that each individual is the best judge of how well off he or she is in a given situation. Each individual's welfare depends not only on that individual's consumption of private goods and of goods and services produced by the government, but also on the quantities and qualities each receives of nonmarket goods and service flows from the resource-environment system—for example, health, visual amenities, and opportunities for outdoor recreation. It follows that the basis for deriving measures of the economic value of changes in resource-environmental systems is their effects on human welfare.[5] The anthropocentric focus of economic valuation does not preclude a concern for the survival and well-being of other species. Individuals can value the survival of other species not only because of the uses people make of them (for food and recreation, for example), but also because of an altruistic or ethical concern. The latter can be the source of existence or nonuse values, a form of economic value discussed in chapter 5.

If society wishes to make the most (in terms of individuals' well-being) of its endowment of all resources, it should compare the values

---

[5]For a discussion of the relationship between economic value and the broader philosophical concepts of value, see Brown (1984).

of what its members receive from any environmental change or use of a resource (that is, the benefits) with the values of what its members give up by taking resources and factor inputs from other uses (that is, the costs). Since the benefits and costs are valued in terms of their effects on individuals' well-being, the terms "economic value" and "welfare change" can be used interchangeably. Society should make changes in environmental and resource allocations only if the results are worth more in terms of individuals' welfare than what is given up by diverting resources and inputs from other uses.

The standard economic theory for measuring changes in individuals' well-being was developed for the purpose of interpreting changes in the prices and quantities of goods purchased in markets. This theory has been extended in the past fifteen years or so to public goods and other nonmarket services such as environmental quality and health.[6] The theory makes the assumption that people have well-defined preferences among alternative bundles of goods, where bundles consist of various quantities of both market and nonmarket goods. The theory also assumes that people know their preferences, and that these preferences have the property of substitutability among the market and nonmarket goods making up the bundles. By substitutability economists mean that if the quantity of one element in an individual's bundle is reduced, it is possible to increase the quantity of some other element so as to leave the individual no worse off because of the change. In other words, the increase in the quantity of the second element substitutes for the decrease in the first element. The property of substitutability is at the core of the economist's concept of value because substitutability establishes trade-off ratios between pairs of goods that matter to people.

The trade-offs that people make as they choose less of one good and substitute more of some other good reveal something about the values people place on these goods. If one of the goods has a monetary value, the revealed values are monetary values. The money price of a market good is just a special case of a trade-off ratio because the money given up to purchase one unit of one element of the bundle is a proxy for the quantities of one or more of the other elements in the bundle that had to be reduced in order to make the purchase.

The value measures based on substitutability can be expressed either in terms of willingness to pay (WTP) or willingness to accept compensation (WAC). WTP and WAC measures can be defined in terms of any other good that the individual is willing to substitute for

---

[6]See Freeman (1979) for an earlier treatment of this theory.

the good being valued. In the following discussion, I use money as the numeraire in which trade-off ratios are expressed, but WTP and WAC could be measured in terms of any other good that mattered to the individual.

WTP is the maximum sum of money the individual would be willing to pay rather than do without an increase in some good such as an environmental amenity. This sum is the amount of money that would make the individual indifferent between paying for and having the improvement and forgoing the improvement while keeping the money to spend on other things. WAC is the minimum sum of money the individual would require to voluntarily forgo an improvement that otherwise would be experienced; it is the amount that would make a person indifferent between having the improvement and forgoing the improvement while getting extra money. Both value measures are based on the assumption of substitutability in preferences, but they adopt different reference points for levels of well-being. WTP takes as its reference point the absence of the improvement, while WAC takes the presence of the improvement as the base level of welfare or utility. In principle WTP and WAC need not be exactly equal. WTP is constrained by the individual's income. But there is no upper limit on what that person could require as compensation for forgoing the improvement. Differences between WTP and WAC measures are discussed in more detail in chapters 3 and 6; the question of which measure is appropriate under various circumstances is addressed in chapter 3.

VALUE, BENEFITS, AND DAMAGES

The terms "benefit," "damage," "environmental cost," and "cost of pollution" are often used interchangeably. This results in an unfortunate potential for confusion. For example, cost also refers to the negative welfare changes that occur when resources are absorbed in the process of reducing pollution or managing the resource. Thus we can speak of the "costs of pollution control" as well as the "costs of pollution." The distinction between benefits and damages or environmental costs lies in the choice of a reference point from which environmental changes are to be measured. The benefits of an environmental improvement are measured by comparing the existing known level of environmental services with some specified hypothetical alternative in which pollution has been reduced or other environmental services have been increased. Benefits are the gain associated with the environmental improvement.

The terms damages and environmental costs are sometimes used to represent the mirror image of benefits, that is, what is lost in moving from the hypothetical "clean" state to the existing level of pollution. In this usage, the "damage" terminology requires that some "clean" state or reference point be identified. But this may be difficult or impossible where there are naturally occurring sources or natural background levels for the pollutant in question. For conventional air and water pollutants, the existing polluted state can be taken as the reference point; and the benefit terminology is appropriate. The term damage also has a specific meaning in the context of CERCLA: it refers to the economic values lost as a consequence of a chemical spill or the release of hazardous wastes into the environment.[7]

ECONOMIC VALUES IN PUBLIC POLICY

There may be potential for substantial gains in economic welfare through better resource management and the judicious use of the principles of resource valuation in some cases, such as those involving the issues described at the beginning of this chapter. If the objective of management is to maximize the net economic values associated with the use of environmental and natural resources, then benefit-cost analysis becomes, in effect, a set of rules for optimum management and a set of definitions and procedures for measuring benefits and costs. Once the objective of maximum net economic value or economic efficiency has been accepted, policy becomes an almost mechanical (but not necessarily easy) process of working out estimates of marginal benefit and marginal cost curves and seeking their point of intersection.

But most current resource and environmental policy is not based solely or even primarily on the efficiency criterion. One reason, of course, is that at the time that many of the basic policy objectives were established, it was not within the capability of analysts to provide the kind of information about values that would be required to implement the efficiency objective. But it is also true that decision makers may have other objectives besides economic efficiency. For example, decision makers may be concerned with equity considerations, intergenerational effects, the sustainability of resource systems, or social risk aversion. Thus it is not particularly useful to advocate

---

[7] For a discussion of the application of economic methods to natural resource damage assessment under CERCLA, see Kopp and Smith (1993).

benefit-cost analysis as a simple decision rule. Rather, it should be considered as a framework and a set of procedures to help organize available information. Viewed in this light, benefit-cost analysis does not dictate choices; nor does it replace the ultimate authority and responsibility of decision makers. It is simply a tool for organizing and expressing certain kinds of information on the range of alternative courses of action. It is in the context of this framework for arraying information that the usefulness of value estimates must be assessed.

Some people may be distrustful of economists' efforts to extend economic measurements to such things as human health and safety, ecology, and aesthetics, and to reduce as many variables as possible to commensurate monetary measures. Some skepticism about the economist's penchant for monetary measurement is no doubt healthy, but it should not be overdone. It is not correct to say that there are some things like human health and safety or the preservation of endangered species that cannot be valued in terms of dollars or some other numeraire. The real world often creates situations where trade-offs between such things as deaths avoided and some other things of value cannot be avoided. The questions really are how the problem of making choices about such trade-offs is to be approached and what information can be gathered to help in the problem of choice.

Consider a hypothetical and highly simplified case of an air pollutant. Assume that the following information is known with certainty. At present levels of emissions, the pollutant causes excess mortality of 1,000 deaths per year in the population at risk. Reducing emissions by 50 percent would cost $500,000 and would reduce the excess mortality to 500 deaths per year. One hundred percent control of emissions would reduce excess mortality to zero, but would cost $1.5 million.

This information can be displayed as follows:

Level of control (percentage)	Costs of control (dollars)	Excess mortality (deaths per year)
0	0	1,000
50	500,000	500
100	1,500,000	0

The problem is clearly one of trade-offs between lives and the value of resources used up in the process of controlling emissions. If the monetary value of saving lives were known, the right-hand column of the table could be converted to dollar measures of benefits, and the appropriate benefit-cost rules could be applied to determine the optimum level of emissions control. But in the absence of some agreed-upon basis for making lifesaving and control costs commensurable in dollar terms, no simple decision rule can be applied to determine the correct choice.

Choices of this sort are made in the political realm by decision makers such as the administrator of the Environmental Protection Agency (EPA). Whatever the choice, there is an implicit value of lifesaving that is consistent with that choice and can be said to have been revealed by that choice.[8] In this example, if the decision maker chooses the 50 percent control level, the value of lifesaving is revealed to be at least $1,000 per death avoided. The 50 percent control level "buys" 500 lives saved at a cost of $500,000. The choice further reveals that the value of lifesaving to the decision maker is less than $2,000, since the decision maker declined the opportunity to "purchase" the additional 500 lives saved that the additional $1 million of control costs would make possible. If the 100 percent control level had been chosen, this would have revealed a value of lifesaving of at least $2,000.[9]

In this example, choice revealed value, rather than value determining the choice. But either way, the problem of valuation cannot be avoided. It can be hidden. But I would argue that in a democratic society, the more open decision makers are about the problems of making choices and the values involved and the more information they have about the implications of their choices, the better their choices are likely to be. Estimates of values in monetary terms are one such source of information.

Most resource and environmental management problems have structures similar to the one just discussed. For example, in the preceding table the first column could be rate of harvest in a forest, the second column the net revenues from the harvest, and the third column the probability of survival of an endangered species requiring old-growth forest habitat. Reducing harvest levels increases the prob-

---

[8]For examples of estimates of the values of reducing risks of death implicit in government policy choices, see Bailey (1980, p. 26); Travis and coauthors (1987); Travis, Pack, and Fisher (1987); and Cropper and coauthors (1992).

[9]In this example with only three data points, the implicit value can only be determined within some range. If control costs and mortality as functions of the level of control were continuous relationships and known with certainty, the choice of a control level would imply a precise value of life. If it is assumed that the control level was established so as to equate marginal benefits and marginal costs, and marginal costs are known, the marginal benefit or value can be inferred.

ability of survival of the species, but at a cost. The rate of harvest actually chosen implies something about the value to the decision makers of increasing the probability of survival. Of course, the second column need not be in dollars. The trade-off could be between having more timber wolves and fewer deer. The fact that there is no monetary measure in that example does not make it at its root any less an economic problem. Whatever choice is made about a population level for wolves implies something about the relative values placed on wolves and deer by the person making the decision. This example is another manifestation of the fundamental economic fact of scarcity, that is, that more of one thing means less of something else that people value.

Because policy choices about resources and environmental quality are made in a political context and are likely to involve comparisons and trade-offs among variables for which there is no agreement about commensurate values, monetary benefit and cost data will not always be the determining factors in decision making. But benefit and cost estimates are an important form of information. Their usefulness lies in the fact that they use easily understood and accepted rules to reduce complex clusters of effects and phenomena to single-valued commensurate magnitudes, that is, to dollars. The value of the benefit-cost framework lies in its ability to organize and simplify certain forms of information into commensurate measures.

CLASSIFICATIONS OF VALUES

In this section I describe some ways of classifying the types of environmental and resource service flows for which value measures might be desired. Of course, any classification system contains a certain element of arbitrariness. The usefulness of any particular classification depends on how well it illuminates important similarities and differences among types of service flows. Just which similarities and differences are important depends upon the particular questions being examined.

Several authors have offered taxonomies of resource service flows or benefits from environmental improvements, including Freeman (1979, 1982), Smith, Desvousges, and McGivney (1983), Fisher and Raucher (1984), Smith, Desvousges, and Freeman (1985), and Mitchell and Carson (1989). Each of these authors has chosen to emphasize one set of features over another. But most of the classification systems I have looked at recognize the following sets of features as being relevant.

I. Types of resource or environmental media. Examples include air, water, a fishery, a forest. The legal and administrative separation of responsibilities for controlling pollution of air and water bodies and the assignment of responsibilities for fisheries, forests, and national parks to different agencies seem to reflect this approach to classification.

II. Types of effects. Another way of classifying environmental and resource service flows is according to whether they impinge directly upon humans, indirectly on humans through their impact on other living organisms, or indirectly through inanimate systems. An example of this approach to classification is adapted here from Freeman (1979):

  A. Direct impacts on humans
    1. Human health—morbidity and mortality effects associated with air and water pollution
    2. Odor, visibility, visual aesthetic
  B. Ecosystem impacts—biological mechanisms
    1. Impacts on the economic productivity of ecological systems
        Agricultural productivity
        Forestry
        Commercial fisheries
    2. Other ecosystem impacts
        Recreational uses of ecosystems—fishing, hunting
        Ecological diversity, stability
  C. Impacts through nonliving systems
    1. Materials damage, soiling, production costs
    2. Weather, climate

III. Economic channels. The environmental and resource service flows can also be classified according to whether they convey their effects through the market system in the form of changes in incomes to producers and changes in the availability of and prices for marketed goods and services to consumers, or through changes in the availability of goods and services not normally purchased through markets—for example, health, environmental amenities such as visibility, and opportunities for water-based recreation. The former have been variously termed ''indirect market'' and ''producers' '' values while the latter have been termed ''direct,'' ''non-marketed,'' and ''individual'' values. The direct and non-marketed values are often further classified as either use values or nonuse (or sometimes intrinsic) values. Nonuse values will be the focus of chapter 5.

EX POST AND EX ANTE ANALYSIS OF VALUES

The decision maker who is trying to allocate scarce resources and is faced with a number of competing goals needs ex ante analyses of the effects of alternative policy actions to guide decision making. Ex ante analysis involves the prediction of the physical and economic consequences of policies on the basis of a model of the physical and economic processes involved. It involves visualizing two alternative states of the world, one with the policy in question and one without, and then comparing these alternative futures in terms of some established criterion such as net economic efficiency. Ex post analysis of a policy involves measuring the actual consequences of the policy by comparing the observed state with a hypothetical alternative, the state of the world without the policy. Ex post analysis, in effect, treats the policy as a controlled laboratory experiment except that the control is hypothetical rather than real. Natural resource damage assessment is an example of ex post evaluation in that the damaged state is observed and must be compared with a hypothetical or counterfactual alternative in which there was no pollution event but all other factors are assumed to have remained unchanged.

Ex post and ex ante analyses are not competitive alternatives; they should be viewed as complementary techniques for improving our knowledge. An ex post analysis of a policy can be viewed as a check on the validity of the ex ante analysis. The ex ante analysis is a prediction of what will happen; the ex post analysis is a check of what actually did happen.[10]

It is particularly important that the economic analysis of environmental and resource policies include ex post analysis. Our knowledge of the physical and economic systems on which present ex ante analyses are based is extremely limited. It is necessary not only to develop more comprehensive models of the physical, biological, and economic aspects of the system but also to devote more effort to verifying these models through ex post comparisons of the predictions with observed results.

It must be emphasized that the ex post verification of the analytical models used in resource valuation is not simply a comparison of actual results with predictions. Ex ante models are based upon some view of

---

[10]For further discussion of this and related points, see Haveman (1972).

the future with projections of economic magnitudes such as population levels, real income, and price levels. Care must be taken in ex post analysis to sort out the effects of unforeseen developments such as war or uncontrolled inflation on the variables in question. For example, if the failure of income levels to rise on the projected path results in a shortfall of recreation benefits at a particular site, this is not a failure of the analytical model so much as a reflection on our inability to perceive the future. The real benefit of ex post analysis is in making the most of the opportunity to improve on the analytical models used.

PREVIEW

The major task of this book is to review and summarize the basic theory of economic welfare measurement and to derive resource evaluation and benefit measurement techniques that are consistent with this underlying theory. In the next chapter, I provide an overview of valuation and welfare measurement methods and discuss the relationship between the economic methods of valuation and the physical and biological relationships that define the resource and environmental systems being valued.

Chapters 3 and 4 constitute the theoretical core of the book. Chapter 3 lays out the basic premises and value judgments that underlie the economic concept of benefits and presents the basic theory of the measurement of economic welfare changes. Chapter 4 introduces the basic methods and models for deriving welfare and value measures from the revealed choices of individuals and from observed changes in market prices. Chapter 5 introduces the concept of nonuse or existence value and shows that the indirect methods of measurement outlined in chapter 4 are not likely to be capable of measuring this type of value. Chapter 6 describes methods for obtaining hypothetical data, rather than observed prices and quantities, for calculating welfare measures. These methods are based on individuals' responses to hypothetical questions; for this reason they are termed "hypothetical methods" in this book. These methods are often referred to as contingent choice methods. They include willingness-to-pay surveys and bidding games, direct referendum questions, and questions about how individuals would rank alternative bundles of environmental and private goods and how their behavior would change under alternative hypothetical circumstances.

Chapter 7 takes up the question of valuation across time and the role of discounting in welfare measurement. Chapter 8 extends the theory of value and welfare change to a situation of risk where people are uncertain about what the actual state of the world will be. Public policies toward the management of environmental resources can effect either the probabilities of alternative outcomes or the magnitudes of environmental services in alternative states of the world. Thus this chapter describes the application of the basic theory of welfare change to evaluating environmentally induced changes in risks.

The remaining chapters describe the application of the various models and methods for welfare measurement to specific situations such as measuring the values of environmental changes affecting producers' costs and productivity (chapter 9), environmentally induced changes in longevity and health (chapter 10), applications of the hedonic price model to housing prices and wage rates (chapters 11 and 12), and the valuation of resources that support recreation activities (chapter 13). Chapter 14 offers some conclusions.

MATHEMATICAL NOTATION

In this book I use the following conventions regarding mathematical notation:

Vectors are represented by boldface uppercase letters; lowercase letters with or without subscripts represent values for individual variables in these vectors—for example, $X = x_1, ..., x_i, ..., x_n$.

The subscript letters i, j, k, m, and n and subscripted numbers index elements of vectors.

The meanings of other subscripted letters are specified when they are first used.

Superscripted letters are used to index such things as utility functions and production functions to specific individuals and firms. For example, $u^j(X)$ gives individual j's utility as a function of that individual's consumption of goods $x_1, ..., x_i, ..., x_n$.

Primes and superscripted numerals represent specific values for variables. For example, M^0 represents the initial value of the variable M. Similarly, $\Delta x = x'' - x'$ means the change in x from x' to x''.

Uppercase letters represent variables expressed as quantities of money. For example, M represents income, and CV is the compensating variation measure of welfare change.

REFERENCES

Bailey, Martin J. 1980. *Reducing Risks to Life: Measurement of the Benefits*. Washington, D.C.: American Enterprise Institute.

Bowes, Michael D., and John V. Krutilla. 1989. *Multiple-Use Management: The Economics of Public Forestlands*. Washington, D.C.: Resources for the Future.

Brown, Thomas C. 1984. "The Concept of Value in Resource Allocation," *Land Economics* vol. 60, no. 3, pp. 231–246.

Chameides, W. L., R. W. Lindsay, J. Richardson, and C. S. King. 1988. "The Role of Biogenic Hydrocarbons in Urban Photochemicals Smog: Atlanta as a Case Study," *Science* vol. 241, pp. 1473–1475.

Cohen, S. D., J. H. Eto, J. Beldock, and G. Crandall. 1990. *A Survey of State Regulatory Commission Activities to Incorporate Environmental Externalities into Electric Utility Planning and Regulation*. Berkeley, Calif.: Lawrence Berkeley Laboratory.

Cropper, Maureen L., William N. Evans, Stephen J. Beradi, Maria M. Ducla-Soares, and Paul R. Portney. 1992. "The Determinants of Pesticide Regulation: A Statistical Analysis of EPA Decision Making," *Journal of Political Economy* vol. 100, no. 1, pp. 175–189.

Fisher, Ann, and Robert Raucher. 1984. "Intrinsic Benefits of Improved Water Quality: Conceptual and Empirical Perspectives," in V. Kerry Smith and Ann Dryden Witte, eds., *Advances in Applied Microeconomics*. Greenwich, Conn.: JAI Press.

Freeman, A. Myrick, III. 1979. *The Benefits of Environmental Improvement: Theory and Practice*. Baltimore, Md.: The Johns Hopkins University Press for Resources for the Future.

Freeman, A. Myrick, III. 1982. *Air and Water Pollution Control: A Benefit-Cost Assessment*. New York: John Wiley.

Freeman, A. Myrick, III, Robert H. Haveman, and Allen V. Kneese. 1973. *The Economics of Environmental Policy*. New York: John Wiley.

Haveman, Robert H. 1972. *The Economic Performance of Public Investments*. Baltimore, Md.: The Johns Hopkins University Press for Resources for the Future.

Kneese, Allen V., Robert U. Ayres, and Ralph C. d'Arge. 1970. *Economics and the Environment: A Materials Balance Approach*. Baltimore, Md.: The Johns Hopkins University Press for Resources for the Future.

Mitchell, Robert C., and Richard T. Carson. 1989. *Using Surveys to Value Public Goods: The Contingent Valuation Method*. Washington, D.C.: Resources for the Future.

Kopp, Raymond J., and V. Kerry Smith. 1993. *Valuing Natural Assets: The Economics of Natural Resource Damage Assessment*. Washington, D.C.: Resources for the Future.

Mohring, Herbert, and J. Hayden Boyd. 1971. "Analyzing Externalities: Direct Interaction vs. Asset Utilization Frameworks," *Economica* vol. 38, no. 152, pp. 347–361.

Smith, V. Kerry. 1988. "Resource Evaluation at the Crossroads," *Resources* no. 90, pp. 2–6.

Smith, V. Kerry, William H. Desvousges, and A. Myrick Freeman III. 1985. "Valuing Changes in Hazardous Waste Risks: A Contingent Valuation

Analysis,'' vol. 1. Draft Interim Report to U.S. Environmental Protection Agency. Nashville, Tenn.: Vanderbilt University.

Smith, V. Kerry, William H. Desvousges, and Matthew P. McGivney. 1983. *A Comparison of Alternative Approaches for Estimating Recreation and Related Benefit of Water Quality Improvements.* Research Triangle Park, N.C.: Research Triangle Institute.

Travis, Curtis C., S. A. Richter Pack, and Ann Fisher. 1987. ''Cost-Effectiveness as a Factor in Cancer Risk Management,'' *Environment International* vol. 13, pp. 469–474.

Travis, Curtis C., S. A. Richter, E. A. C. Crouch, R. Wilson, and E. Klema. 1987. ''Cancer Risk Management,'' *Environmental Science and Technology* vol. 21, no. 5, pp. 415–420.

2

Measuring Values and Benefits: An Overview

In the first part of this chapter, I present a simple general equilibrium model of an economy with nonmarket environmental and resource service flows to show how the values of these flows emerge as shadow prices from the solution to a welfare maximization problem. In the second part, I describe in general terms the major types of methods for estimating values and benefits. In the third, I present a simple model to illustrate the relationships between physical and biological changes in environmental and resource systems and the changes in well-being and values realized by the people affected by these changes. These models help to make clear the kinds of data required to carry out resource evaluations, and to show the roles of economists and physical and biological scientists in the evaluation process.

RESOURCE VALUES AS SHADOW PRICES

Economic values can only be defined in terms of some underlying criterion that identifies what is to be considered good. As discussed in chapter 1, in neoclassical welfare economics good is defined in terms of the well-being of individuals. Here, I assume that an individual's well-being can be represented by an ordinal utility function. Of course, this assumption is only a first step in defining good because it does not deal with interpersonal comparisons. Specifically, it does not answer the question of whether it is good when one individual's utility increases while another individual's utility decreases.

The concept of Pareto optimality is the route I choose for dealing with this set of problems. An allocation of resources, goods, and services in an economy is Pareto optimal if there is no feasible real-location that can increase any one person's utility without decreasing someone else's utility. Of course, there is an infinite number of Pareto optimum allocations for an economy, each with a different distribution of utilities across individuals. In order to rank the allocations it is necessary to have a social welfare function that aggregates the utilities of the individuals, perhaps by assigning social welfare weights. If such a social welfare function exists, Pareto optimality is a necessary but not a sufficient condition for maximizing that function. Hence, despite its limitations, Pareto optimality has some usefulness in evaluating economic outcomes.

Pareto optimality is the solution to a constrained maximization problem in which some of the constraints are the exogenously determined environmental and resource service flows. The shadow prices on these constraints are the economic values of these service flows. In the general equilibrium model I use here, it is clear that the values assigned to the environmental and resource services flows are not fixed parameters but are determined by their roles in enhancing individuals' well-being and arise from their scarcity or limited availability.[1]

In this model there is a vector of m market goods (resources), each of which is divisible with well-defined and enforceable property rights. X represents the vector of goods (resources) being consumed; Y represents the vector of goods (resources) being produced; and S represents the vector of natural endowments of goods (resources) to the economy. More specifically, let

x_{ji} = the amount of good (resource) j consumed by individual i, ($j = 1, ..., m$) ($i = 1, ..., n$),

y_{jk} = the amount of good (resource) j produced (used) by firm k, ($j = 1, ..., m$) ($k = 1, ..., h$),

s_j = the total of endowment of good (resource) j available to the economy ($j = 1, ..., m$).

For each good (resource), the initial endowment plus the quantities produced by firms must just equal the sum of the quantities used and consumed by firms and individuals. This requirement is expressed in

---

[1] This model is a variation of the general equilibrium externalities model presented in Baumol and Oates (1975, 1988). Mäler (1985) has also used a general equilibrium model of an economy with environmental and resource service flows and pollution to analyze the values of nonmarket service flows.

a set of production-consumption constraints:

$$s_j + \sum_{k=1}^{h} y_{jk} - \sum_{i=1}^{n} x_{ji} = 0 \qquad (j = 1, ..., m)$$

In addition, let there be a resource service flow r, that is nondivisible so that it is, in effect, a public good. The endowment of r is determined exogenously. For simplicity, I assume that r does not vary across space. This means that r takes the same value for all individuals. Finally, let d be some environmental quality parameter, say the concentration of a pollutant at a specific point in space. The level of d will depend on the discharges of pollutants by firms and will determine the quantities of pollution experienced by each individual. These relationships will be specified below.

Individuals' preferences can be represented by utility functions:

$$u^i = u^i (x_{ji}, r, d_i) \qquad (i = 1, ..., n) \, (j = 1, ..., m)$$

Let $\partial u^i / \partial x_{ji} \geq 0$, $\partial u^i / \partial r \geq 0$, and $\partial u^i / \partial d_i < 0$, where d_i is the concentration of the pollutant to which the ith individual is exposed.

The production side of the economy can be represented by a set of production functions for multiproduct firms:

$$f^k (y_{jk}, r, d_k) = 0 \qquad (k = 1, ..., h)$$

where y_{jk} is the production of good j by firm k, r is the level of the nonmarket resource service flow, and d_k is the firm's contribution to the environmental quality parameter d as represented by, for example, emissions of a pollutant. The variable d is some function of the emissions of all firms. For simplicity, let $d = \sum_k d_k$. Also, assume that $d_i = \alpha_i \cdot d$, where α_i relates each individual's exposure to the pollutant to the aggregate level of pollution, d.[2]

The Pareto optimum conditions for this economy can be found by determining the conditions that maximize each individual's utility, subject to the production function and resource constraints and the constraint that all other individuals' utilities are held constant at some level representing the status quo. This procedure makes it clear that any Pareto optimum allocation is only as good as our judgment about

---

[2]For a perfectly mixed unavoidable pollutant, $\alpha_i = 1$ for all i. It would also be possible to represent averting behavior by making α_i a choice variable, perhaps a function of some x_{ji}.

the associated distribution of utilities across individuals. For each individual one, the problem is to choose the values for x_{ji} and d_i that

$$\max L_1 = u^1(\cdot) + \sum_{i=2}^{n} \lambda_i \left[u^{i*} - u^i(\cdot) \right] - \sum_{k=1}^{h} \mu_k f^k(\cdot)$$

$$+ \sum_{j=1}^{m} \rho_j \left(s_j + \sum_{k=1}^{h} y_{jh} - \sum_{i=1}^{n} x_{ji} \right) + \gamma \left(r^* - r \right)$$

where r^* is the exogenously determined quantity of r. The first-order conditions for a Pareto optimum are:[3]

$$\lambda_i \frac{\partial u^i}{\partial x_{ji}} - \rho_j = 0 \qquad (i = 1, \ldots, n)\,(j = 1, \ldots, m) \quad (1)$$

$$-\mu_k \frac{\partial f^k}{\partial y_{jk}} + \rho_j = 0 \qquad (k = 1, \ldots, h)\,(j = 1, \ldots, m) \quad (2)$$

$$\sum_{i=1}^{n} \lambda_i \frac{\partial u^i}{\partial r} - \sum_{k=1}^{h} \mu_k \frac{\partial f^k}{\partial r} - \gamma = 0 \qquad (i = 1, \ldots, n)\,(k = 1, \ldots, h) \quad (3)$$

$$\sum_{i=1}^{n} \alpha^i \lambda_i \frac{\partial u^i}{\partial d_i} + \mu_k \frac{\partial f^k}{\partial d_k} = 0 \qquad (i = 1, \ldots, n)\,(k = 1, \ldots, h) \quad (4)$$

where $\partial u_i / \partial d_i$, μ_k, and $\partial f^k / \partial d_k < 0$.

Conditions (1) and (2) generate the marginal rates of substitution and marginal rates of transformation for the marketed goods. They can be interpreted as saying either that each individual's marginal valuation of each good must equal the marginal cost to firms of its production, or that each individual's marginal rate of substitution between any pair of goods must equal each firm's marginal rate of transformation between that pair of goods.

In condition (3) the Lagrangian term γ is the shadow value for r^*, that is, it gives the increase in the objective function for a marginal increase in the constrained resource service r. Condition (3) also shows that the shadow price depends on the utility and production functions. Specifically, the shadow price is equal to the sum of the marginal values attached to r by all individuals and producers in the constrained solution.

---

[3]For notational simplicity, I assume an interior solution.

The first term in (4) is the value to individuals of reducing firm k's emissions by one unit. Condition (4) says, in effect, that Pareto optimality requires that the aggregate value of reducing the emissions of firm k must be just equal to the marginal cost of that reduction (the second term in the equation). The second term in (4) can also be interpreted as the marginal value to the kth firm of being able to emit one unit to the environment—that is, it is the marginal value of the waste receptor services of the environment.

Conditions (3) and (4) give the *marginal* values for changes in r and d_k. They also imply the existence of aggregate marginal willingness-to-pay curves that give the marginal willingnesses to pay (WTP) for the service flows as functions of the quantities of the flows being supplied. As is shown in chapter 3, willingness to pay can be taken to be equal to the area under such a marginal willingness-to-pay curve. Value estimation, then, involves determining directly or indirectly the shapes of these marginal willingness-to-pay curves for environmental services.

If the services of the environment could be purchased in a perfectly functioning market, estimating the marginal willingness-to-pay curves would be a fairly straightforward econometric problem, and environmental and resource management would not be an important public policy matter. But environmental and resource service flows typically have characteristics such as nonexcludability and nonrivalry in consumption, which make it difficult or impossible for markets for these services to function well. Often individuals are not free to vary independently the level of the services that they consume. The public good character of environmental services then leads to market failure. And without a market, there are no price and quantity data from which the demand relationships can be estimated.

APPROACHES TO MEASURING VALUES

Mitchell and Carson (1989, pp. 74–87) have offered a classification of methods for estimating values that is based on two characteristics of the methods. The first characteristic is whether the data come from observations of people acting in real-world settings where people live with the consequences of their choices, or come from people's responses to hypothetical questions of the form "what would you do if . . . ?" or "would you be willing to pay . . . ?" The second characteristic is whether the method yields monetary values directly or whether monetary values must be inferred through some indirect technique based on a model of individual behavior and choice.

Table 2-1.   Methods for Estimating Values

	OBSERVED BEHAVIOR	HYPOTHETICAL
DIRECT	*Direct Observed*	*Direct Hypothetical*
	Competitive market price	Bidding games
	Simulated markets	Willingness-to-pay questions
INDIRECT	*Indirect Observed*	*Indirect Hypothetical*
	Travel cost	Contingent ranking
	Hedonic property values	Contingent activity
	Avoidance expenditures	Contingent referendum
	Referendum voting	

Source: Adapted from Mitchell and Carson (1989), p. 75.

On the basis of these two methodological characteristics, any method for estimating environmental and resource values can be placed in one of four possible categories—Direct Observed, Indirect Observed, Indirect Hypothetical, and Direct Hypothetical. These are shown in table 2-1. *Direct Observed* methods include the use of competitive market prices and the use of results from simulated markets set up specifically to learn about individual values.[4] With Direct Observed methods, the observations are based on the actual choices made by people who are maximizing their utility, subject to the relevant constraints, and who are free to choose the quantity of the good at a given price. The data reveal values directly in monetary units since the choices are made on the basis of prices.

Indirect Observed methods are also based on actual behavior reflecting utility maximization. One type of indirect method is based on observed choices in a referendum setting. If an individual is offered a fixed quantity of a good at a given price on a take-it-or-leave-it or a yes-no basis (as in a referendum), observation of the choice reveals only whether the value of the offered good to the individual was greater or less than the offering price. A family of discrete choice and random utility models has been developed for the purpose of deriving exact value measures from these choices. In other instances the resource-environmental service does not have an offering price, but its quantity does affect the choices people make about other things such as quantities of market goods. In these cases, the value of the

---

[4]An example of a simulated market is Bishop and Heberlein's offer to purchase goose hunting permits from Wisconsin hunters at stated prices. See Bishop and Heberlein (1979).

resource-environmental service must be inferred through the application of some model of the relationship between market goods and the environmental service. Most such models are based on the assumption of some kind of substitute or complementary relationship between the environmental service and marketed goods and services. Examples of these models include the hedonic property value and hedonic wage models, models of household spending on cleaning and on repair of materials damaged by air pollution, and the travel cost demand model for visits to a recreation site.

Where environmental changes affect producers of market goods, values can also be observed indirectly through examination of changes in product and factor prices and in producers' quasi-rents. Indirect Observed methods involve a kind of detective work, in which clues about the values individuals place on environmental services are pieced together from the evidence that people leave behind as they respond to prices and other economic signals. Models for implementing Indirect Observed methods are described in chapter 4.

The principal difference between Indirect Observed and *Indirect Hypothetical* methods is that the latter draw their data from people's responses to hypothetical questions rather than from observations of real-world choices. The models and techniques used to draw inferences about values from these data are often the same as those used in the Indirect Observed methods. McConnell (1986), for example, applied a recreation travel cost demand model to questions of the form "how often would you visit these beaches if they were free of PCBs?" in order to estimate the damages (resource values lost) from the pollution of the waters of New Bedford harbor, Massachusetts, with polychlorinated biphenyls (PCBs).

Also in the Indirect Hypothetical category is the contingent ranking approach, in which individuals are given a set of cards, with each card depicting a different situation with respect to the level of the environmental service flow in question and other attributes of choice (number of visits to a site, congestion level, and admission fee, for example). People are asked to place their cards in order of preference. Values for environmental services can be inferred from these rankings. Uses of this approach include a study of the value of visibility in national parks (Charles River Associates, 1981) and a study of water quality benefits in a river basin (Smith and Desvousges, 1986).

The fourth category, *Direct Hypothetical* methods, involves asking people directly about the values they place on environmental services by, in effect, creating hypothetical markets. For example, people could be asked what value they place on a specified change in environmental services, or how much of an environmental service they would "purchase" at a given price. Direct and Indirect Hypothetical methods are discussed in chapter 6.

THE METHODOLOGY OF INDIRECT MEASUREMENT

Since the indirect methods for measuring values use data on behavior (either observed or hypothetical), some theoretical framework must be developed to model this behavior and to relate the behavior to the desired monetary measures of value and welfare change. A key element in the theoretical framework is the model of individual optimizing behavior that relates the individual's choices to the relevant prices and constraints, including the level of environmental or resource quality q. If a behavioral relationship between observable choice variables and q can be specified and estimated, this relationship can be used to calculate the marginal rate of substitution between q and some observed choice variable with a money price tag.

More specifically, welfare measurement using these models involves three steps. The first is to derive the expression for willingness to pay as a function of the environmental variable, usually either from the indirect utility function or the expenditure function. This expression gives the compensating change in income that holds utility constant for the change in the environmental parameter. The second step is to develop a model of individual utility maximizing behavior that relates the individual's choices to the relevant prices and constraints, including the level of environmental quality. The first-order conditions for optimization are derived. These first-order conditions involve equating measures of marginal value to price or equating some marginal rate of substitution or marginal rate of transformation to some price ratio. The third step is to examine the model to see whether the first-order conditions include a relationship between the desired marginal value for the environmental change and some observable variable. If they do, then the observable variable can be taken as a measure of the marginal change in welfare. Several categories of models based on this methodology are presented in chapter 4.

The measures of value and welfare change derived from optimization models often produce results that are quite different from those of the naive models employed in the early literature on value and benefit measurement. The early models were often based on what has come to be known as the "damage function" approach. The damage function approach involves estimating some physical relationship between a measure of environmental quality (or its converse, pollution) and some physical measure of damage or loss (such as number of workdays lost to sickness in the case of health, or percentage of crop lost in the case of effects of air pollution on agricultural productivity). Then some unit price is applied to the physical measure to convert it to monetary terms. In some studies, for example, lost wages and medical costs were used to determine the value of avoiding one day of illness induced by air pollution.[5] Similarly, the market price of a crop was often used to determine the value of lost productivity. The benefit of a pollution control program would be estimated as the reduction in the damages calculated according to this damage function approach.

I have characterized the damage function approach as being naive because this approach implies a model of the world in which behavioral and market responses to changes in q are ruled out. Farmers can, in fact, adjust to changes in air pollution by changing cultivation practices, shifting to more resistant cultivars of the same crop, or even changing to entirely different crops that are less sensitive to pollution. Furthermore, the prices of agricultural crops may change because of changes in crop supplies. It may be that the changes in prices are of greater significance to human welfare than the changes in physical yields of crops.[6] Similarly, people can choose defensive or mitigating activities in response to air pollution that affects health. These behavioral changes result in welfare consequences that have a monetary dimension, which should be taken into account in calculating values. I show in chapter 4 that the conditions under which a naive damage function model will yield conceptually correct measures of value and welfare change are quite restrictive.

A MODEL OF ENVIRONMENTAL AND RESOURCE VALUES

Although measuring values involves the use of economic theory and technique, value measures must be based on other types of knowledge. For example, estimates of the value of a salt marsh in sustaining a marine fishery must be based on knowledge of the biological and ecological links between the marsh and the exploited fish species (Lynne, Conroy, and Prochaka, 1981; Ellis and Fisher, 1987). Esti-

---

[5]See Freeman (1982) for a discussion of some of these studies.

[6]Examples of studies of the benefits of air pollution costs that model these adjustments and take into account the effects of price changes include Adams, Hamilton, and McCarl (1984), Kopp and coauthors (1985), and Kopp and Krupnick (1987).

mates of the health benefits from air pollution control must be based on scientific knowledge of the relationship between pollutant concentrations and human health (Lave and Seskin, 1977). And estimates of the recreational fishing benefits stemming from water pollution control require knowledge of the relationships among pollutant levels, biological productivity, and anglers' activities. Lack of knowledge of these relationships may, in some instances, be a major barrier to empirical measurement of values. In this section, I lay out a simple model for examining the relationship between the economic concept of value and the physical and biological dimensions of the resource system being valued. The model helps to make clear the dependence of economic technique on some knowledge of the underlying physical and biological relationships that determine the quantity and quality of environmental and resource service flows.

The economic values of the service flows from a resource-environmental system can be viewed as the product of three sets of functional relationships. The first relates some measure of environmental or resource quality to the human interventions that affect it. Let q represent a qualitative or quantitative measure of some environmental or resource attribute. Examples include the biomass of some species of fish of commercial or recreational interest, the stock of standing timber in a forest, or the concentration of some pollutant in the atmosphere. Two kinds of human intervention need to be specified. One involves the unregulated activities of the market economy (for example, the commercial exploitation of a fishery or the discharge of pollutants into the air); these will be left implicit in the relationships presented here. The other kind of intervention is government actions taken to prevent or ameliorate the adverse impacts of unregulated market activities or to protect or enhance the value of market and nonmarket services provided by the environment. Let S represent the government interventions. For example, if q represents the population of water fowl, S could be increases in protected habitat and breeding grounds. S could represent a set of regulations designed to attain a stated ambient air quality standard. Alternatively, S could be a management plan for a national forest. Let us represent the relationship between q and S as

$$q = q(S) \tag{5}$$

As discussed below, this relationship could be quite complex in its spatial and temporal dimensions.

Where the government regulates private activities that influence q, the effect of a change in S on q can depend in complex ways on the responses of private decision makers to the public regulation. The most obvious example of this dependence is the question of compliance with pollution control regulations. For any given S, q increases as the degree of compliance with the regulation increases. Some public regulations by their very nature are only indirectly linked to the relevant environmental quality measure. An example is the automotive emissions standards set under the Clean Air Act, which regulate emissions of certain pollutants in grams per mile traveled by the automobile. In this case the effect of a change in S on q depends also on how automobile use is affected. To take account of these complexities, it might be more appropriate to write q as a function of both S and some measure of private responses to the government regulation, $R(S)$:

$$q = q[S, R(S)] \tag{5'}$$

The second set of functional relationships involves the human uses of the environment or resource and their dependence on q. Let X represent the levels of some set of activities involving use of the environment or resource. For example, X could be days of recreational activity on some water body, tons of fish caught from some commercially exploited fishery, and, where human health depends on the level of environmental quality, some measure of health status. Typically, the level of X will depend not only on q but also on the inputs of labor, capital, and other materials and resources, including time; and these will also depend on q. Let Y represent these other inputs into the production of environmental services or activities based on the resource. The second functional relationship can be written as

$$X = X[q, Y(q)] \tag{6}$$

The third set of functional relationships gives the economic value of the uses of the environment. Let V represent the money value of the flow of services or activities based on the environment or resource. The relationship

$$V = V(X) \tag{7}$$

embodies whatever value judgments society has adopted for economic welfare purposes. Here I assume that the value function is a simple aggregation of individuals' values; but $V(X)$ could also incorporate social welfare weights that reflect some social equity goals. Alternatively, $V(\cdot)$ could incorporate concepts of environmental ethics or social norms.

By substitution of (5'), (6), and (7), we have

$$V = f\{S, R(S), Y[S, R(S)]\} \tag{8}$$

The marginal value of the change in S can be found by taking the total derivative of equation (8) to reflect the private adjustments of R and Y to the public intervention. The benefits of a nonmarginal change in policy intervention that increases q are given by

$$B \equiv \Delta V = f\{S^2, R(S^2), Y[S^2, R(S^2)]\} - f\{S^1, R(S^1), Y[S^1, R(S^1)]\}$$

where S^1 and S^2 are the pre- and post-policy levels of intervention and R and Y are optimally adjusted to the change in S.

The set of relationships represented by equation (5') is almost totally noneconomic in nature because it involves a variety of physical and biological processes. The relationship reflected in equation (7) is wholly within the realm of economics because it involves the theory of economic welfare and the use of economic data. The set of relationships reflected in equation (6) represents the interface between the natural science and social science disciplines. Some aspects of these relationships—for example, how recreation use varies with changes in water quality—are primarily behavioral or social. Other aspects are almost wholly physical or biological, as in the effects of air pollution on human health and mortality. But even here, to the extent that people can "defend themselves" against the adverse effects of air pollution (say, by purchasing home air purifiers) or mitigate the symptoms of illness induced by air pollution, behavioral relationships are embedded in equation (6). The effect of an air pollutant on a particular type of vegetation is also a biological question; but if farmers alter crop patterns as a way of adapting to changes in air pollution, then the behavioral and biological aspects of the relationship must be considered together.

Figure 2-1 illustrates the three sets of relationships for the case of the benefits associated with an improvement in ambient water quality. A variety of substances are discharged into water bodies. Reductions in the discharges can affect physical, chemical, and biological indicators of water quality such as dissolved oxygen, temperature, algae levels, and fish populations (stage 1). Changes in the indicators can be predicted with water quality models. The resulting water quality, as measured by the indicators, can in turn affect human uses of the water body (stage 2). These include withdrawal uses—for example, for industrial or municipal water supply and irrigation— or instream uses—for example, for fishery production or

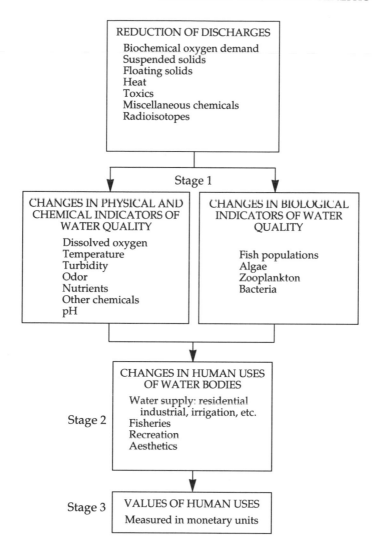

Figure 2-1.   The production of benefits from improved ambient water quality

recreation. The major difficulty at stage 2 arises from the fact that only rarely is the level of use a simple function of a single water-quality indicator like dissolved oxygen. Rather, some uses (as in the case of commercial fisheries and recreation) depend in complex ways on the whole range of physical, chemical, and biological water-quality indicators.

In our illustration, estimating water-quality benefits involves determining the monetary values that people place on such things as improved recreational opportunities, increases in fish production, and the availability of particular species of fish (stage 3). Regarding the analysis of this stage, there is a well-developed theory of economic value. As discussed earlier, there are a number of approaches for estimating these values under different circumstances.

Figure 2-2 illustrates a similar application of the model to the abatement of air pollution. Understanding stage 1 is the task of atmospheric scientists who study photochemistry, atmospheric dispersion and transport, and the like. They must provide the air quality models that relate changes in emissions to changes in the ambient concentrations of the substances of interest.

Understanding the effects of changes in air quality on agricultural productivity and materials damages is a technical task for

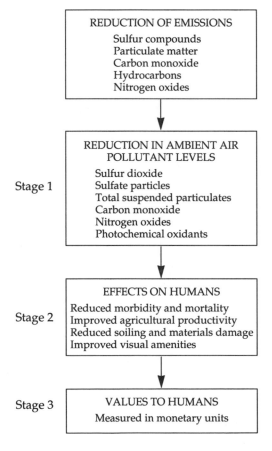

Figure 2-2.   The production of benefits from improved ambient air quality

botanists, plant physiologists, and materials engineers. But economic models are also necessary because the effects and their valuation also depend on human responses, such as changes in cropping patterns, materials substitution, and a variety of mitigating activities. These must be analyzed properly if an accurate picture of the effects of air pollution on human activity and welfare is to be obtained. Similarly, we must look to the biomedical sciences for information on the health effects of air pollution. But a full understanding of the relationship between air pollution and the health status of the population requires a comprehensive epidemiological analysis that controls for socioeconomic influences and other confounding variables, such as diet, life style, and occupational exposures to harmful substances.

THE NONECONOMIC FOUNDATIONS OF RESOURCE VALUATION

In a book that is about determining the economic values attached to the services affected by environmental change, the discussion cannot proceed far without acknowledging the importance of the relationship between environmental and resource quality and the uses of the environment. For example, the value of a recreation user-day at a lake is affected by fish populations and species distribution, algae levels, the number and type of bacteria present, temperature, smell, turbidity, and concentration of toxic substances. To further complicate matters, an increase in the magnitude of one characteristic may affect one use favorably while affecting an alternative use in a negative way. Higher water temperatures, for example, may make for better swimming while adversely affecting trout and salmon populations. Industrial discharges of acids may adversely affect recreation and fisheries while improving the value of water for industrial uses because of retarded algae growth.

The difficulties in tracing out the effects of the discharge of a pollutant on the many parameters of environmental quality and, in turn, their effects on man's uses of the environment substantially limit our ability to do careful benefit-cost analyses of environmental quality and improvements. This is not fully appreciated by many advocates of greater use of benefit-cost analysis in this field. It was in 1968 that Allen Kneese wrote,

> I believe that our limited ability to evaluate the recreational losses associated with poor quality water, or conversely, the

benefits of water improvement, is an extremely important barrier to rational water quality management. . . . The first [complexity] is the relationship between the level of various water quality parameters and the recreational attractiveness of the water resource. This relationship can be viewed as being composed of two linkages: a natural one and a human one. I think these are both about equally ill-understood. It is my impression . . . that the biological sciences are almost never able to tell us specifically what difference a change in measured parameters of water quality will make in those biological characteristics of the water that contribute to its recreational value. . . . Perhaps the undeveloped state of forecasting is a result of the fact that biologists have seldom been confronted with the types of questions we would now like them to answer. . . . There is also a human linkage that is ill-understood. What quality characteristics of water do human beings find attractive for recreation? This is still largely an area of ignorance. (Kneese, 1968, pp. 180–181)

What Kneese said about water quality and recreation applies as well to the other effects of water pollution and to air pollution. And it is as true in 1993 as it was more than twenty years ago.

Describing Resource and Environmental Quality

An analysis of the value of a resource or of the benefits of an environmental or resource policy change must begin with a description of the resource flow or some measure of environmental quality. This description requires choices about what attributes or characteristics of the resource-environmental service are important. Suppose the question is, what will be the benefits of achieving the auto emissions standards mandated by federal law? In order to answer this question it is necessary to determine what things that matter to people are adversely affected by automotive emissions, and then to trace out the links between emissions, and those things that are valued. It is now understood that automobile emissions and the subsequent products of their photochemical reactions adversely affect human health, visibility, and agricultural productivity, among other things. So, in order to estimate values it is necessary to determine what specific measures of air quality are linked to these effects and how these measures of air quality are affected by the mandated standards. The measures used to characterize effects on visibility may be quite different from those relevant to effects on human health.

Two kinds of problems arise in this stage of the analysis. The first concerns the choice of parameters for describing the resource or environmental quality. The second involves determining the functional relationship between the policy instrument and the resource service flow or environmental quality measure.

Consider the case of water quality. A single effluent discharge can contain many substances that affect water quality—for example, oxygen-demanding organic wastes (biochemical oxygen demand), suspended solids, waste heat, and toxic chemicals. When these substances enter the waterway they affect—in sometimes simple and sometimes complex ways—such measurable components of water quality as dissolved oxygen, temperature, and concentrations of chemicals. A nondegradable chemical substance, for example, will simply be diluted, and its concentration in the water body will be a calculable fraction of its concentration in the effluent stream. In contrast, organic wastes affect water quality parameters in a more complicated way. As they are degraded by bacteria, they reduce dissolved oxygen levels to an extent and at a rate that depends on water temperature, wind, rates of river flow, and other physical and biological characteristics of the receiving water.

Some of the physical measures of water quality, such as turbidity and smell, affect man's uses of the water directly. In addition, these and other physical parameters affect the stream ecology in complex and not always well-understood ways. The populations and species distributions of fish, algae, zooplankton, and bacteria may also be affected, and not necessarily in the same direction, by changes in the physical and chemical parameters of water quality.

Even providing a descriptive characterization of this first stage is a formidable task. Water quality cannot be represented by a single number on some scale, but rather is an n-dimensional vector of the relevant parameters. Which subsets of these parameters are most important in influencing the uses of a water body (commercial fishing, boating, or swimming, for example) is still a major question for research.

Developing predictive models for these parameters is also a major research priority. The most commonly used water quality models relate dissolved oxygen to the biochemical oxygen demand of discharges of organic wastes. But dissolved oxygen levels are only crudely related to the suitability of a water body for fishery production or recreational use.

In the case of air quality, the choice of parameters is somewhat easier, but not without pitfalls. It is only recently that attention has been turned from measurement of sulfur dioxide to its transformation products, sulfate particles. The latter measure of air quality is now thought to be a more important air pollutant indicator than sulfur dioxide because of its effects on human health and ecology (U.S. Environmental Protection Agency, 1986a, 1986b).

Temporal and Spatial Aggregation

Typically, measures of environmental quality vary over time and space. The dissolved oxygen level at any point in a river rises and falls with changes in streamflow, discharge rates, water temperature, and the like, and is different at different locations. Air pollution readings vary with the time of day, day of the week, over the year, and from one place to another. One problem in empirical research on effects of pollution is how to define a variable or set of variables in a way that adequately reflects the temporal and spatial variations in environmental quality while still being manageable.

To put the problem in a concrete setting, consider an attempt to estimate the relationship between the level of an air pollutant and some health effect, say the occurrence of an asthma attack. The air pollution level at any particular point in an urban area is an instantaneous variable that fluctuates over time. The true exposure of an individual located at that point is measured by a trace of the time path of that instantaneous variable over the relevant time period. But the individual may also move from one point to another in the urban space. The published data on air pollution levels that are used to generate exposure variables for empirical research involve various approaches to summarizing the instantaneous time paths at the locations of air pollutant measurement devices. One common measure is the annual mean, either arithmetic or geometric. Averages are also struck over shorter time periods—for example, a 24-hour average for particulates, and 8-hour and 1-hour averages for other pollutants. Also, readings taken at one or two points in the space must be used to represent the space as the whole. Inevitably, summarizing involves loss of information. No one or two of these temporal and spatial summaries can completely represent the true exposure of any individual. Research on health effects is hampered by our inability to characterize accurately the exposure of individuals.

SUMMARY

Economists seek measures of values that are based on the preferences of individuals. When value measures are derived via indirect methods based on models of behavior, these models should be internally consistent and be based on accepted theories of preferences, choice, and economic interactions. If empirical observations are taken without benefit of an underlying theoretical model, researchers may be led to make faulty or erroneous interpretations of the data.[7] But equally important is the need for a sound understanding of the underlying biological and physical processes by which environmental and resource service flows are generated.

REFERENCES

Adams, Richard M., Scott A. Hamilton, and Bruce A. McCarl. 1984. *The Economic Effects of Ozone on Agriculture.* Corvallis, Ore.: U.S. Environmental Protection Agency.

Baumol, William J., and Wallace E. Oates. 1975. *The Theory of Environmental Policy.* Englewood Cliffs, N.J.: Prentice-Hall.

Baumol, William J., and Wallace E. Oates. 1988. *The Theory of Environmental Policy.* 2d ed. Cambridge: Cambridge University Press.

Bishop, Richard C., and Thomas A. Heberlein. 1979. "Measuring Values of Extra Market Goods: Are Indirect Measures Biased?" *American Journal of Agricultural Economics* vol. 61, no. 5, pp. 926–930.

Charles River Associates, Inc. 1981. *Benefits of Improving Visibility at Mesa Verde National Park.* Boston, Mass.: Charles River Associates, Inc.

Ellis, Gregory M., and Anthony C. Fisher. 1987. "Valuing the Environment as Input," *Journal of Environmental Management* vol. 25, no. 2, pp. 149–156.

Freeman, A. Myrick, III. 1982. *Air and Water Pollution Control: A Benefit-Cost Assessment.* New York: John Wiley.

Kneese, Allen V. 1968. "Economics and the Quality of the Environment: Some Empirical Experiences," in Morris Garnsey and James Gibbs, eds., *Social Sciences and the Environment.* Boulder, Colo.: University of Colorado Press.

---

[7]An important example is found in studies of the land value–air pollution relationship. Early researchers discovered that land value and air pollution levels were inversely related in urban areas, other things being equal. They then assumed that changes in welfare associated with reduced pollution would be accurately measured by the associated increases in land values as predicted by the regression equation relating land values at a point in time to air pollution. Subsequent research based on theoretical models of urban land markets has shown that this assumption is not true in general. The relationships among air pollution, land values, and measures of welfare change are discussed in chapter 11.

Kopp, Raymond J., and Alan J. Krupnick. 1987. "Agricultural Policy and the Benefits of Ozone Control," *American Journal of Agricultural Economics* vol. 69, no. 5, pp. 956–962.

Kopp, Raymond J., William J. Vaughan, Michael Hazilla, and Richard T. Carson. 1985. "Implications of Environmental Policy for U.S. Agriculture: The Case of Ambient Ozone Standards," *Journal of Environmental Management* vol. 20, no. 4, pp. 321–331.

Lave, Lester B., and Eugene P. Seskin. 1977. *Air Pollution and Human Health.* Baltimore, Md.: The Johns Hopkins University Press for Resources for the Future.

Lynne, Gary D., Patricia Conroy, and Frederick J. Prochaka. 1981. "Economic Valuation of Marsh Areas for Marine Production Processes," *Journal of Environmental Economics and Management* vol. 8, no. 2, pp. 175–186.

Mäler, Karl-Göran. 1985. "Welfare Economics and the Environment," in Allen V. Kneese and James L. Sweeney, eds., *Handbook of Natural Resource and Energy Economics,* vol. 1. Amsterdam, The Netherlands: Elsevier Science Publishers.

McConnell, Kenneth E. 1986. "The Damages to Recreational Activities from PCBs in New Bedford Harbor." Report prepared by Industrial Economics, Inc., Cambridge, Mass., for the Ocean Assessment Division, National Oceanic and Atmospheric Administration.

Mitchell, Robert Cameron, and Richard T. Carson. 1989. *Using Surveys to Value Public Goods: The Contingent Valuation Method.* Washington, D.C.: Resources for the Future.

Smith, V. Kerry, and William H. Desvousges. 1986. *Measuring Water Quality Benefits.* Norwell, Mass.: Kluwer-Nijhoff.

U.S. Environmental Protection Agency. 1986a. "The Addendum to the OAQPS Staff Paper." Research Triangle Park, N.C.: Environmental Protection Agency.

U.S. Environmental Protection Agency. 1986b. *The Second Addendum to Air Quality Criteria for Particulate Matter and Sulfur Oxides.* Research Triangle Park, N.C.: Environmental Protection Agency.

3

Defining and Measuring Welfare Changes: Basic Theory

The theory of the measurement of welfare change has been discussed by others both at the most rigorous levels of abstraction and in pragmatic, practical terms of application.[1] Most of this discussion has focused on the welfare effects of changes in the prices people pay for the goods they consume. My major purpose in adding to this literature is to provide a systematic development of the definition and measurement of welfare effects of changes in quantities or qualities of environmental and resource service flows.

Changes in environmental quality can affect individuals' welfares through any of the following four channels: changes in the prices they pay for goods bought in markets; changes in the prices they receive for their factors of production; changes in the quantities or qualities of nonmarketed goods (for example, public goods such as air quality); and changes in the risks individuals face. The first three of these channels are the focus of this chapter. After a brief review of the theory of individual preferences and demand, I review the principles of welfare measurement for price changes. These principles are relevant because some forms of environmental change affect people only indirectly through their effects on prices. Also, a review of these principles will provide a good foundation for the treatment of quantity and quality changes that follows. The extension of these principles to the valuation of changes in risk—the fourth channel—raises some interesting questions, which will be left to chapter 8.

---

[1]A thorough exposition of the theory is presented in Boadway and Bruce (1984). See also Johansson (1987). For a treatment designed more for practitioners, see Just, Hueth, and Schmitz (1982).

The principles and measures developed in this chapter apply equally to decreases and increases in individuals' welfare. It is a basic principle of welfare economics that all costs ultimately take the form of reductions in the utility welfare of individuals. This principle applies equally to the costs of public policies (for example, investment in resource development and the regulation of private activities) and to the costs of private uses of the environment (for example, harvesting from a common property resource and using the waste receptor services of the environment). Hence, the welfare measures developed here provide a foundation for the analysis of both the benefits and the costs of environmental change. I return to this point in chapter 14.

In this chapter, I consider three sets of questions in some detail. The first set concerns how to define an acceptable monetary measure of changes in economic welfare for an individual. The answer to this question hinges in part on what the measure would be used for. As Hallam (1988) suggests, a welfare measure should answer the questions posed by policymakers. But policymakers can ask different kinds of questions. For example, suppose that policymakers wish to evaluate proposed policy changes in terms of an aggregate social welfare function that placed different weights on individuals' changes in utility depending on their positions in the income distribution (Bergson, 1966). In that case, the welfare measure that answers the policymakers' question must be a money metric of utility changes. Alternatively, if policymakers wish to select policies on the basis of the potential Pareto improvement criterion, they will want measures of required compensation and willingness to make compensating payments. In the concluding section of this chapter, I return to the question of choosing from among the alternative measures described in the chapter.

The second set of questions concerns how changes in welfare would be measured both in theory and in practice. Theory suggests several alternative ways of calculating either exact or approximate welfare measures, using data on observed behavior of individuals—for example, their demand functions for market goods. These alternatives will be described and evaluated, especially from the practical perspective of implementation.

The third set of questions concerns how any measure of welfare changes for individuals might be used to make judgments about social policies affecting many individuals. For example, is it possible to speak of a measure of aggregate welfare for the society as a whole? If so, what significance can be attached to changes in such a measure? Measures of welfare change for an individual can be defined and analyzed without reference to the notions of efficiency and equity. In this sense the concept is objective; that is, one can define and measure a monetary equivalent of an individual's welfare change without being committed to any particular set of value judgments concerning aggregation across individuals or the role of such welfare measures in social choice.[2] It is in answering the third set of questions that value judgments about the relative deservingness of individuals, the meaning of efficiency, and the objectives of public policy come into play. Some of these issues are discussed in the section headed Aggregation and Social Welfare.

In the following section I review some of the basic terminology and theory involving individual preferences and demand. I then take up the standard case where utility depends only on the consumption of market goods. I examine the theory of measuring the welfare value of changes in the prices of these goods and the relationships among the Marshallian and Hicksian surplus measures of welfare change. There are two reasons for choosing this order of presentation. First, it parallels the historical evolution of the theory of welfare change. And second, it makes for an easier exposition of the basic principles. The section concludes with a review of methods for obtaining exact measures of and approximations to the desired Hicksian surpluses. In the third section I briefly describe how the theory of welfare value for changes in prices of goods can be applied to changes in factor prices. I then take up, in the fourth section, the case of the welfare effects of changes in the quantities of nonmarket goods where individuals are not free to adjust these quantities in response to price changes. In the final two sections, I review some of the issues involved in aggregating measures of individual welfare change for public policy decision making and in selecting the appropriate welfare measure.

INDIVIDUAL PREFERENCES AND DEMAND

Before introducing the various possible welfare measures, it will be useful to review briefly the basic theory of individual preferences and

---

[2]Of course, in a more fundamental sense focusing attention on changes in income and consumption to the exclusion of other aspects of well-being, such as social relationships and sense of self-worth, involves a kind of value judgment. But this judgment is implicit in the division of labor between economists and others.

the demand for goods as it relates to welfare theory.[3] This theory starts with the premises that individuals are their own best judge of their welfares and that inferences about welfare can be drawn for each individual by observing that individual's choices among alternative bundles of goods and services. If an individual prefers bundle A over bundle B, then bundle A must convey a higher level of welfare.

What things are to be included in the bundles (such as A and B) among which individuals are assumed to have preferences? There is little controversy over the inclusion of all the goods and services that can be bought or sold in markets—consumer goods, the services of household assets such as a house or a car, and consumer durables. Since time can be used in leisure activities or sold at some wage rate in the labor market, individuals must also have preferences among alternative uses of time such as reading, outdoor recreation, and working at some wage rate. And since government and the environment both provide a variety of services that enhance the welfares of individuals, these services should also be included in the bundles among which people have preferences. Environmental services include those provided by cleaner air, cleaner water, and scenic amenities.

If we assume that individuals can rank the alternative bundles according to their preferences, what properties will the resulting ordering of bundles display? For our purpose, two are important. The first is nonsatiation, or the "more-is-better" property. This means that a bundle with a larger quantity of an element will be preferred to a bundle with a smaller quantity of that element, other things being equal.[4] The second property is substitutability among the components of bundles. This means that if the quantity of one element of a bundle, say x_i, is decreased, it is possible to increase the quantity of another element, say x_j, sufficiently to make the individual indifferent between the two bundles. More formally, suppose that X' consists of $x'_1, ..., x'_i, ..., x'_j, ..., x'_n$ and X'' consists of $x'_1, ..., x''_i, ..., x'_j, ..., x'_n$, with $x''_i < x'_i$. Substitutability means that there is another bundle X^* consisting of $x'_1, ..., x''_i, ..., x^*_j, ..., x'_n$ with $x^*_j > x'_j$ such that the individual is indifferent as to X' and X^*. In other words, X' and X^* lie on the same indifference surface.[5]

---

[3]For alternative treatments of this and related topics, the reader may wish to consult other texts, such as Just, Hueth, and Schmitz (1982), Boadway and Bruce (1984), Varian (1984), and Johansson (1987).

[4]Formally, if X' consists of $x'_1, ..., x'_i, ..., x'_n$ and X'' consists of $x''_1, ..., x''_i, ..., x''_n$, and $x'_i > x''_i$, then this individual will prefer X' to X''.

[5]Two other important properties are transitivity and quasi-concavity. If

The property of substitutability is at the core of the economist's concept of value. This is because substitutability establishes trade-off ratios between pairs of goods that matter to people.[6] The money price of a market good is just a special case of a trade-off ratio, because the money given up to purchase one unit of one element of the bundle is a proxy for the quantities of one or more of the other elements in the bundle that had to be reduced in order to make the purchase.

If the preference ordering has the properties described here, it can be represented by an ordinal preference function or utility function that assigns a number to each bundle as a function of the quantities of each element of the bundle. Specifically,

$$u = u(X, Q, T)$$

where X is a vector of the quantities of market goods, Q is a vector of public goods and environmental and resource services whose quantities or qualities are fixed for the individual, and T is a vector of the times spent in various activities that yield utility to the individual. This utility function is assumed to be increasing in all of its arguments and unique up to a monotonic transformation.[7] This preference function is not the same thing as the cardinal utility function of the classical utilitarians. Since there is no unit of measurement for this ordinal utility, it is not possible to add or otherwise compare the utilities of different individuals.

To simplify the exposition and notation, let us now consider an individual whose utility is a function only of private goods that can be bought and sold in markets. Assume that tastes and preferences (that is, the utility function) are given and do not change. The individual faces a set of given prices for these goods and is assumed to choose the quantities of the goods so as to maximize his utility, given the

---

there are three bundles X', X'', and X^*, and the individual prefers X' over X'' and X'' over X^*, then transitivity is satisfied if the individual prefers X' over X^*. For more on the axiomatic description of these properties of preference ordering, see Boadway and Bruce (1984) or Varian (1984).

[6]In the formulation in the text above, the trade-off ratio is $(x_j^* - x_j')/(x_i' - x_i'')$ or $|\Delta x_j / \Delta x_i|$. In the limit for infinitesimally small changes, this reduces to $|dx_j / dx_i|$, which is the definition of the marginal rate of substitution between x_i and x_j or the slope of the two-dimensional indifference curve between these two elements.

[7]For purposes of mathematical modeling and analysis, it is convenient also to assume that this function is continuous, convex, and twice differentiable.

constraints of prices and a fixed money income M. The maximization problem can be expressed as

$$\text{maximize } u = u(X)$$

$$\text{subject to } \sum_i p_i \cdot x_i = M$$

where X is the vector of quantities ($X = x_1, ..., x_i, ..., x_n$). The solution to this problem leads to a set of ordinary demand functions

$$x_i = x_i(P, M)$$

where P is the vector of prices ($P = p_1, ..., p_i, ..., p_n$).

Substituting the expressions for x_i as functions of P and M into the direct utility function gives the indirect utility function, that is, utility as a function of prices and income, assuming optimal choices of goods:

$$u = v(P, M)$$

According to Roy's Identity, the demand functions can also be expressed in terms of derivatives of the indirect utility function,

$$x_i(P, M) = -\frac{\partial v / \partial p_i}{\partial v / \partial M}$$

The expenditure function represents a useful perspective on the problem of individual choice.[8] It is derived by formulating the dual of the utility maximization problem. The individual is assumed to minimize total expenditure,

$$e = \sum_i p_i \cdot x_i$$

---

[8]For a general discussion of the expenditure function, see Diamond and McFadden (1974). See also Moss (1976) for a correction of one of the derivations. Mäler (1974) develops the concept of the expenditure function and relates it specifically to the welfare effects of changes in environmental quality.

subject to a constraint on the level of utility attained,

$$u(X) = u^0$$

where u^0 is the maximum utility attained with the solution to the primal problem. Just as the solution to the utility maximization problem yields a set of ordinary demand curves conditional on prices and money income, the solution of the expenditure minimization problem yields a set of functions giving optimal quantities for given prices and utility. These are Hicks-compensated demand functions that show the quantities consumed at various prices assuming that income is adjusted (compensated), so that utility is held constant at u^0. Substituting these demand functions into the expression for total expenditure yields the expenditure function. This expression gives the minimum dollar expenditure necessary to achieve a specified utility level, given market prices. In functional notation:

$$e = e(P, u^0)$$

where e is dollar expenditure and u^0 is the specified utility level. The compensated demand functions can also be found by differentiating the expenditure function with respect to each of the prices:

$$\partial e / \partial p_i = h_i(P, u^0) \tag{1}$$

Now consider the set of ordinary demand functions derived from the utility maximization problem. In order to determine the functional form and parameters of these demand functions, it is necessary to know the underlying utility function, and this may not be directly observable. Suppose instead that we observe an individual's behavior and estimate the demand functions that describe the individual's responses to changes in prices and income. These functions should contain all of the same information as the underlying preferences. This is assured provided that the demand functions satisfy the so-called integrability conditions. These conditions require that the Slutsky matrix of substitution terms,

$$\frac{\partial x_i(P, M)}{\partial p_j} + \frac{\partial x_i(P, M)}{\partial M} x_j$$

be symmetric and negative semidefinite.[9] If these conditions are satis-

---

[9] See, for example, Varian (1984). The original reference is Hurwicz and Uzawa (1971).

fied, the system of demand functions can be integrated to yield the expenditure function, which in turn can be used to derive the indirect and direct utility functions. If the integrability conditions are not satisfied, the implication is that the observed demand functions are not consistent with the maximization of a well-behaved utility function. As I explain below, if the integrability conditions are satisfied, it may be possible to utilize empirically derived descriptions of demand behavior to obtain a complete description of the underlying preferences, as well as exact measures of welfare change for a wide range of postulated changes in economic circumstances.

WELFARE MEASURES FOR CHANGES IN PRICES

An Overview

In order to introduce the alternative welfare measures, consider first the simplest case of only two goods and the welfare gain associated with a nonmarginal decrease in the price of one of these goods. Five alternative measures of this welfare change have been identified in the literature. The first is the change in ordinary consumer's surplus, a concept whose origin can be traced back through Alfred Marshall to Dupuit.[10] As Marshall explained it,

> [The individual] derives from a purchase a surplus of satisfaction. The excess of the price which he would be willing to pay rather than go without the thing, over that which he actually does pay is the economic measure of this surplus of satisfaction. It may be called consumer's surplus. (1920, p. 124)

Ordinary consumer's surplus is measured by the area under a Marshallian ordinary demand curve but above the horizontal price line. As we will see, the consumer surplus measure cannot be defined in terms of the underlying utility function. And, in general, it will be different from the other four measures defined below.

The other four measures of welfare change are theoretical refinements of the ordinary consumer's surplus (Hicks, 1943). Each can be defined in terms of the underlying individual preference mapping. Figure 3-1 shows two indifference curves for an individual. Assume

---

[10]Mishan (1960) and Currie, Murphy, and Schmitz (1971) provide useful discussions of the history and evolution of the concept of consumer's surplus.

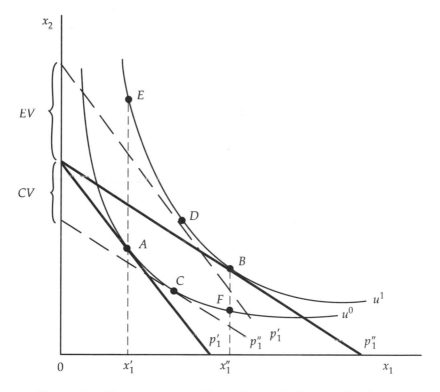

Figure 3-1.   Four measures of the welfare gain from a price decrease

that an environmental improvement reduces the cost of producing x_1 so that its price drops from p_1' to p_1''. In response to the price reduction, the individual shifts from the consumption bundle marked A at utility level u^0 to consumption bundle B at utility level u^1. What is the welfare benefit of the price reduction to this individual? Four alternative welfare measures can be defined in terms of good x_2, which is taken to be the numeraire. The units of x_2 are chosen so that the price of x_2 is equal to one. Thus, x_2 can be taken to represent income.[11]

---

[11]In the many-good case, x_2 is a composite good that can be treated as an index of the consumption levels of all other goods except x_1. The aggregation of all other goods into a composite good for graphical representation is valid so long as the prices of all of the goods are assumed to move in the same proportion, that is, that there are no changes in the relative prices of components of the composite good bundle. This assumption can be maintained, since we are analyzing only the consequences of the change in the price of x_1.

The four remaining measures of welfare changes are:

1. Compensating Variation (CV)—This measure asks what compensating payment (that is, an offsetting change in income) is necessary to make the individual indifferent between the original situation (A in figure 3-1) and the new price set. Given the new price set with consumption point B, the individual's income could be reduced by the amount of CV and that person would still be as well off at point C as at point A with the original price set and money income. The measure CV is often interpreted as the maximum amount that the individual would be willing to pay for the opportunity to consume at the new price set. However, for a price increase, CV measures what must be paid to the individual to make that person indifferent to the price change. For price decreases the CV cannot be greater than the individual's income; but for a price increase, the CV could exceed income.

2. Equivalent Variation (EV)—This measure asks what change in income (given the original prices) would lead to the same utility change as the change in the price of x_1. As shown in figure 3-1, given the original prices, the individual could reach utility level u^1 at point D with an income increase equal to EV. EV is the income change equivalent to the welfare gain due to the price change. The EV measure has also been described as the minimum lump sum payment the individual would have to receive to induce that person to voluntarily forgo the opportunity to purchase at the new price set. For a price increase, EV is the maximum amount the individual would be willing to pay to avoid the change in prices.

Note that both the EV and CV measures allow the individual to adjust the quantities consumed of both goods in response to both changes in relative prices and income levels. The remaining two measures are defined so as to place restrictions on the individual's adjustment of the consumption bundles.

3. Compensating Surplus (CS)—This measure asks what compensating payment will make the individual indifferent as to the original situation and the opportunity to purchase the new quantity x_1'' of the good whose price has changed. The CS measure is the vertical distance between the indifference curves at the new quantity x_1''. This is the distance from B to F in figure 3-1. This measure is closely related to the CV measure, the only difference being the restriction on adjusting the purchases of x_1 in response to the compensating change in income.

4. Equivalent Surplus (ES)—This measure asks what change in income is required, given the old prices and consumption level of x_1, in order to make the individual as well off as that person would be with the new price set and consumption point B. In figure 3-1 the ES

measure is the vertical distance between the two indifference curves, holding the consumption of good x_1 at the original level, that is, the vertical distance from A to E. The ES measure is closely related to the EV measure, the only difference being in the restriction on the adjustment of the consumption of x_1 in the former case. The ES measure is larger than the EV measure for price decreases because it must include an income equivalent to the individual's welfare loss stemming from the inability to adjust the consumption of x_1 so as to equate the marginal rate of substitution with the price ratio.

Each of these concepts measures something different and has a different meaning. The next subsection is devoted to a comparison and evaluation of the compensating and equivalent variations and their relationship to the ordinary consumer surplus. The compensating and equivalent surplus measures for price changes do not answer very useful questions since they both arbitrarily restrict the individual to consuming a specific quantity of the good whose price has changed.

A Closer Look at the Welfare Measures

In this section I first present the basic welfare measure for a marginal change in one price. Then I give more rigorous derivations of the consumer surplus, compensating variation, and equivalent variation measures of welfare change for the case of changes in price.[12] For a marginal change in, say p_1, the basic welfare measure is the change in expenditure necessary to hold utility constant. Using equation (1) from the previous section, we have

$$w_{p_1} = \frac{\partial e\,(P,\,u)}{\partial p_1} = -\,h_1(P,\,u^0) \tag{2}$$

where w_{p_1} is the marginal welfare measure. This result also follows from the indirect utility function and Roy's Identity:

$$\frac{\partial v / \partial p_1}{\partial v / \partial M} = -\,x_1$$

or

$$\frac{dM}{dp_1} = -\,x_1$$

---

[12]For more detailed treatment of these topics, see Silberberg (1972, 1978), Just, Heuth, and Schmitz (1982, Appendix B), Varian (1984), and Johansson (1987).

In the left-hand side of the first expression, the marginal utility of the price change is converted to monetary units by dividing by the marginal utility of income. The second expression says that the change in income required to hold utility constant is equal to the change in price multiplied by the quantity of the good being purchased.

MARSHALLIAN CONSUMER SURPLUS

In figure 3-2, panel A shows one individual's preference mapping in the simple two-good case. Suppose that the price of good x_1 falls from p_1' to p_1''. The individual responds by moving from the original equilibrium at point A to point B on the new budget line. In panel B of figure 3-2, these equilibrium positions are plotted in the price and quantity plane. Points A and B are on the ordinary demand curve holding the price of good x_2 and money income constant. Since the Marshallian surplus associated with the consumption of a good at a given price is the area under the demand curve, the change in surplus for a change in the good's price is the geometric area $p_1'ABp_1''$ in panel B of figure 3-2. In mathematical form,

$$S = \int_{p_1''}^{p_1'} x_1(P, M)\, dp_1 \tag{3}$$

where S is the change in surplus.

The conditions under which S can be interpreted as an indicator of utility change can be seen by employing Roy's Identity:

$$x_1(P, M) = -\frac{\partial v(P, M)/\partial p_1}{\partial v(P, M)/\partial M}$$

and substituting this into equation (3) to obtain

$$S = -\int_{p_1''}^{p_1'} \frac{\partial v/\partial p_1}{\partial v/\partial M}\, dp_1$$

If the marginal utility of income is constant over the range of the price change, this can be written as (see Varian, 1984, pp. 208–209)

$$S = \frac{[v(p_1'', p_2, M) - v(p_1', p_2, M)]}{\partial v/\partial M}$$

This expression shows that the Marshallian surplus can be interpreted as the utility change converted to monetary units by a weight-

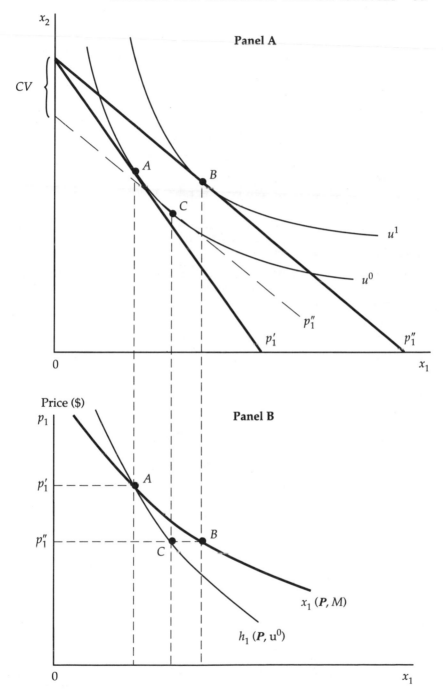

Figure 3-2.   The compensating variation and the Hicks-compensated demand curve

ing factor, the marginal utility of income. If the marginal utility of income is constant, then S can be said to be proportional to the change in utility for any price change. But the constancy of the marginal utility of income as any one price changes is a restrictive condition. The marginal utility of income cannot simultaneously be invariant with respect to income and to changes in all of the prices.[13]

Alternatively, as Silberberg explains (1978, pp. 350–361), the integral of equation (3) can be viewed as the sum of a series of small steps from an initial price and income vector of (p_1', p_2', M') to (p_1'', p_2', M') following a path on which p_2 and M are held constant. But there are other paths over which (3) can be integrated involving changes away from the initial values for p_2 and/or M as long as the terminal point is (p_1'', p_2', M'). And the other paths will not in general lead to the same solution value for the integral. In other words, the integral will not in general be path-independent.

A similar problem arises when the Marshallian surplus measure is generalized to simultaneous changes in all prices. In this case, S is defined as a line integral. This integral will be independent of the path of integration (that is, the order in which prices and/or incomes are assumed to change) only if the income elasticities of demand for all goods are equal. The income elasticities of all goods can be equal to each other only if they are all equal to one, in other words, if preferences are homothetic. Finally, if the prices of only a subset of all goods change, a unique S exists if the marginal utility of income is constant with respect to only those prices that are changed. (See Just, Hueth, and Schmitz [1982] for more details.)

COMPENSATING VARIATION

Suppose now that as the price of good x_1 is decreased, income is taken away from the individual so that he remains at the initial utility level and indifference curve u^0. Given the price change and the compensating income change, the individual would be in equilibrium at point C in panel A of figure 3-2. Point C is also plotted in panel B of figure 3-2. Points A and C are on the Hicks-compensated demand curve, a demand curve that reflects only the substitution effect on the change in relative prices. The income effect of the price change has been eliminated by the device of compensating withdrawals of money income. Since x_1 is a normal good by assumption, that is, it has an income elasticity greater than zero, the Hicks-compensated demand curve is less price-elastic than the ordinary demand curve.

---

[13]For a full treatment of this issue, see Samuelson (1942).

The difference between the Hicks-compensated and the ordinary demand functions is one of the main considerations in the comparison of EV, CV, and consumer surplus measures of welfare change.

Panel A of figure 3-2 shows the compensating variation measure of the welfare change associated with the price decrease, that is, the reduction in income needed to hold the individual on the original indifference curve. In terms of the indirect utility function, CV is the solution to

$$v(P', M) = v(P'', M - CV) = u^0$$

The CV can also be defined in terms of the expenditure function. It is the difference between the expenditures required to sustain utility level u^0, at the two price sets:

$$CV = e(p_1', p_2, u^0) - e(p_1'', p_2, u^0) > 0$$
$$= M - e(p_1'', p_2, u^0) \qquad (4)$$

Since CV is defined as the difference between two levels of expenditure, it can also be written as the integral of the marginal welfare measure (equation [2]) over the relevant range. Specifically,

$$CV = \int_{p_1''}^{p_1'} \frac{\partial e(P, u^0)}{\partial p_1} \, dp_1 = \int_{p_1''}^{p_1'} h_1(P, u^0) \, dp_1$$

Since spending M at the new price set yields a higher level of utility, we can also write

$$M = e(p_1'', p_2, u^1)$$

and by substitution

$$CV = e(p_1'', p_2, u^1) - e(p_1'', p_2, u^0) \qquad (5)$$

In other words, although the CV is *defined* in terms of u^0, it also measures the amount of money required to raise utility from u^0 to u^1 at the new set of prices.

The CV is equal to the area to the left of the Hicks-compensated demand curve between the two prices, that is, the area $p_1'' CA p_1'$. The partial derivative of the expenditure function with respect to p_1 gives the change in expenditure (income) necessary to keep the individual on u^0 for small changes in p_1. As shown above, this derivative gives the Hicks-compensated demand curve, that is, it gives the optimal quantity for x_1, holding utility constant. For finite changes, the integral of this derivative is the area to the left of the Hicks-compensated demand curve, that is, the CV. In other words,

$$CV = \int_{p_1''}^{p_1'} h_1(P, u^0)\, dp_1$$

Unlike the Marshallian measure of surplus given by equation (3), this measure does not rely on any assumption about the constancy of the marginal utility of income. This is because this measure integrates along a constant utility indifference curve at u^0. In the many-good case, when several prices change, the CV of the price changes taken together is the integral of the set of compensated demand functions evaluated by taking each price change successively. The order in which the price changes are evaluated is irrelevant.[14]

EQUIVALENT VARIATION

The equivalent variation can also be derived through the expenditure function. Panel A of figure 3-3 shows the same preference mapping and price change for an individual. With a price decrease, the EV is defined as the additional expenditure (income) necessary to reach utility level u^1, given the initial set of prices. In terms of the indirect utility function, EV is the solution to

$$v(P', M + EV) = v(P'', M) = u^1$$

In figure 3-3, the EV is the additional expenditure necessary to sustain point C' over point A at the initial prices, or

$$EV = e(p_1', p_2, u^1) - e(p_1', p_2, u^0) > 0$$
$$= e(p_1', p_2, u^1) - M \tag{6}$$

Since the money expenditure levels are the same at point A and point B, that is, $e(p_1', p_2, u^0) = e(p_1'', p_2, u^1)$, this can also be written as

$$EV = e(p_1', p_2, u^1) - E(p_1'', p_2, u^1) \tag{7}$$

---

[14]This follows from the symmetry of the cross price substitution terms, that is, $\partial x_i/\partial p_j = \partial x_j/\partial p_i$ along the indifference curve.

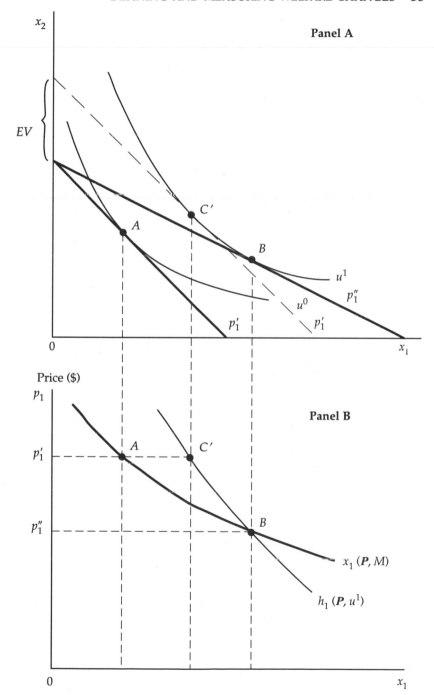

Figure 3-3.   The equivalent variation and the Hicks-compensated demand curve

In other words, although the *EV* is defined in terms of the monetary equivalent of a change from u^0 to u^1, it can also be measured by the change in expenditure associated with price changes given utility level u^1.

The *EV* can also be written as the integral of the marginal value measure (equation [2]):

$$EV = \int_{p_1''}^{p_1'} \frac{\partial e(P, u^1)}{\partial p_1} \, dp_1$$

The price derivative of the expenditure function (this time holding utility constant at u^1) generates another Hicks-compensated demand curve through point *B* in panel B of figure 3-3. The area to the left of this Hicks-compensated demand curve between the two prices is the equivalent variation welfare measure. In other words,

$$EV = \int_{p_1''}^{p_1'} h_1(P, u^1) \, dp_1$$

As in the case of the *CV*, this measure does not require any assumption about the constancy of the marginal utility of income. And the measure for multiple price changes is path-independent.

All of this discussion has been in terms of the welfare gain due to a price decrease. The derivation of the welfare cost of a price increase can be worked out in a symmetrical fashion. In general, for any price change, the *CV* welfare measure is the area to the left of the Hicks-compensated demand curve that passes through the initial position. The *EV* measure of the welfare change is the area to the left of the Hicks-compensated demand curve that passes through the final position.

A COMPARISON OF THE THREE MEASURES

Although the Marshallian consumer surplus has some intuitive appeal as a welfare indicator, it does not measure any of the theoretical definitions of welfare change developed here. It is neither an index of utility change, except under special conditions, nor a measure of gain or loss that can be employed in a potential compensation test. The Marshallian surplus does lie between the *CV* and the *EV*. This opens the question of whether it can be a useful approximation to either of these other measures, a question that I take up below in the subsection ''Consumer's Surplus Without Apology.''

In contrast, the *CV* and the *EV* do represent welfare relevant measures. The EV is the monetary equivalent of a price change. It can be interpreted as an index of utility in the sense that it imputes the same monetary value to all changes from an initial position that result in the same final utility level.[15] This is because it evaluates all changes from an initial position at the same set of prices. The *CV* cannot be interpreted as an index of utility. Rather, it measures the offsetting income change necessary to "prevent" a utility change. As Silberberg put it, "the [*EV*] imputes a dollar evaluation to a change in utility levels for a particular path of price changes, while the [*CV*] derives dollar values necessary to hold utility constant when prices change" (1972, p. 948).

The two measures *EV* and *CV* will be the same if the income elasticity of demand for good x_1 is zero. In this case, the ordinary and Hicks-compensated demand curves are identical. With a positive income elasticity, the *EV* exceeds the *CV* for price decreases, but the *CV* exceeds the *EV* when price increases are considered. The difference between points *C* and *B* in figure 3-2 and between points *A* and *C'* in figure 3-3 is one of income level. If the income elasticity of demand for x_1 were zero, the income differences would have no effect on the purchase of x_1. The *CV* and the *EV* would be exactly equal; and they could both be measured by the area under the ordinary demand curve. The higher the income elasticity of demand for x_1, the larger is the difference between the *EV* and the *CV*, and the larger is the difference between either measure and the ordinary consumer surplus.

There is a symmetry between the *CV* and the *EV* measures. This can be seen by comparing figures 3-2 and 3-3 and by comparing equation (5) with equation (6) and equation (4) with equation (7). For simplicity, let I represent the initial price set (with p_1') and let II represent the second price set (with p_1''). The *CV* for moving from I to II with u^0 as the reference utility level is exactly equal to the *EV* of moving from II to I with u^1 as the reference utility level. The *CV* is a welfare measure for the move from *A* to *B* via point *C*; the *EV* starts at point *B* and measures the reduction in income necessary to get to point *C* and therefore u^0. Similarly, the *EV* for the move from I to II is just equal to the *CV* starting at II and u^1 and moving to I.

---

[15]This is an ordinal utility index. For example, suppose a change from initial position *A* to position *B* has an *EV* of $10, while a change from *A* to *C* has an *EV* of $20. It cannot be inferred that the second change conveys twice as much extra utility as the first change. For a discussion of cardinal versus ordinal indexes of welfare change, see Morey (1984).

This symmetry relates to the interpretation of *CV* and *EV* as measures of willingness to pay (WTP) and willingness to accept compensation (WAC). The *CV* is sometimes described as the maximum willingness to pay for the right to purchase the good at the new price level, that is, that lump sum payment that the individual would be willing to make which would just exhaust the potential for welfare gain from the new price. This description is accurate only for a price decrease. For a price increase, the *CV* defines the minimum payment to the individual sufficient to prevent a utility decrease—in other words, it defines a WAC measure. Similarly, the *EV* defines a WAC measure for a price decrease, that is, the sum of money the individual would require to voluntarily forgo a proposed price decrease. But for a proposed price increase, the *EV* is a WTP measure, that is, the maximum sum of money that could be taken away from the individual, yielding a loss of utility equivalent to that caused by the price change. Whatever the direction of the price change, the *CV* takes the initial utility as the reference point.

These two measures can also be interpreted in terms of the implied rights and obligations associated with alternative price sets. The *CV* carries with it implicitly the presumption that the individual has no right to make purchases at a new set of lower prices, but does have a right to the original price set in the case of price increases. In contrast, the *EV* contains the presumption that the individual has a right to (an obligation to accept) the new lower (higher) price set and must be compensated (make a payment) if the new price set is not to be attained. Based on this interpretation of the two measures, some economists have argued that the choice between them is basically an ethical one, that is, one that depends on a value judgment as to which underlying distribution of property rights is more equitable.[16] All of this can be summarized as follows: .

Welfare measure	*Price increase*	*Price decrease*
EV—Implied property right in the change	WTP to avoid	WAC to forgo
CV—Implied property right in the status quo	WAC to accept	WTP to obtain

For two alternative price changes, the welfare measures should be the same if both changes place the individual on the same higher indifference curve. But if the two price changes place the individual on different indifference curves, the welfare measure should correctly

---

[16]See Krutilla (1967) and Mishan (1976).

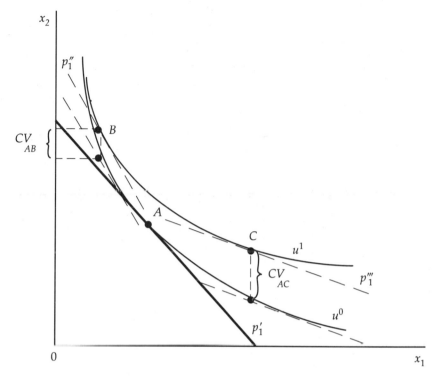

Figure 3-4. The compensating variation incorrectly ranks two alternative policies

indicate the preference ranking of the two alternatives. The *EV* measure always provides a consistent ranking in this sense, but the *CV* measure does not.

Figure 3-4 illustrates why this is the case. It shows an individual in equilibrium at point *A*, given prices and money income. Suppose that one policy proposal would increase the price of x_1 and decrease the price of x_2 simultaneously. The individual would achieve a new equilibrium at point *B*. The *CV* measure of the welfare change is shown as CV_{AB}. The second policy alternative would decrease the price of x_1 while increasing the price of x_2. This would lead to a new consumer's equilibrium at point *C*. Point *C* has been drawn on the same indifference curve as point *B*. Therefore, the measure of welfare change should be the same for the two policy alternatives. But as can be seen by inspection, the *CV* for the second policy, CV_{AC}, is larger. The *CV* measure would indicate a preference for the second policy while the individual is in fact indifferent between the two policies.[17] The *EV* gives the same welfare measure for the two policy alternatives. This is because the *EV* measure bases its comparison on a point on the indifference curve passing through the new equilibrium, but with the old prices. If two policies are on the same new indifference curve, the *EV* measure picks the same point for measuring the welfare effects for both policies.

If the question being asked by policymakers is, Does the proposed change pass the Kaldor potential compensation test? then *CV* is the measure to use. The Kaldor potential compensation test is one form of potential Pareto improvement test that asks whether it is possible for the winners to fully compensate all of the losers from the proposed policy change and still leave someone better off. For each person, the *CV* gives the compensating income change required to maintain that person at his or her initial utility level. If the sum of what could be collected from all gainers exceeds the sum of the required compensations for losers, the proposal passes this form of the potential Pareto improvement test. The fact that the *CV* cannot rank consistently two or more policy changes is no obstacle to its use in this manner. This is because the potential Pareto criterion itself provides no basis for ranking two or more proposed policy changes. If two proposed changes both pass the Kaldor potential compensation test, the potential Pareto improvement criterion provides no basis for choosing between them.

On the other hand, if the question being asked by policymakers is, Does the policy pass the Hicks version of the potential compensation test? then *EV* is the appropriate measure. The Hicksian test asks whether it is possible for the losers to bribe the gainers to obtain their consent to forgo the proposed policy change. The potential gainers would accept a bribe only if it were large enough to raise their utility by the same amount as the proposed project would have. The offered bribe would have to be as large as each individual's *EV* measure of welfare gain. And the maximum bribe that would be offered by the potential losers would be their *EV* measure of loss. Thus if the sum of the *EV* of all gainers exceeded the sum of the *EVs* of all losers, the proposal would pass the Hicks form of the potential compensation test. Also, since the Hicks form of the compensation test is based on the *EV* measure, it will consistently rank two or more policy changes, provided that society is indifferent as to the distribution of gains and losses across individuals.

---

[17]See Hause (1975). Chipman and Moore (1980) discuss the conditions under which *CV* will provide a consistent ranking of alternatives.

Measurement

Simply put, the problem posed for applied welfare economics is that the desired welfare measures, the *CV* or the *EV,* are based on the unobservable Hicks-compensated demand functions, while the one measure based on the observed Marshallian demand functions is flawed as a welfare indicator. The typical practice had been to use the Marshallian surplus anyway, and to offer such justifications as "income effects are likely to be small"; "with only one price change, path dependence is not an issue"; and "it is the only measure we have and it is better than nothing." Then Robert Willig (1976), in a widely cited article, provided a justification for using the Marshallian surplus by examining the magnitude of the differences between *S* and *CV* or *EV* under different conditions. He argued that "in most applications the error of approximation will be very small. In fact the error will often be overshadowed by the errors involved in estimating the demand curve" (1976, p. 589). More recently several authors have developed methods for direct calculation of the *CV* and *EV* from information contained in the ordinary demand function, either through a Taylor's series approximation (McKenzie and Pearce, 1982, and McKenzie, 1983) or, as exact measures, through integration to obtain the indirect utility function and the expenditure function (see, for example, Hausman, 1981). In this subsection I review these recent developments in applied welfare analysis and assess their significance for valuing environmental services.

"CONSUMER'S SURPLUS WITHOUT APOLOGY"

Willig (1976) has offered rigorous derivations of expressions relating *CV*, *S*, and *EV*. These expressions provide a way of calculating the magnitude of the differences among the three measures for given prices, quantities, and income. The differences among the three measures depend on the income elasticity of demand for the good in question and consumer surplus as a percentage of income. The differences among the measures appear to be small and almost trivial for most realistic cases. The differences are probably smaller than the errors in the estimation of the parameters of demand functions by econometric methods.

Willig's bounds for the approximation errors are based on the fact that the differences between *S* and *CV* or *EV* arise from an income effect on the quantity demanded; and the size of that effect depends on the change in real income brought about by the price change and on the income elasticity of demand for the good. This can be shown in

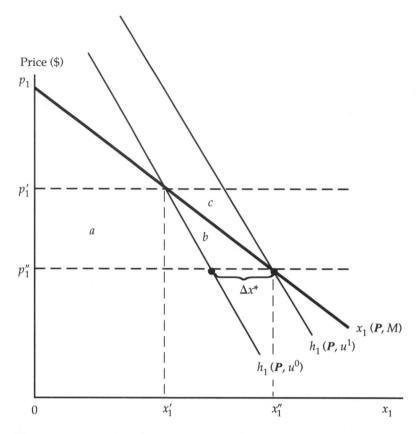

Figure 3-5. Deriving the Willig bounds for S as an approximation to CV

a nonrigorous way for the case of one price change with the help of figure 3-5.[18] In this figure the ordinary and compensated demand curves are assumed to be linear. Let S represent the area $a + b$, so:

$$CV = a$$
$$= S - b$$

and
$$EV = a + b + c$$
$$= S + c$$

___

[18]Although this exposition applies to the case of only one price change, the Willig expressions can be generalized to accommodate multiple price changes (Willig, 1979), provided that a specific path of integration is chosen.

The errors in using S to approximate CV and EV are equal to the areas b and c respectively. For a price change from p' to p'', the factors influencing the size of the approximation error can be seen by examining the determinants of the area b:

$$CV - S = -b$$
$$= 1/2\, \Delta p \cdot \Delta x^* \qquad (8)$$

where Δx^* is the income effect on the quantity demanded of x associated with reducing income sufficiently to hold utility at u^0. Let ΔM^* represent this income change. By definition, ΔM^* is CV. The definition of income elasticity of demand is

$$E_M = \frac{\Delta x}{\Delta M} \cdot \frac{M}{x}$$

Rearranging this expression gives

$$\Delta x^* = E_M \cdot x \cdot \frac{\Delta M^*}{M}$$

Substituting this into equation (8), we obtain

$$CV - S = -\frac{\Delta p \cdot x \cdot E_M \cdot CV}{2M}$$

In general, for small changes in p, $\Delta p \cdot x \simeq -S$. This is strictly true for the linear demand curve when x is evaluated at the mid point between x' and x''. Finally, dividing both sides by CV to express the error in percentage terms gives

$$\frac{CV - S}{CV} \simeq -\frac{E_M}{2} \cdot \frac{S}{M}$$

This is similar to the Willig expression for the approximation error.[19] It says that the error is proportional to the income elasticity of demand and consumer surplus as a percentage of income. A similar line of reasoning can be used to derive the relationship between EV and S.

Willig's analysis is more rigorous than this in that it takes into account the possibility that for finite changes in price and quantity,

---

[19]The principal difference is that it expresses the error as a percentage of CV, while Willig's term makes the error a percentage of S.

the income elasticity of demand may vary over the range of the price change. He derives rules of thumb for calculating the maximum error in using S as an approximation for EV or CV. The rules of thumb are applicable if the following conditions are met:

$$\frac{S}{M} \cdot \frac{\underline{E}_M}{2} \leq .05$$

$$\frac{S}{M} \cdot \frac{\overline{E}_M}{2} \leq .05 \tag{9}$$

and
$$\frac{S}{M} \leq .9 \tag{10}$$

where $\underline{E}_M$ = the smallest value of income elasticity of demand for the good in the region under consideration

$\overline{E}_M$ = the largest value of income elasticity of demand for the good in the region under consideration.

Given these conditions, the rule of thumb for CV is

$$\frac{S}{M} \cdot \frac{\underline{E}_M}{2} \leq \left|\frac{CV - S}{S}\right| \leq \frac{S}{M} \cdot \frac{\overline{E}_M}{2} \tag{11}$$

and the rule of thumb for EV is

$$\frac{S}{M} \cdot \frac{\underline{E}_M}{2} \leq \left|\frac{S - EV}{S}\right| \leq \frac{S}{M} \cdot \frac{\overline{E}_M}{2} \tag{12}$$

The first thing to note is the conditions under which these rules of thumb are valid. Consider equation (10) first. The change in consumer surplus as a percentage of income depends on the size of the price change, the price elasticity of demand, and expenditure on this good as a percentage of total income. The smaller the price change and the smaller the proportion of income spent on the good, the smaller is S/M. It can readily be shown that:[20]

$$\frac{S}{M} \leq \left|\frac{\Delta p}{p}\right| \frac{p \cdot x}{M} \tag{13}$$

---

[20]From a given initial situation, S is largest when the demand curve is perfectly inelastic. Then $S = |x \cdot \Delta p|$. Condition (12) follows readily.

This shows that, for example, for a good absorbing 50 percent of total income and for a 100 percent price change, S/M cannot exceed 0.5, while for a 10 percent price change for a good absorbing 10 percent of income, S/M will be less than 0.1. Thus condition (10) is likely to be satisfied except for very large price increases for goods with low price elasticities that also absorb a large proportion of the total budget.

As for the first condition, the smaller that consumer surplus is as a percentage of income, and the smaller that the income elasticity of demand is, the more likely (9) is to be satisfied. For example, if consumer surplus is 5 percent of income, the income elasticity of demand can be as high as 2.0 and still satisfy (9). If S/M just barely satisfies condition (10), the income elasticity cannot exceed .11 to satisfy (8).

Assuming that conditions (9) and (10) hold, let us turn to the rules of thumb. First, according to (9), the maximum error involved in using S as an approximation for either CV or EV is 5 percent. Second, the smaller the change in income elasticity over the range being considered, the more precise are (11) and (12) as statements of the error involved in using S rather than CV or EV. If the income elasticity of demand does not change over the range being considered ($\underline{E}_M = \bar{E}_M$), the left-hand and right-hand sides of (11) and (12) are equal to each other and the errors are zero, as discussed above. Finally, as the income elasticity of demand for the good decreases, the differences among ordinary consumer surplus, CV, and EV decrease, disappearing as E_M goes to zero.

Willig's analysis has been interpreted as providing a justification for using consumer surplus as an approximation of the CV or the EV. But there are two reasons why one should be cautious about adopting the Willig approach to welfare measurement. The first has to do with limitations on the applicability of the Willig conditions to some kinds of problems of welfare measurement, including some of specific interest to environmental and resource economists. The second arises because of the recent development of new methods for obtaining exact measures of CV and EV from the same information that is required to use the Willig approximation.

The Willig conditions for valid approximation were developed for changes in S as a result of changes in the price of some market good. Many environmental and resource policy issues require information on the total value of some environmental service as a measure of what would be lost if the resource were destroyed or diverted to some other use. For example, the economic cost of damming a river that provides whitewater canoeing and trout fishing would be measured by the total areas under the Hicks-compensated demand curves for these activities. This is equivalent to measuring the change in consumer surplus for a price increase from the present price to the vertical intercept of the Hicks-compensated demand curve. Bockstael and McConnell (1980) point out that for the linear demand function, the income elasticity of demand goes to infinity as the price approaches the vertical intercept. Thus the approximation error cannot be calculated.[21]

For some questions, the variable of interest to policymakers is not CV but some fraction of CV—for example, the dead weight loss associated with a tax on a commodity. Suppose an excise tax raises the price of a good from p_1' to p_1'' , as shown in figure 3-6. The consumer's loss as measured by CV is the area $a + b + c$, but only $b + c$ is an efficiency loss, since a is a revenue transfer to the government. If the ordinary demand curve is used to approximate the consumer loss, the area c is the error. If the Willig conditions are satisfied, c is an acceptably small percentage of S and CV; but it can be an unacceptably large percentage of the true dead weight loss.[22]

The second reason for being cautious about using the Willig approximation is that better methods of welfare measurement now exist. If the demand functions being used to calculate S reflect utility maximizing behavior on the part of individuals, they should satisfy the integrability conditions. And if this is the case, as I explain in the next section, it is possible to calculate CV and EV directly without approximation. On the other hand, if the demand functions do not satisfy the integrability conditions, then it is inappropriate to use the Willig approximations, since their derivation was also based on the assumption of utility-maximizing behavior.

EXACT WELFARE MEASUREMENT

It is a basic proposition in the theory of demand that if the ordinary demand functions reflect utility-maximizing behavior, they contain all of the information on preferences that is necessary to reconstruct an ordinal index of utility. Recognition of this dual relationship between

---

[21]In a comment on Bockstael and McConnell (1980), Hanemann (1980) showed that if the parameters of the Marshallian demand function were known, it was unnecessary to compute the Willig approximation error, since the CV could be calculated directly. In this, Hanemann apparently anticipated the analysis of Hausman (1981), to be discussed below.

[22]See Hausman (1981).

Price ($)

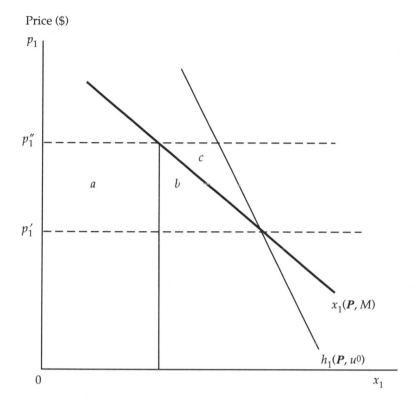

Figure 3-6. The Willig approximation and the error in estimating dead weight losses

preferences and observable demand has resulted in a search for practical procedures for recovering the utility function and *CV* and *EV* welfare measures directly from estimated demand functions. One path that has been followed is reflected in the work of Reaume (1973), McKenzie and Pearce (1976, 1982), and McKenzie (1976, 1983). This method involves Taylor's series expansions of the indirect utility function based on terms that are derivatives of the ordinary demand functions. This approach has been criticized as being cumbersome (Vartia, 1983, for example), potentially inexact because successive terms of a convergent Taylor's series are not necessarily monotonically decreasing (Willig, 1979, p. 463, for example), and unnecessary.

More recently, several authors have presented operational procedures for recovering *CV* and *EV* exactly or as arbitrarily close approximations.[23] For example, Hausman's (1981) procedure, which was developed for the case of only one price change, involves four steps. The first involves combining the ordinary demand function and Roy's Identity to obtain a partial differential equation:

$$x_1(P, M) = - \frac{\partial v(P, M)/\partial p_1}{\partial v(P, M)/\partial M}$$

If the utility function is separable so that the demand function contains only its own price argument, and if the demand function is linear, this becomes:

$$a - b \cdot p_1 + c \cdot M = - \frac{\partial v(P, M)/\partial p_1}{\partial v(P, M)/\partial M} \tag{14}$$

where the parameters a, b, and c are estimated econometrically and where p_1 and M are deflated by an appropriate index of the other prices. Changes in p_1 and M that involve moving along an indifference curve must satisfy

$$\frac{\partial v(\cdot)}{\partial p_1(t)} \cdot \frac{dp_1(t)}{dt} + \frac{\partial v(\cdot)}{\partial M(t)} \cdot \frac{dM(t)}{dt} = 0$$

where t defines a path of price changes. Rearranging this expression, substituting into (14), and using the implicit function theorem gives

$$\frac{dM(p_1)}{dp_1} = a - b \cdot p_1 + c \cdot M$$

the solution of which is

$$M\left(p_1\right) = k \cdot e^{c \cdot p_1} - \frac{1}{c}\left(a - b \cdot p_1 - \frac{b}{c}\right)$$

where k is the constant of integration, which depends on the initial level of utility. If units are arbitrarily chosen so that k is the initial utility level, the quasi-indirect utility function and quasi-expenditure

---

[23]See Hausman (1981) and Bergland and Randall (1984) for examples of exact measurement. Vartia (1983) presents an algorithm for approximating in welfare measures to an arbitrarily small error.

function follow directly:

$$u = k^0 \cdot e^{-c \cdot p_1} \left[\left(M + \frac{1}{c} \right) \left(a - b \cdot p_1 - \frac{b}{c} \right) \right]$$

and,

$$e = k^0 \cdot e^{c \cdot p_1} - \frac{1}{c} \left(a - b \cdot p_1 - \frac{b}{c} \right)$$

These expressions are termed "quasi-" functions because they do not contain information about the effects of the prices of other goods on utility or expenditure.[24] As can be seen, this method depends on the ability to solve the differential equation that is obtained from Roy's Identity. Hausman (1981) has shown a method of solution for the case when only one price changes, and has discussed in general terms the solution in the case of multiple price changes.

Conclusions

Selection of a welfare measure has long involved questions both of appropriateness and practicality. The Marshallian surplus measure was frequently chosen on the grounds of practicality, even though it was recognized that the measure was inappropriate in that it did not answer any specific well-formed welfare question. Willig's development of the bounds for the errors of approximation in using S gave encouragement to this practice. But quickly on its heels have come new approaches to exact welfare measurement that offer the opportunity to calculate the more appropriate CV and EV measures directly.

One question related to practicality remains, however. Do we know enough about the functional form of the utility function to implement the exact measurement methods? Assuming a functional form for the system of demand functions for purposes of estimation is equivalent to assuming the functional form of the underlying utility function. One approach is to assume a specific functional form for the utility function or indirect utility function and to derive the demand functions for estimation. If this is the approach taken, then plugging the estimated parameters back into the utility function to calculate welfare changes is straightforward, provided that the parameter estimates of the demand function satisfy the integrability conditions. Since researchers have been reluctant to specify the functional form of the utility function, one alternative has been to specify

---

[24]For further analysis of exact welfare measurement when demand information is available for only some of the goods being purchased, see LaFrance and Hanemann (1989) and Hanemann and Morey (1992).

so-called flexible forms for the indirect or direct utility function.[25] Again, if the integrability conditions are satisfied, deriving "exact" welfare measures from the "approximate" flexible functional form of the utility function is straightforward.[26] The alternative is to seek guidance from the data by selecting the functional form for the demand functions on the basis of goodness of fit and consistency with the restrictions imposed by theory.

WELFARE MEASURES FOR CHANGES IN FACTOR PRICES

If the quantity of the factor that the individual supplies to the market is unaffected by changes in its price, the price change affects only income. In this case, the income change is an exact measure of the welfare effect of the price change. But if the change in factor price affects the quantity supplied—for example, by altering the choice between labor and leisure—then the welfare measure must reflect both changes in income and changes in the utility derived from the quantity of the factor withheld from the market. CV and EV measures of welfare change can be defined in a manner consistent with the preceding analysis of product price changes. The relationship between these measures and the area above the individual's factor supply curve is similar to the relationships among CV, EV, and S for product price changes.[27]

Consider the supply of labor where the individual allocates time between income-producing work and utility-yielding leisure. The individual's preferences are described by $u(X, f)$ where f is time spent in leisure (fun). The individual maximizes utility subject to the budget and time constraints:

$$M^* + p_w \cdot w - P \cdot X = 0$$

$$t^* - w - f = 0$$

or
$$M^* + p_w (t^* - f) - P \cdot X = 0$$

where M^* is exogenous income, p_w and w are the wage rate and hours worked, and t^* is the total endowment of time. The individual's labor supply function, $w = w(p_w, P, M^*)$ is part of the solution to this maxim-

---

[25]See, for example, Deaton and Muellbauer (1980).

[26]But see, for example, Deaton and Muellbauer (1980, chapter 3) for a discussion of some of the evidence for satisfying the restrictions of theory in empirical estimation of demand systems.

[27]See Mishan (1959) for a discussion of the CV and EV measures, consumer surplus, and economic rent.

ization problem. As is well known, the labor supply curve can be back-ward bending, at least over some range, if the positive income effect on the demand for leisure outweighs the negative substitution effect of more expensive leisure. See figure 3-7.

Following the usual procedure, the indirect utility function, $v(P, p_w, M^*)$, and the pseudo-expenditure function, $e^*(P, p_w, u^0)$, can be derived (Just, Hueth, and Schmitz, 1982). The pseudo-expenditure function gives the amount of exogenous income (not expenditure) required to sustain any utility level as a function of prices. The compensated labor supply curve shows the quantities supplied at various prices, assuming that exogenous income has been adjusted to maintain the individual at that person's initial utility level, say u^0. It is derived by differentiating the pseudo-expenditure function with respect to p_w. Because the compensated supply curve represents only the substitution effect between work and leisure, it is always upward sloping and more elastic than the ordinary supply curve.

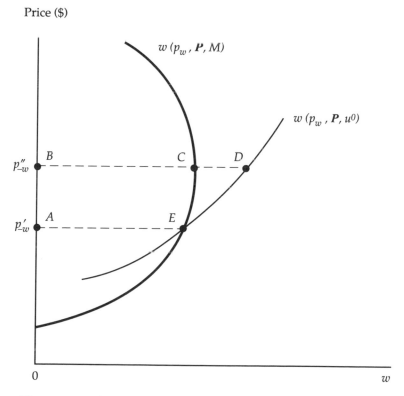

Figure 3-7.   The compensating variation for a factor price increase

If p_w increases from, say, p'_w to p''_w, the CV measure of welfare gain is the integral of the derivative of the pseudo-expenditure function with respect to p_w. Geometrically this is the area to the left of the compensated supply curve between the two price levels, that is, $ABDE$ in figure 3-7. This is greater than the area between the prices bounded by the ordinary supply curve. There is also another compensated supply curve defined for the new utility level after the price change. It goes through point C. The EV measure of welfare change is the area to the left of this curve (not shown in figure 3-7) between the price levels. Unless the income elasticity of demand for leisure is zero, CV is greater than EV for wage increases, while EV is greater than CV for wage decreases. Given an estimate of the ordinary labor supply function, exact calculation of EV and CV measures is possible (Hausman, 1981).[28]

WELFARE MEASURES FOR QUANTITY CHANGES

Many environmental policy proposals involve changes in either the quantities or the qualities of nonmarket environmental goods and services rather than changes in the price of a marketable good. From the individual's point of view, the most important characteristic of some environmental goods is that they are available only in fixed, unalterable quantities. These quantities act as constraints on each individual's choice of a consumption bundle. The analysis of this class of problems is often referred to as the theory of choice and welfare under quantity constraints (Johansson, 1987) or under imposed quantities (Lankford, 1988). The imposition of quantity constraints raises some new issues in the theory of choice and welfare measurement. The analysis of these problems has evolved out of the theory of rationing.[29]

In this section, I briefly describe the model of individual preferences and choice under imposed quantity constraints. I then derive measures of welfare change for changes in the quantities of imposed goods. These measures are essentially similar to the compensating and equivalent surplus measures for price changes discussed earlier. Finally, I turn to the question of measurement, describing what is

---

[28]See Haveman, Gabay, and Andreoni (1987) for a comparison of the exact CV and Marshallian approximation of the welfare cost of a tax on labor income.

[29]See, for example, Tobin and Houthakker (1950/1) and Neary and Roberts (1980).

necessary to obtain exact measures by solving for the expenditure function or indirect utility function, and the potential errors involved in approximating these exact measures by areas under uncompensated inverse demand (bid) curves.

The Basic Model

Consider an individual whose utility function has the following form:

$$u = u(X, Q)$$

where X is the vector of private goods quantities ($X = x_1, ..., x_i, ..., x_n$) and Q is a vector of environmental and resource service flows ($Q = q_1, ..., q_2, ..., q_m$) that is given to the individual. It is possible that for at least some of the elements in Q there is a positive price charged. Let $R \geq 0$ be the vector of these prices, and P be the vector of prices for X. The individual maximizes utility subject to a budget constraint

$$P \cdot X + R \cdot Q = M$$

where M is money income. This yields a set of conditional demand functions for the marketed goods:

$$x_i = x_i(P, M - R \cdot Q, Q)$$

In general, Q will be an argument in these conditional demand functions, along with the prices of both market and nonmarket goods. The term "conditional" refers to the fact that these functions are conditioned upon the imposed Q.

Inserting the conditional demand functions into the utility function gives the conditional indirect utility function

$$v = v(P, M - R \cdot Q, Q)$$

Inverting the conditional indirect utility function for the term $M - R \cdot Q$ yields a conditional expenditure function that gives the minimum expenditure on market goods required to produce utility level u, given P and Q. This is

$$e^* = M - R \cdot Q - e^*(\Gamma, Q, u)$$

The dual to the utility maximization problem can be stated as follows: minimize expenditure subject to the constraint that utility equal or exceed some stated level, say u^0. The solution to this problem gives the restricted expenditure function

$$e = e(P, R, Q, u^0)$$

which is the total expenditure on all goods necessary to achieve u^0 given P, R, and Q. The conditional and restricted expenditure functions are related by the following expression:

$$e = e^* + R \cdot Q \tag{15}$$

Welfare measures can be defined in terms of either the conditional or restricted expenditure function. These two forms of expenditure function coincide in the special case where $R = 0$.

Welfare Measures

Welfare measures can be defined for either changes in Q or changes in R, or both. Changes in R have the same effect as lump sum changes in money income equal to $-\Delta R \cdot Q$, so they pose no special problem for welfare measurement. For simplicity, in what follows I will consider welfare measures only for changes in Q; and I will assume that Q consists of only one element, q, with a price of r.[30] I will first present welfare measures for the general case where $r \geq 0$. I will also show how these measures can be simplified when $r = 0$. So that graphic presentations of some of the key points can be made, I will assume that X is the numeraire represented as x. Finally, I will assume that at the given prices and income, the individual would choose more of q if given the option. The case where more q is a bad is taken up below.

To begin with, the marginal value of a small increase in q is the reduction in income that is just sufficient to maintain utility at its original level. If w_q is the marginal value of a change in q, it is given by the derivative of the restricted expenditure function with respect to q or

$$w_q = -\frac{\partial e}{\partial q} \tag{16}$$

The right-hand side of this expression is also equal (in absolute value) to the slope of the indifference curve through the point at which

---

[30]Lankford (1988) has derived measures for the general case with simultaneous changes in P, Q, R, and M.

the welfare change is being evaluated. Alternatively, from equation (15):

$$w_q = -\frac{\partial e^*}{\partial q} - r$$

The second term, $-r$, reflects the fact that the additional q imposes an obligation on the individual to increase the lump sum payment given by $r \cdot q$ whenever r is positive.[31]

The quantity-constrained nature of the problem makes it impossible for an individual to adjust the quantity of q to satisfy the conventional optimizing conditions of equality of marginal rates of substitution and price ratios. Therefore, compensating surplus (CS) and equivalent surplus (ES) are the relevant measures of welfare change. And the choice between them depends on the same considerations applying to the choice between CV and EV measures for price changes. There are several ways to present CS and ES for changes in quantity-constrained goods.

In the first way, from the conditional indirect utility function the CS and ES measures are defined implicitly as the solutions to the following expressions:

CS is the solution to

$$v(P, M - r \cdot q^0, q^0) = v(P, M - r \cdot q^1 - CS, q^1)$$

and ES is the solution to

$$v(P, M - r \cdot q^0 + ES, q^0) = v(P, M - r \cdot q^1, q^1)$$

These two measures can also be defined in terms of the expenditure function. Consider first the CS. Using the restricted expenditure function, CS is

$$CS = e(P, r, q^0, u^0) - e(P, r, q^1, u^0)$$
$$= M - e(P, r, q^1, u^0)$$

This case is shown in figure 3-8. The initial position is at point A, where the individual consumes q^0 and x^0 and achieves u^0 in utility.

---

[31]Lankford (1988) calls this the "income value" of the change in q. An increase in q has a negative income value.

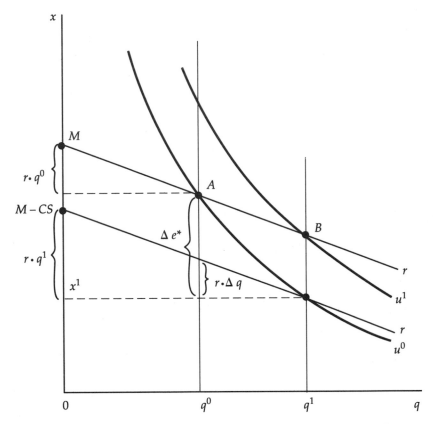

Figure 3-8.   The compensating surplus measure for an increase in q

The increase in q enables the individual to reach u^1 at point B. If income is reduced by CS, the individual is pushed back to u^0 at point C.

The conditional expenditure function can be used to show that this welfare measure consists of two components. Recalling that the conditional expenditure function gives the expenditure on privately chosen market goods necessary to achieve a specified utility level given the level of q, we have:

$$e^*(P, q^0, u^0) = e(P, r, q^0, u^0) - r \cdot q^0$$

Thus: $$CS = e^*(P, q^0, u^0) + r \cdot q^0$$

$$- e(P, q^1, u^0) - r \cdot q^1$$

$$= e^*(P, q^0, u^0) - e(P, q^1, u^0) - r(q^1 - q^0) \qquad (17)$$

The last term is the "income value" term reflecting the extra expenditure on q given r. As shown in figure 3-8, the reduction in the expenditures on x that is given by the first two terms in equation (17) overstates the welfare gain from the increase in q when r is positive. The true welfare gain is found by deducting from Δe^* an amount equal to the unavoidable increase in expenditure on q, $r(q^1 - q^0)$.

The ES measure is given by

$$ES = e(P, r, q^0, u^0) - e(P, r, q^0, u^0)$$
$$= e(P, r, q^0, u^1) - M$$

This is shown in figure 3-9. As before, the increase in q enables the individual to reach point B with utility equal to u^1. Alternatively, if income were increased by ES while holding q constant, the individual could achieve u^1 at point C.

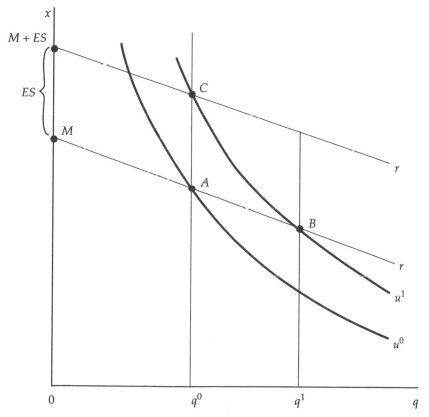

Figure 3-9. The equivalent surplus measure for an increase in q

The second way to present *CS* and *ES* measures is based on the conditional expenditure function

$$ES = e^*(P, q^0, u^1) + r \cdot q^0$$
$$- e^*(P, q^0, u^0) - r \cdot q^0$$

Since the welfare measure is defined in terms of income changes holding q constant at q^0, there is no need to take account of the income value of a change in q.

If q is a pure public good with $r = 0$, then there is no need to take into account the income value term in either of the welfare measures. The budget lines are horizontal, as shown in figure 3-10. The full and conditional expenditure functions are the same. And the *CS* and *ES* measures are as shown in the diagram.

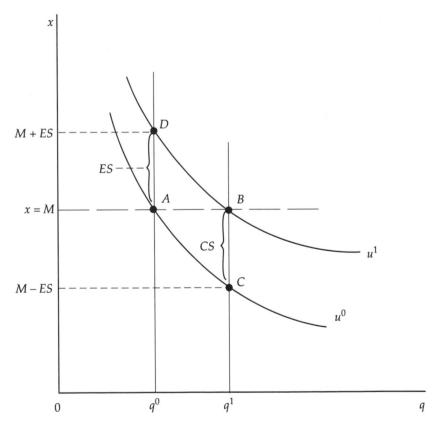

Figure 3-10.   Compensating and equivalent surpluses for a change in q when $r = 0$

The third way to derive the *ES* and *CS* measures is to make use of equation (16). The value of a nonmarginal change in *q* is the integral of (16) taken over the relevant range, or

$$W_q = - \int_{q^0}^{q^1} \partial e(P, r, q, u^t)/\partial q \cdot dq$$

This is either a *CS* or an *ES* measure, depending on whether *t* = 0 or 1.

Another way of looking at the welfare measurement problem is to ask what income levels and prices for *q* would be necessary to support each quantity-constrained bundle as a freely chosen utility-maximizing bundle. For example, consider the initial position shown in figure 3-11, with q^0 and the individual at point *A* realizing utility

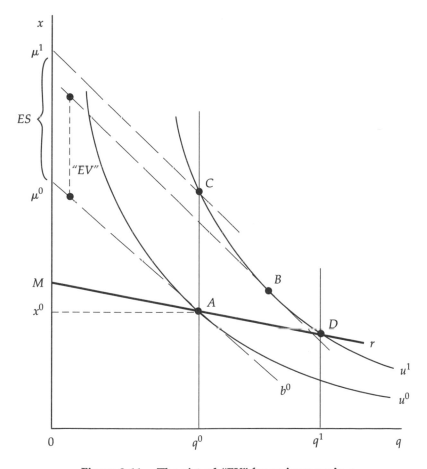

Figure 3-11.   The virtual "*EV*" for an increase in *q*

level u^0. A "virtual" budget line through point A and tangent to u^0 has a slope equal to $-b^0$ and an intercept equal to μ^0. We can interpret b^0 as the virtual price of q and μ^0 as the virtual income associated with point A; b and μ are functions of q, u, and the prices of marketed goods. The function $b(\cdot)$ is the compensated inverse demand function for q. It might also be called either the marginal willingness-to-pay function or the virtual price function. Also, clearly,

$$b = -\frac{\partial e}{\partial q} \tag{18}$$

These terms provide yet another basis for defining ES. It is increase in virtual income at initial virtual prices and q that is required to attain the same utility level as can be attained with the increase in q to q^1:

$$ES = \mu^1\left(P,\, b^0(P,\, q^0,\, u^0),\, q^0,\, u^1\right) - \mu^0(P,\, b^0(P,\, q^0,\, u^0),\, q^0,\, u^0)$$

This is measured by the vertical distance AC, which is the same as the vertical distance AC that measures ES in figure 3-9. Similarly, CS is the decrease in virtual income that maintains u^0 given the increase in q, assuming b^1; that is:

$$CS = \mu^0(P,\, b^1(P,\, q^0,\, u^1),\, u^0)$$
$$- \mu^1(P,\, b^1(P,\, q^1,\, u^1),\, u^0)$$

The virtual incomes and prices can also be used to define a "virtual" equivalent variation labeled "EV." This is the increase in income that has the same effect as the postulated increase in q, assuming, however, that the individual can choose an optimal quantity of q. A virtual budget line is tangent to the indifference curve representing u^1 at point B in figure 3-11. The increase in virtual income necessary to achieve point B is "EV." "EV" is less than ES, showing that less additional income is necessary if the individual can optimally adjust compared to what is necessary in the quantity constrained regime. Similar reasoning can be used to derive a "virtual" CV, denoted as "CV," and to show that "CV" is larger than CS for the same change in q.[32] This is because more income can be taken away from the individual when that person can optimally adjust q than in the quantity-constrained regime.

---

[32]The virtual budget line and price, b, used to derive "CV" is tangent to u^1 at point D with q^1.

The virtual price functions can also be used to define welfare measures. Since the virtual price is equal to the slope of the indifference curve at that point, it is also equal to $\partial e/\partial q$ evaluated at that point. Thus:

$$W_q = \int_{q^0}^{q^1} b(P, q, u^t)\, dq$$

This is either a *CS* or an *ES* measure, depending on whether $t = 0$ or 1.

Exact and Approximate Welfare Measures

The question to be considered now is whether there is some relationship describing the individual's behavior with respect to choices of q that is, at least in principle, observable and that can be used to solve for the indirect utility function or expenditure function in a manner similar to the Hausman (1981) method for exact welfare measurement for price changes. Lankford (1988) and Bergland and Randall (1984) have described techniques for recovering the expenditure function or indirect utility function when the individual faces a quantity constraint. But these techniques require information that is not often available, especially in the case of public goods and nonmarketed environmental and resource service flows.

What all of these techniques have in common is the requirement for information that reflects the choices of the individual who is maximizing utility subject only to price and income constraints. Simply put, in the absence of quantity constraints and when price ratios are known, marginal rates of substitution can be inferred from utility-maximizing choices. But as is clear from figures 3-8 through 3-11, the presence of quantity constraints means that marginal rates of substitution are not in general equal to price ratios; so marginal rates of substitution are not directly revealed.

The Lankford and the Bergland and Randall techniques require some knowledge of what the individual would choose to do if he or she could freely choose q at exogenous prices. The required data might be available, for example, in the case of wartime rationing in which the quantity-constrained goods were previously available in unconstrained markets. But these data are not likely to be available for environmental and resource services.

A related question is whether there is an observable counterpart to the Marshallian measure for price changes and, if so, what its relationship is to the exact *ES* and *CS* measures for quantity changes. Randall and Stoll (1980) have analyzed the difference between *ES* and

CS measures and their relationship to a Marshallian measure. However, their Marshallian surplus is somewhat unusual, and it is not directly observable. In order to explain this, we need to define two other functions. The first is a bid function conditional on M and r. Assume that the individual receives q and pays r per unit. This bid function gives the maximum willingness to pay for one additional unit of q as a function of q, while holding constant the other prices and the income available to allocate between additional q and X, $(M - r \cdot q)$. This bid function can be written as

$$b^* = b^*(P, q, M - r \cdot q) \tag{19}$$

This gives the slope of the indifference curve passing through the point defined by q, $M - r \cdot q$. At point A in figure 3-11, $b^* = b$, but at other levels of q, $b^* \neq b$. For example, at q^1 in figure 3-11, $b^*(\cdot)$ gives the slope of u^1 at point D, while $b(\cdot)$ as given by equation (18) gives the slope of u^0 at that level of q; and $b^* > b$. This function cannot be inverted to get the optimal choice of q as a function of b_1^* except in the special case where $b^* = r$. For example, at q^1 this function would yield the slope of u^1 at point D; but this point would not be freely chosen given M and r and the preferences described by these indifference curves.

The second function to be defined also determines a bid, say b^{**}, but makes this bid depend on q and virtual income, $\mu = M + (b - r)q$. This income-compensated bid function is given by

$$b^{**} = b^{**}[P, q, M + (b^{**} - r)q] \tag{20}$$

Given q_0, figure 3-11 shows that $\mu = \mu^0$ and $b^{**} = b^0$ at point A. If q increases to q^1, the compensated bid function gives the new b^{**} and virtual income necessary to support point D as an optimal choice. In other words, at point D, $b^{**} = b^1$, the slope of u^1 at point D.

In principle this function could be inverted to find the optimal choice of q given b^{**}. But this would not be a true Marshallian demand function because it assumes compensation for the income value of the Δq. Also, it cannot be estimated from observable data because only M, q, and r are observed, not $M + (b^{**} - r)q$.

It is interesting to note that although these two forms of bid function are defined differently and have different forms, they will compute the same bid for given q, M, and preferences. They both compute the slope of the indifference curve passing through the intersection of the $M - r \cdot q$ budget line and the given q. Yet neither function can be given a strictly Marshallian interpretation.

The relationship between the Hicks-compensated inverse demand function (equation 17) and the bid function $b(\cdot)$ is shown in figure 3-12. At q^0, $b^0(\cdot)$ and $b^*(\cdot)$ coincide. But as q increases, $b^0(\cdot)$ diverges from $b^*(\cdot)$. This is because the first curve holds utility constant while the second holds income (M) constant.

As Lankford (1988, pp. 56–61) shows, if the income-compensated bid function (20) is known, it is possible to solve for the conditional expenditure function and through that to obtain the CS or the ES welfare measures exactly. CS is the area under $b^0(\cdot)$ between q^0 and q^1. ES is the area under $b^1(\cdot)$ over the same interval.

Randall and Stoll (1980) employed a line of reasoning similar to Willig's (1976) to analyze the errors involved in using the area under the $b^{**}(\cdot)$ function as an approximation to either CS or ES and the likely magnitude of the difference between CS and ES. As I have done in this section, they assumed that there is only one quantity-constrained good. Referring to figure 3-12, for a move from q^0 to q^1 the three alternative welfare measures are given by the following geometric areas:

$$CS = a$$
$$S\ \ = a + b$$
$$ES = a + b + c$$

Randall and Stoll show that when

$$\frac{S}{M} \cdot \frac{E_q}{2} \le 0.05,$$

the error bounds are

$$\frac{S}{M} \cdot \frac{E_q}{2} \le \frac{ES - S}{S} \le \frac{S}{M} \cdot \frac{\overline{E}_q}{2}$$

and

$$\frac{S}{M} \cdot \frac{\underline{E}_q}{2} \le \frac{S - CS}{S} \le \frac{S}{M} \cdot \frac{\overline{E}_q}{2}$$

where S is the area under $b^*(\cdot)$ and E_q is the price flexibility of income, defined as

$$E_q \equiv \frac{\partial b^*(P, q, M)}{\partial M} \cdot \frac{M}{b^*}$$

The lower and upper bars denote the smallest and largest values of E_q

Price ($)

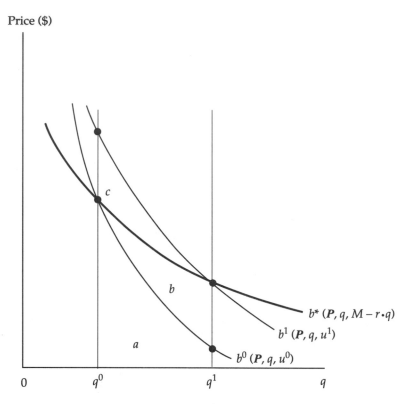

Figure 3-12. The compensated inverse demand curves and the bid curve for q

when E_q is not constant over the relevant range. General expressions for the error bounds are also presented. This result implies that using the area under $b^*(\cdot)$ to approximate CS or ES is an acceptable procedure. But recall that $b^*(\cdot)$ is not observable; and if it is known somehow, it should be possible to calculate CS or ES directly, making the approximation unnecessary.

The result also seems to imply that CS (or willingness to pay for a change) and ES (or willingness to accept compensation to forgo a change) will be approximately equal in many circumstances. Hanemann (1991) has shown that this is not the case, again when there is only one quantity-constrained good. He has shown that the price flexibility of income can be expressed as the ratio of two other terms:

$$E_q = \frac{E_M}{\sigma_q}$$

where σ_q is the aggregate Allen-Uzawa elasticity of substitution between q and the composite commodity X and E_M is the income elasticity of demand for q. If there are no close substitutes for q among the private goods in X, σ_q can be close to zero, leading to a high value for E_q and a large difference between CS and ES.

When q Is a Bad

There are two ways in which more q could be a bad, rather than a good, for an individual. The first way is when r is greater than zero and the individual would prefer to have less than the quantity being imposed given that price. This signifies that $r > b > 0$. The welfare measures ES and CS are still defined in the same way, but now they are negative for increases in q and positive for decreases in q.

The second way in which q can be a bad is the more fundamental one; it is when the marginal utility of q is negative. Even at a zero price, the individual would prefer to receive a smaller quantity. This means that b is negative. In both cases, the welfare measures ES and CS are defined in the same way, and again they are negative for increases in q and positive for decreases in q. Also, all of the discussion of exact welfare measurement techniques and approximations carries over with appropriate changes to the case of q as a bad.

AGGREGATION AND SOCIAL WELFARE

Assume now that we have obtained measures of the welfare changes, either plus or minus, for all individuals. How can we use that information to make choices about public policy alternatives? To put the question in its most profound sense, what is the appropriate relationship between the welfare of individuals and the social welfare? In the literature on welfare economics there are basically four ways to approach this question.[33]

The first approach to the question is the so-called Pareto criterion. Only policy changes that make at least one person better off (that is, an individual experiences a positive welfare change) and make no individual worse off (that is, no individual experiences a

---

[33]What follows is a brief review of alternative social welfare criteria. Since the main concern of this book is with measurement, the question of social welfare criteria—that is, how to use the measures—is off the main track. For a more extensive discussion of the problem, see Mishan (1960), especially Section III. Boadway and Bruce (1984) provide a more up-to-date treatment.

negative welfare change) pass this criterion. This criterion deliberately rules out any attempts to add up or otherwise make commensurable the welfare measures of different individuals. Since virtually all actual public policy proposals impose net costs on at least some individuals, most policy actions by the state could not be accepted under this criterion. This would be particularly true in the environmental area, where environmental management costs are often channeled through the production sector while benefits accrue to households in the form of increased levels of environmental services. It is highly unlikely that this would result in a pattern of incidence of benefits and costs in which no one would lose.[34] The restrictive features of the Pareto criterion have stimulated an ongoing search for a welfare criterion that would justify the state doing certain things that at least some people feel it should be able to do.[35]

The second approach to the question was proposed in slightly different forms by Kaldor (1939) and Hicks (1939). These are the two different forms of a potential compensation test discussed earlier, in the section on Welfare Measures for Changes in Prices. Let us review these tests in the present context of aggregation and social welfare.

As noted earlier, the Kaldor version of the test asks whether those who gain by the policy can fully compensate for the welfare losses of those who lose by the policy. The Kaldor version of the test would be satisfied if the sum of all individual CV and CS measures of welfare changes were greater than zero. The criterion is essentially one of potential Pareto improvement, since if the compensation were actually paid, no one would lose from the policy.

The Hicks version of the potential compensation test asks whether those who lose from the policy could compensate the gainers for a decision not to proceed with the policy. If the answer is yes, the policy should be rejected according to the Hicks criterion. If the policy were rejected and compensation were actually paid, those who would have gained from the policy would be just as well off as if the policy had been adopted, and those who would have lost are at least as well off as they would have been with the policy. The Hicks version of the test takes acceptance of the project as its reference point. In effect, it is a decision to forgo the project that creates the gains and losses that are relevant to the Hicksian version of potential Pareto improvement criterion.

---

[34]See, for example, Gianessi, Peskin, and Wolff (1979) and Peskin (1978).

[35]However, for a strong defense of the Pareto criterion on classical liberal grounds, see Peacock and Rowley (1975).

In either the Kaldor case or the Hicks case, should the compensation actually be paid? If one thinks the answer should be yes, then the compensation test is transformed into a variation of the Pareto criterion in which the state serves to enforce the taxes and transfer payments that are necessary to ensure that no one actually experiences a welfare loss, assuming that such taxes and transfers would be costless. If one thinks that the answer should be no, this is equivalent to, in effect, assuming that all individual welfare changes are commensurate and can be summed together into an aggregate measure of welfare change. This is the efficiency criterion of the new welfare economics. According to the efficiency criterion, the objective of social policy is to maximize the aggregate value of all of the goods and services people receive, including environmental and resource services.[36]

Alternatively, one might believe that whether compensation should be paid depends upon who has to pay and who gets the benefits. This requires consideration of the equity (fairness) in the distribution of income as an element in the evaluation of social policy. The third approach to the question of social welfare criteria, proposed by Little (1957), makes explicit the concern for equality. He proposed a twofold test. First, does the policy pass the Kaldor test? And second, does the resulting change improve the distribution of income? The Little criterion legitimizes a concern with the distributional effects of changes in resource allocation, but it does not resolve the question of what constitutes an improvement.

The fourth approach to the question involves an attempt to make specific social judgments regarding equity and to introduce equity considerations systematically into the evaluation of social policy. The most common proposal calls for the establishment of a social welfare function that gives different weights to individual welfare changes according to the relative deservingness of the different individuals.[37] Of course, the main problem with the social welfare weight approach is the determination of the weighting function.[38]

---

[36]One justification for the Hicks-Kaldor potential compensation test is that a large number of efficient projects will spread benefits sufficiently widely so that everyone is a net gainer from the set of projects taken as a whole, even though some might be losers on individual projects. See Polinsky (1972) for an interesting development of this line of reasoning.

[37]See, for example, Eckstein (1961). For a more extended discussion of this approach, see Haveman and Weisbrod (1975, pp. 61–64) and references therein.

[38]See, for example, Freeman (1971).

Nevertheless, willingness to make explicit value judgments about equity makes it possible to consider a wider range of policy choices. For example, if one opts for the Pareto criterion or the potential compensation version of the Hicks-Kaldor test, one rules out the possibility of accepting a project that has a sum of individual welfare changes that is less than zero but that would substantially improve the distribution of income. An example of such a policy would be one that imposes a welfare loss of $1,000 on a millionaire while bringing benefits of $99 to each of ten impoverished orphans. A welfare-weighting function could approve negative sum policies like this provided that the weights given to the beneficiaries were sufficiently greater than the welfare weights of the losers. In addition, neither of these criteria would reject a project that imposes costs on no one but distributes benefits only to the richest in our society. Some might make the value judgment that that, in itself, is undesirable. A social welfare function that included some measure of inequality of the aggregate distribution as an argument might reject inequality-creating projects like this. And it would also be likely to accept negative sum projects that reduced inequality.

SUMMARY

In this chapter, I have explained the derivation of the compensating and equivalent measures of individual welfare change for changes in prices and imposed quantities. The compensating and equivalent measures answer different kinds of policy-relevant questions because they make different implicit assumptions about the relevant status quo. It is interesting to examine some hypothetical examples.

Suppose that the question is whether to locate a landfill in a particular neighborhood. The neighbors are likely to oppose this proposal. Suppose that it is accepted that the neighbors have a right to an undisturbed neighborhood. Then the relevant measure of the harm for locating the landfill in their neighborhood would be the sum of their compensating measures of loss (CV and CS). The appropriate measure of the gain to those who would use the landfill would be their willingness to pay to locate it in this neighborhood—also a compensating measure. Alternatively, if it is argued that the larger society has a right to locate the landfill anywhere, then what is relevant is the neighbors' willingness to pay to keep it out their neighborhood. This is an equivalent measure of the potential loss (EV and ES). And for the users of the landfill, the value of locating the landfill in this neighborhood is what its users would require to compensate them for locating it in a less desirable place, an equivalent measure of benefit.

Suppose, instead, that the offending facility is a polluting factory that has been in the neighborhood for a long time. If the neighbors are deemed to have a right to a clean neighborhood, then the appropriate reference point for welfare measurement is their utility levels after the factory has stopped its pollution. This implies an equivalent measure of welfare change (*EV* and *ES*). Specifically, this is the compensation that the neighbors would require to forgo having the pollution stopped and the factory owners' willingness to pay to continue to pollute. Alternatively, if it is the factory that has a right to pollute, compensating measures of the gain from stopping the pollution are appropriate (*CV* and *CS*).

In each case, the appropriate welfare measure can be found by examining the nature of the social transaction that is implied by the policy decision at hand and by the implicit rights to the services of the environment presumed to be held by the various parties to the transaction. The results for the examples discussed here can be summarized as follows:

Implicit "rights"	Policy question	Gainers	Losers
To the present pollutor	Require clean up?	Neighbors—compensating measure (WTP)	Pollutor—compensating measure (WAC)
To the potential pollutor	Allow pollution?	Pollutor—equivalent measure (WAC)	Neighbors—equivalent measure (WTP)
To the neighbors	Require clean up?	Neighbors—equivalent measure (WAC)	Pollutor—equivalent measure (WTP)
To the neighbors	Allow pollution?	Pollutor—compensating measure (WTP)	Neighbors—compensating measure (WAC)

REFERENCES

Bergland, Olvar, and Alan Randall. 1984. "Operational Techniques for Calculating the Exact Hicksian Variations from Observable Data," Staff Paper 177. Lexington, Ky.: Department of Agricultural Economics, University of Kentucky.

Bergson, Abram. 1966. *Essays in Normative Economics*. Cambridge, Mass.: Harvard University Press.

Boadway, Robin, and Neil Bruce. 1984. *Welfare Economics*. Oxford, England: Basil Blackwell.

Bockstael, Nancy E., and Kenneth E. McConnell. 1980. "Calculating Equivalent and Compensating Variation for Natural Resource Facilities," *Land Economics* vol. 56, no. 1, pp. 56–63.

Chipman, John S., and James C. Moore. 1980. "Compensating Variation, Consumer's Surplus, and Welfare," *American Economic Review* vol. 70, no. 5, pp. 933–949.

Currie, John M., John A. Murphy, and Andrew Schmitz. 1971. "The Concept of Economic Surplus," *Economic Journal* vol. 81, no. 324, pp. 741–799.

Deaton, Angus, and John Muellbauer. 1980. *Economics and Consumer Behavior.* Cambridge: Cambridge University Press.

Diamond, Peter A., and Daniel McFadden. 1974. "Some Uses of the Expenditure Function in Public Finance," *Journal of Public Economics* vol. 3, no. 1, pp. 3–21.

Eckstein, Otto. 1961. "A Survey of the Theory of Public Expenditures Criteria," in National Bureau of Economic Research, *Public Finances: Needs, Sources, and Utilization*. Princeton, N.J.: Princeton University Press.

Freeman, A. Myrick, III. 1971. "Project Design and Evaluation with Multiple Objectives," in R. Haveman and J. Margolis, eds., *Public Expenditures and Policy Analysis*. Chicago, Ill.: Markham.

Gianessi, Leonard P., Henry M. Peskin, and Edward Wolff. 1979. "The Distributional Effects of the Uniform Air Pollution Policy in the United States," *Quarterly Journal of Economics* vol. 93, no. 2, pp. 281–301.

Hallam, Arne. 1988. "Measuring Economic Welfare: Is Theory a Cookbook for Empirical Analysis?" *American Journal of Agricultural Economics* vol. 70, no. 5, pp. 442–447.

Hanemann, W. Michael. 1980. "Measuring the Worth of Natural Resource Facilities: Comment," *Land Economics* vol. 56, no. 4, pp. 482–490.

Hanemann, W. Michael. 1991. "Willingness to Pay and Willingness to Accept: How Much Can They Differ?" *American Economic Review* vol. 81, no. 3, pp. 635–647.

Hanemann, W. Michael, and Edward Morey. 1992. "Separability, Partial Demand Systems, and Consumer's Surplus Measures," *Journal of Environmental Economics and Management* vol. 22, no. 3, pp. 241–258.

Hause, John C. 1975. "The Theory of Welfare Cost Measurement," *Journal of Political Economy* vol. 83, no. 6, pp. 1145–1182.

Hausman, Jerry A. 1981. "Exact Consumer's Surplus and Dead Weight Loss," *American Economic Review* vol. 71, no. 4, pp. 662–676.

Haveman, Robert H., Mary Gabay, and James Andreoni. 1987. "Exact Consumer's Surplus and Deadweight Loss: A Correction," *American Economic Review* vol. 77, no. 3, pp. 494–495.

Haveman, Robert H., and Burton A. Weisbrod. 1975. "The Concept of Benefits in Cost-Benefit Analysis: With Emphasis on Water Pollution Activities,'' in Henry M. Peskin and Eugene P. Seskin, eds., *Cost-Benefit Analysis and Water Pollution Policy.* Washington, D.C.: The Urban Institute.

Hicks, John R. 1939. ''The Foundations of Welfare Economics,'' *Economic Journal* vol. 49, no. 196, pp. 696–712.

Hicks, John R. 1943. ''The Four Consumer Surpluses,'' *Review of Economic Studies,* vol. 11, no. 1, pp. 31–41.

Hurwicz, Leonid, and Hirofumi Uzawa. 1971. ''On the Integrability of Demand Functions,'' in John Chipman and others, eds., *Preferences, Utility and Demand.* New York: Harcourt Brace Jovanovich.

Johansson, Per-Olov. 1987. *The Economic Theory and Measurement of Environmental Benefits.* Cambridge: Cambridge University Press.

Just, Richard E., Darrell L. Hueth, and Andrew Schmitz. 1982. *Applied Welfare Economics and Public Policy.* Englewood Cliffs, N.J.: Prentice-Hall.

Kaldor, Nicholas. 1939. ''Welfare Propositions of Economics and Interpersonal Comparisons of Utility,'' *Economic Journal* vol. 49, pp. 549–552.

Krutilla, John V. 1967. ''Conservation Reconsidered,'' *American Economic Review* vol. 57, no. 4 pp. 777–786.

LaFrance, Jeffery T., and W. Michael Hanemann. 1989. ''The Dual Structure of Incomplete Demand Systems,'' *American Journal of Agricultural Economics* vol. 71, no. 2, pp. 262–274.

Lankford, R. Hamilton. 1988. ''Measuring Welfare Changes in Settings With Imposed Quantities,'' *Journal of Environmental Economics and Management* vol. 15, no. 1, pp. 45–63.

Little, I. M. D. 1957. *A Critique of Welfare Economics.* 2d ed. Oxford, England: Clarendon Press.

Mäler, Karl-Göran. 1974. *Environmental Economics: A Theoretical Inquiry.* Baltimore, Md.: The Johns Hopkins University Press for Resources for the Future.

Marshall, Alfred. 1920. *Principles of Economics: An Introductory Volume.* 8th ed. London: Macmillan.

McKenzie, George W. 1976. ''Measuring Gains and Losses,'' *Journal of Political Economy* vol. 84, no. 3, pp. 641–646.

McKenzie, George W. 1983. *Measuring Economic Welfare: New Methods.* Cambridge: Cambridge University Press.

McKenzie, George W., and I. F. Pearce. 1976. ''Exact Measures of Welfare and the Cost of Living,'' *Review of Economic Studies* vol. 43, no. 4, pp. 465–468.

McKenzie, George W., and I. F. Pearce. 1982. ''Welfare Measurement—A Synthesis,'' *American Economic Review* vol. 72, no. 4, pp. 669–682.

Mishan, Ezra J. 1959. ''Rent as a Measure of Welfare Change,'' *American Economic Review* vol. 49, no. 3, pp. 386–394.

Mishan, Ezra J. 1960. ''A Survey of Welfare Economics, 1939–1959,'' *Economic Journal* vol. 70, no. 278, pp. 197–256.

Mishan, Ezra J. 1976. ''The Use of Compensating and Equivalent Variations in Cost-Benefit Analysis,'' *Economica* vol. 43, no. 170, pp. 185–197.

Morey, Edward R. 1984. "Confuser Surplus," *American Economic Review* vol. 74, no. 1, pp. 163–173.

Moss, William G. 1976. "Some Uses of the Expenditure Function in Public Finance: A Comment," *Journal of Public Economics* vol. 5, no. 3, pp. 373–379.

Neary, J. P., and K. W. S. Roberts. 1980. "The Theory of Household Behavior Under Rationing," *European Economic Review* vol. 13, no. 2, pp. 25–42.

Peacock, Alan T., and Charles K. Rowley. 1975. *Welfare Economics: A Liberal Restatement*. London: M. Robertson.

Peskin, Henry M. 1978. "Environmental Policy and the Distribution of Benefits and Costs," in Paul R. Portney, ed., *Current Issues in U.S. Environmental Policy*. Baltimore, Md.: The Johns Hopkins University Press for Resources for the Future.

Polinsky, A. Mitchell. 1972. "Probabilistic Compensation Criteria," *Quarterly Journal of Economics* vol. 86, no. 3, pp. 407–425.

Randall, Alan, and John R. Stoll. 1980. "Consumer's Surplus in Commodity Space," *American Economic Review* vol. 70, no. 3, pp. 449–455.

Reaume, David M. 1973. "Cost-Benefit Techniques and Consumer Surplus: A Clarificatory Analysis," *Public Finance* vol. 28, no. 2, pp. 196–211.

Samuelson, Paul A. 1942. "Constancy of the Marginal Utility of Income," in Oscar Lange, Francis McIntyre, and Theodore O. Yntema, eds., *Studies in Mathematical Economics and Econometrics, in Memory of Henry Schultz* (Chicago, Ill.: University of Chicago Press); reprinted in Joseph E. Stiglitz, ed., *The Collected Scientific Papers of Paul A. Samuelson*, vol. 1. Cambridge, Mass.: MIT Press, 1966.

Silberberg, Eugene. 1972. "Duality and the Many Consumer's Surpluses," *American Economic Review* vol. 62, no. 5, pp. 942–952.

Silberberg, Eugene. 1978. *The Structure of Economics: A Mathematical Analysis*. New York: McGraw-Hill.

Tobin, James, and Hendrik S. Houthakker. 1950/1. "The Effects of Rationing on Demand Elasticities," *Review of Economic Studies* vol. 18, no. 3, pp. 140–153.

Varian, Hal R. 1984. *Microeconomic Analysis*. 2d ed. New York: Norton.

Vartia, Yrjö O. 1983. "Efficient Methods of Measuring Welfare Change and Compensated Income in Terms of Ordinary Demand Functions," *Econometrica* vol. 51, no. 1, pp. 79–98.

Willig, Robert D. 1976. "Consumers' Surplus Without Apology," *American Economic Review* vol. 66, no. 4, pp. 589–597.

Willig, Robert D. 1979. "Consumer's Surplus Without Apology: Reply," *American Economic Review* vol. 69, no. 3, pp. 471–474.

4

Models for Indirect
Benefit Estimation:
Basic Theory

W elfare changes have been defined in terms of the area under
the appropriate Hicks-compensated demand curve for a market good or marginal willingness-to-pay curve for a nonmarket good
or service. The marginal willingness-to-pay curves exist even for public goods and nonmarketed goods such as the services of the environment; but they cannot be estimated from direct observations of
transactions in these goods. Given the absence of markets for public
goods and environmental goods, how can information on demand
and benefits be obtained?

As described in chapter 2, there are basically four approaches to
obtaining demand and value information. Two approaches involve
the direct estimation of value from observations of market behavior or
from responses to hypothetical valuation questions. These approaches require observations of the prices of the goods being valued
or expressions of willingness to pay or to accept compensation. The
other two approaches use indirect or inferential methods based on
some hypothesized relationship between the observable demands for
marketed goods and the unobservable demands for environmental
services and public goods. The indirect observed methods are based
on actual observations of behavior, while the indirect hypothetical
methods are based on responses to hypothetical questions about
choices and activities under alternative circumstances.

In this chapter we shall explore some of the possible relationships between demands for private goods and demands for environmental services in an effort to determine under what circumstances
the demands for environmental services can be inferred from information on market transactions for the related private good. We will

use q to denote some parameter of environmental or resource quality. The problem is to estimate in monetary terms the changes in individuals' welfares associated with changes in q. The basic thesis of this chapter is that the degree to which inferences about the benefits of increases in q can be drawn from market observations, and the appropriate techniques to be used in drawing these inferences, both depend on the way in which q enters individual utility functions. Broadly speaking, there are three ways in which q can affect an individual's utility:

1. q can produce utility indirectly as a factor input in the production of a marketed good that yields utility;
2. q can be an input in the household production of utility-yielding commodities; or
3. q can produce utility directly by being an argument in an individual's utility function.

In the third case, there are a variety of ways in which q can interact with one or more market goods in the individual's preference structure. For example, there may be a substitution or complementarity relationship between q and some private good. If the nature of the household production process or the forms of interaction between q and private goods can be specified, it may be possible to infer the value of q to the individual from observations of choices of the related market goods.

The next section of this chapter explores the case where q is a factor of production for a market good. The remainder of the chapter is devoted to examining models of the ways in which q can affect utility more directly. The exploitation of possible relationships between environmental goods and private goods leads to several empirical techniques for estimating environmental and resource values. These techniques have the following characteristics: (1) they are consistent with the basic theory of demand and consumer preferences; (2) they provide a means for estimating the indirect utility function, the expenditure function, or the compensated demand function for the environmental service; and (3) they are practical in the sense of imposing realistic data and computational requirements.

ENVIRONMENTAL QUALITY AS A FACTOR INPUT

When q is a factor of production, changes in q lead to changes in production costs which in turn affect the price and quantity of output or the returns to other factor inputs, or both. The benefits of changes in q can be inferred from these changes in observable market data. There are several examples where q can be interpreted as a factor input. The quality of river water diverted for irrigation affects the agricultural productivity of irrigated land. The quality of intake water may influence the costs of treating domestic water supplies and the costs of production in industrial operations that utilize water for processing purposes. Agricultural productivity is impaired by some forms of air pollution. And to the extent that air pollution causes materials damages, it can affect the costs of production for a wide variety of goods and services.

Assume that good x is produced with a production function

$$x = x(k, w, ..., q)$$

where k and w are capital and labor (work) respectively, and where the marginal product of q is positive. With given factor prices, and assuming cost-minimizing behavior, there is a cost function

$$C = C(p_w, p_k, x, q)$$

Since q affects the production and supply of a marketable good, the benefits of changes in q can be defined and measured in terms of changes in market variables related to the x industry. A change in q will cause shifts in both cost curves and factor demand curves. The consequences of these shifts depend on conditions in factor and product markets. There are two channels through which changes in q can produce benefits. The first is through changes in the price of x to consumers. The second is through changes in the incomes received by owners of factor inputs used in x production.

To illustrate the first channel, assume that x is produced in a competitive industry under conditions of constant cost, that is, factor supplies to this industry are infinitely elastic. Assume that the change in q affects the cost curves of a significant proportion of producers in the market. As a result, the supply curve shifts downward, causing a fall in the price and an increase in total quantity. The benefit of the price reduction accrues to consumers and can be measured by the methods described in chapter 3.

To illustrate the second channel—changes in the incomes received by factors of production—consider only one producer who is a price taker in all markets. If the change in q affects only this producer, output price will not be changed. Since the change in q affects the marginal costs of production, the firm's marginal cost and supply curves are shifted down. In this case the benefit is equal to the increase in quasi-rents to the firm. This benefit will accrue to the owner of a fixed factor—land, for example—or to the residual income claimant as profit. In either case, benefits can be measured by changes in profits and fixed factor incomes. However, if the producers affected by changes in q face less than perfectly elastic factor supply curves, at least some of the benefits will be passed on to factors through changes in factor prices and incomes. The factors' shares of benefits can be approximated by the areas to the left of factor supply curves.

The effects of these two channels are combined in figure 4-1. When the supply curve of the industry is shifted down to S'', price decreases to p''. The benefit to consumers of x is approximated by the change in consumers' surplus, the area $p'BCp''$. Part of this benefit, $p'BFp''$, is at the expense of a reduction in producer and factor sur-

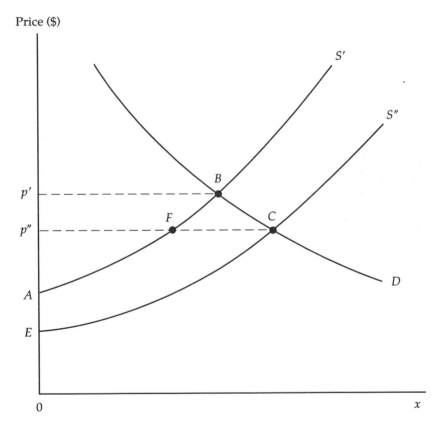

Figure 4-1.   The welfare measure when q affects the production of x

pluses; so the net gain from the lower price is BCF. The lower supply curve results in factor surpluses and quasi-rents equal to $p''CE$. The net increase to producers and factors is $AFCE$. So total benefits are equal to $ABCE$.

Implementation of these measures requires knowledge of the effects of changes in q on the cost of production, the supply conditions for output, the demand curve for good x, and factor supplies. Two special cases make the estimation of benefits relatively straightforward.

The first is the case where q is a perfect substitute for other inputs in the production of a good. An increase in q leads to a reduction in factor input costs. If the substitution relationship is known, the decrease in per-unit production costs is readily calculated. For example, if water quality improvement results in a decrease in chlorination requirements for drinking water supplies, the decrease in chlorination costs per unit of output can be readily calculated. Where the change in total cost does not affect marginal cost and output, the cost saving is a true measure of the benefits of q. If the change in q affects marginal cost, the benefits should include the effect of the lower cost on output and price. However, if the percentage reduction in marginal costs is small, or the marginal cost curve is inelastic, or both, the corresponding increase in output would be relatively small. Thus, the decrease in total cost could still be used to provide a rough approximation of true benefits.[1] This approach, sometimes referred to as the "damage function" approach, has been the basis of a number of estimates of the materials, household-cleaning, and agricultural crop-loss benefits of air pollution control, and of the benefits to municipalities, industries, and households of reduced contamination of intake water supplies (Freeman, 1982).

The second case that makes the estimation of benefits relatively straightforward is where knowledge of cost, demand, and market structure suggests that the benefits of a change in q will accrue to producers. Then benefits may be estimated from observed or predicted changes in the net income of certain factor inputs. If the production unit in question is small relative to the market for the final product and for variable factors, it can be assumed that product and variable factor prices will remain fixed after the change in q. The

---

[1]Unfortunately, it cannot be determined on a priori grounds whether this approximation is biased upward or downward. The direction and magnitude of the error depend on the shape of the cost curve, how it shifts (whether the shift is parallel or a rotation, for example), and the elasticity of the demand for output.

increased productivity then accrues to the fixed factors of production in the form of profit or quasi-rent.

More generally, however, estimates of the value of q require knowledge of the cost and demand functions. In some studies it has been possible to use econometric methods to estimate a cost function that includes an environmental quality variable (Mathtech, Inc., 1982; Mjelde and coauthors, 1984). Other studies have used various simulation approaches to model the behavior of producers and their responses to changes in an environmental variable. Models and techniques for valuing the effects of q on production are discussed in more detail in chapter 9.

AN INDIVIDUAL'S DEMAND FOR ENVIRONMENTAL QUALITY

In order to analyze those cases where q affects individuals directly I first review the basic model of individual preference and demand, with environmental quality included as an argument in the utility function. I then consider the implications of different forms of utility functions for estimation of the demand for q.

Consider a single individual who has a utility function

$$u = u(X, q)$$

where X is a vector of private goods quantities ($X = x_1, ..., x_i, ..., x_n$). By entering environmental quality as an argument in the utility function, we are assuming that the individual perceives at least the effects of changes in environmental quality. For example, if high ozone levels cause respiratory irritation, the individual must be aware of the irritation, so that he feels "better" when it is reduced. He need not know the cause of the irritation or the actual levels of air pollution. If the individual is not aware of the effects of changes in q, the indirect methods of benefit estimation cannot be applied. For example, individuals may not perceive the effects of long-term exposure to air pollutants on their probability of chronic illness or death. If that is the case, changes in q will not affect their behavior; and observations of market behavior will yield no information about the value of reducing risks to health.

Assume that the individual maximizes utility subject to a budget constraint

$$\sum_i p_i \cdot x_i = M$$

where M is money income. The individual takes q as given and does not have to pay a price for this "imposed" quantity. The solution to this problem yields a set of ordinary demand functions

$$x_i = x_i (P, M, q)$$

where P is the vector of private goods prices ($P = p_1, ..., p_i, ..., p_n$). Note that in general q could be an argument in all private goods demand functions.

The dual to the utility maximization problem can be stated as follows: minimize expenditure ($\Sigma_i p_i \cdot x_i$) subject to the constraint that utility equal or exceed some stated level, say u^0. The solution to this problem gives the expenditure function

$$e(P, q, u^0) = M \tag{1}$$

The expenditure function has a number of useful properties for applied welfare analysis. First, as was shown in chapter 3, the derivative of the expenditure function with respect to any price gives the Hicks-compensated demand function for that good, that is,

$$\partial e / \partial p_i = h_i (P, q, u^0)$$

Similarly, the derivative of (1) with respect to q (with the appropriate change of sign) gives the Hicks-compensated inverse demand function or marginal willingness to pay for changes in q. Let w_q be the marginal willingness to pay or marginal demand price for q. Then

$$w_q = -\partial e(P, q, u^0)/\partial q \tag{2}$$

An alternative expression for the marginal willingness to pay can be obtained by setting the total differential of the indirect utility function equal to zero and solving for the compensating change in income associated with the change in q. Specifically,

$$u = v(M, P, q)$$

$$du = \frac{\partial v}{\partial M} dM + \frac{\partial v}{\partial q} dq = 0$$

and

$$\frac{dM}{dq} = -\frac{\partial v/\partial q}{\partial v/\partial M}$$

where dP is zero by assumption.

If the value of the derivative of the right-hand side of equation (2) can be inferred from observed data, then we have a point estimate of the marginal willingness to pay for q. If this derivative can be estimated as a function of q, then we have the marginal willingness to pay function for q. Let W represent the benefit to the individual of a nonmarginal increase in the supply of q. W is the integral of this function, or

$$W = -\int_{q'}^{q''} e_q(P, q, u)dq$$

$$= e(P, q', u) - e(P, q'', u) \qquad (3)$$

This is either a compensating surplus (*CS*) or equivalent surplus (*ES*) measure of welfare change, depending on the level of utility at which (3) is evaluated. The question to be discussed in the next section is whether there are any circumstances in which information about (2) or (3) can be derived from observations of market prices and quantities for private goods.

THE STRUCTURE OF PREFERENCES AND MEASURES OF VALUE

There are several ways in which additional structure can be given to the general model of preferences and choice discussed above. Each of these alternatives involves some kind of assumption about the structure of preferences and/or the constraints on individual choice. Each of the assumptions implies some kind of connection between observable demands for market goods and the values of environmental services and public goods. And each assumption provides a basis for inferring the marginal willingness to pay for q from observations of the relationships between q and the demands for market goods. For each of the alternatives, I describe the specific restrictions and discuss how inferences about value can be drawn from observations of individual choices.

Several of the alternatives involve making some kind of assumption about the separability of the utility function. Thus it will be useful to first review the concept of separability and the implications of various forms of separability for observable demands for market goods.[2]

---

[2]See Goldman and Uzawa (1964) and Katzner (1970).

Separability refers to the possible effect of partitioning the goods entering into the utility function into subsets, and the relationships among these subsets. Suppose now that there are three types of market goods, so that we can write

$$u = u(X, Y, Z)$$

where $X = x_1, ..., x_i, ..., x_m$, $Y = y_1, ..., y_i, ..., y_n$, and $Z = z_1, ..., z_i, ..., z_p$. A utility function is *weakly separable* if the marginal rate of substitution between any pair of goods within the same subset is independent of the quantities of goods in any other subset. The following utility function is weakly separable:

$$u = v[u^1(X), u^2(Y), u^3(Z)]$$

This means that the demand functions for goods in one subset can be written as depending only on the prices of goods in that subset and the expenditure share of that subset. Thus, if separability can be assumed, it simplifies the task of econometric estimation of demand functions.[3]

A utility function is *strongly separable* if the marginal rate of substitution (MRS) between two goods in different subsets is independent of the quantity of any good in any other subset. Specifically, with strong separability the $MRS_{x_i y_i}$ is independent of z_i. Any additive utility function is strongly separable; and any strongly separable utility function is additive in some monotone transform (Deaton and Muellbauer, 1980). The following utility function is strongly separable:

$$u = v[u^1(X) + u^2(Y) + u^3(Z)]$$

If the utility function is not separable, there are no restrictions on the terms that are arguments in marginal rates of substitution or in the demand functions for individual goods. The marginal rate of substitution between any pair of goods depends on the quantities of all goods. And the demand function for any good depends upon the prices of all goods as well as income.

In the rest of this section, I examine three major forms of relationships between q and market goods. These are some form of complementary relationship between q and one or more market goods, some

---

[3]See Katzner (1970), p. 80, or Deaton and Muellbauer (1980), pp. 220–226.

form of substitution relationship, and the case of differentiated goods where the amount of q embodied in or attached to a market good is one of its differentiating characteristics. Since some of these relationships can be implicitly or explicitly modeled as involving some form of household production, I begin with a brief overview of the household production framework for modeling individual preferences and choice.

The Household Production Framework

The household production function model provides a framework for examining interactions between demands for market goods and the availability of a public good such as environmental quality. This framework is based on the assumption that there is a set of technical relationships among goods used by households in the implicit production of utility-yielding final services.[4] Examining the household production technology is one approach to gaining knowledge of the relationship between demands for market goods and the value of environmental quality change.

In the household production framework, utility is a function of the level of final service flows:

$$u = u(Z) = u(z_1, \ldots, z_j, \ldots, z_m) \tag{4}$$

Z is produced according to a technology common to all households and assumed to be known:

$$z_j = z_j(X, Q), \text{ for } j = 1, \ldots, m \tag{5}$$

where X represents a vector of market goods available at prices P_x, and Q represents a vector of environmental quality attributes.

Formally, the individual choice problem is to maximize (4) subject to the constraints provided by (5), the given level of Q and the budget constraint $\sum_i^n P_x \cdot X \le M$. This problem can be solved by a two-step procedure in which the first step is to combine market goods and Q so as to minimize the costs of producing the z_j. This determines the marginal costs of final services P_z. These are essentially implicit

---

[4]In the household production function literature, the utility-yielding final services are often termed "commodities," while market goods are simply "goods." The terminology used in this chapter is more descriptive of the relationship between goods being bought in the market and what finally yields utility to individuals.

prices; but unlike market prices, they are not parametric to the individual unless marginal costs are constant. The second step is to maximize (4) subject to the budget constraint $C(Z) = C(X, P_x, Q) \leq M$. The observable manifestation of the solution to this problem is a set of derived market goods demands:

$$X = X(P_x, Q, M)$$

These demand functions will reflect both the role of Q in the household production technology (equation 5) and the preferences over the final service flows. So it is important to examine under what conditions information on these values can be extracted from the demand functions. I return to this question in a more formal way in the section headed The Household Production Model Again. In the meantime, as we consider various forms of complementarity and substitutability relationships, I will show how they can also be interpreted as applications of the household production function model.

Perfect Complements

Suppose that the single environmental service q is a perfect complement to a market good, say x_1. By perfect complementarity I mean that x_1 and q must be consumed in fixed proportions, for example $x_1/q = a$. This means that to utilize effectively the services of one unit of q, the individual must purchase $1/a$ units of x_1. In the context of the household production model, this could be interpreted as implying a production function for z_1 of the following form:

$$z_1 = \min (x_1/a, q)$$

If this is the case, there are conditions under which the benefits of changes in q can be estimated from knowledge of the demand function for the complementary market good. As long as p_1 is less than some critical value, say p_1^*, the quantity demanded of x_1 will be independent of its price and determined solely by the availability of q. If q is increased by one unit, the individual will purchase $1/a$ additional units of x_1 and experience a higher level of utility. Thus, the marginal willingness to pay for q will be positive. But if p_1 is greater than p_1^*, the individual will purchase fewer units of x_1 than are required to fully utilize the available q. Thus, at p_1 greater than p_1^*, the marginal utility and marginal willingness to pay for q are zero.

Mäler (1974, pp. 180–183) has shown that if the demand functions for x_1 and for other market goods are known, it is possible to compute the expenditure function and the demand price for q when p is less than p_1^*. The exact expression for the demand price for q depends upon the specification of the true demand curves. There is no simple generalization of the technique. Also, as Mäler appears to realize, it is difficult to imagine examples of perfect complementarity between an environmental service and a market good. But as we will now see, there is a less restrictive form of complementarity that also makes possible the calculation of the marginal willingness to pay for q and that appears to have real-world applications.

Weak Complementarity

Suppose that q enhances the enjoyment the individual derives from consuming x_1 and that an increase in q increases the quantity demanded of x_1, other things being equal. One reasonable interpretation of these assumptions is that q is an exogenously determined characteristic of x_1. Examples of environmental services that might fit these assumptions include water quality as a characteristic associated with visits to a lake, and the number of fish caught as a characteristic of fishing trips to a stream. In this section I show that where the "enjoyment" of q requires the purchase of a market good or where q can be treated as a characteristic of a market good, it is possible to identify a measure of the value of a change in q that is based on the demand for the market good. I describe this measure and the conditions under which it is a valid welfare measure. Then I discuss some issues concerning implementation of the measure.

Let us assume that x_1 and q are Hicksian complements, that is, that the compensated demand function

$$h_1 = h_1(P, q, u^0)$$

is characterized by $\partial h_1/\partial q > 0$. This implies that the Marshallian demand for x_1,

$$x_1 = x_1(P, q, M), \tag{6}$$

also has $\partial x_1/\partial q > 0$.

Suppose that the complete system of demand equations for this individual has been estimated econometrically and that the system satisfies the Slutsky Conditions for integrability. In order to derive a valid welfare measure from this system of demand functions, one must be able to integrate this system to solve for the underlying utility function and expenditure function. But Mäler (1974, pp. 183–189) has shown that in general it is not possible to solve completely for the utility and expenditure functions with the information given. Mathematically, the result of the integration contains unknown terms that are themselves functions of q and the constants of integration. It is necessary to impose additional conditions on the problem in order to solve for the unknown terms and determine the constants of integration. The additional conditions involve what Mäler called weak complementarity.

THE CONDITIONS

Weak complementarity requires that the marginal utility or marginal demand price of q be zero when the quantity demanded of the complementary private good x_1 is zero. Mathematically, weak complementarity involves two conditions. The first is that x_1 be nonessential, in the sense that there be a price p_1^* for x_1 such that the compensated demand

$$h_1(p_1^*, q, u^0) = 0$$

This price is often called the choke price. The choke price will usually be an increasing function of q. Graphically, the compensated demand curve has a vertical intercept; and there is some level of expenditure on other goods that will sustain u^0 even when x_1 is zero.

The second condition is that for values of p_1 at or above p_1^*, the derivative of the expenditure function

$$e = e(p_1^*, q, u^0)$$

be zero, that is,

$$\partial e / \partial q = 0$$

where the term for all other prices has been omitted for simplicity. This condition means that at or above the choke price, the marginal utility or marginal willingness to pay for q is zero. Changes in q have no welfare significance unless the price of x_1 is low enough so that the compensated demand would be positive.

These two conditions together establish an initial position for the individual that can be used to determine the constants of integration.[5]

---

[5] For proof and a demonstration with a numerical example, see Mäler (1974), pp. 183–189. Results similar to Mäler's were obtained by Bradford and Hildebrandt (1977).

It is important to note that these conditions describe a position for the individual that satisfies the first-order conditions of a utility maximization problem in which the individual can choose the quantities of all goods, including q, at given prices, including a zero price for q. This is analogous to defining a position in which the virtual prices for imposed quantities (discussed in chapter 3) are just equal to the actual prices.

THE WELFARE MEASURE

Given the conditions for weak complementarity, the compensating surplus for a change in q can be measured by the area between the two compensated demand curves for x_1. The intuition for this measure can be illustrated graphically. Assume that the compensated demand curve for x_1 has been found for environmental quality level q'. In figure 4-2 this demand curve is labeled $h_1(q')$. Assume that the price of x_1 is given at p_1' and does not change throughout the analysis.

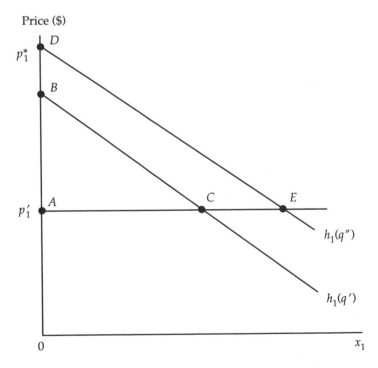

Figure 4-2.   The welfare measure for an increase in q when q and x_1 are weak complements

The compensating surplus associated with the use of x_1 is the area ABC under the demand curve. Now assume that quality is improved to q''. The increase in the quality associated with the use of x_1 is assumed to increase the demand for x_1, thus shifting the demand curve outward to $h_1(q'')$. The calculation of the benefit associated with this change is straightforward and can be divided into three steps.

1. Given the initial demand curve $h_1(q')$, postulate a hypothetical increase in price from p_1' to p_1^*, the choke price. In order to leave the individual no worse off, that person must be compensated by the area ABC.

2. Now postulate the improvement in quality and the shift in the demand curve to $h_1(q'')$. Given the weak complementarity assumption, utility is unaffected since the consumption of x_1 is zero. Therefore there is no need for compensation, either positive or negative.

3. Now postulate a return to the old price of p_1'. The individual is made better off by the area ADE. In order to restore the individual to the original welfare position, that person must be taxed by this amount. The net effect of these changes is a gain to the individual (in the absence of the hypothetical compensating payments) of the area $BCED$ $(= ADE - ABC)$. This is the benefit of the change in q.

The role of the two conditions defining weak complementarity can now be made clear. If there were no choke price, there would be no finite compensating variations (CVs) in steps 1 and 3 above. And if the derivative of the expenditure function were not zero at the choke price, there would be a welfare change in step 2 above (presumably positive) associated with the increase in q, even though the quantity demanded of x_1 were zero.[6] In this case, the area $BCED$ would be an underestimate of the benefits of increasing q.

This result can also be established more rigorously. Recall from chapter 3 that the welfare measure for a change in q is

$$CS_q = W_q = e(p_1, q', u^0) - e(p_1, q'', u^0) \qquad (7)$$

The area between the two compensated demand curves is $CV_{q''} - CV_{q'}$ where these terms are the CVs associated with the ability to

---

[6]If the consumption of x_1 is interpreted as use of an environmental resource (for example, trips to a recreation site), then whatever welfare change is associated with step 2 must be a nonuse value. Nonuse values are the topic of chapter 5.

purchase x_1 at different levels of q. Specifically,

$$CV_{q'} = e\,(p_1^*, q', u^0) - e\,(p_1, q', u^0)$$

and $$CV_{q''} = e\,(p_1^*, q'', u^0) - e\,(p_1, q'', u^0)$$

Using these expressions to calculate the area between the compensated demand curves gives

$$e(p_1^*, q'', u^0) - e(p_1, q'', u^0) - e(p_1^*, q', u^0) + e(p_1, q', u^0) \qquad (8)$$

If the second condition for weak complementarity is satisfied, the first and third terms in (8) sum to zero, and we have

$$W_q = e\,(p_1, q', u^0) - e\,(p_1, q'', u^0)$$

WEAK COMPLEMENTARITY WITH MORE THAN ONE GOOD

So far, we have assumed that q is complementary to only one market good. It is possible for q to be complementary with several goods at the same time. For example, if q is water quality in a lake, it could be complementary to several market activities such as fishing, boating, and swimming. As Bockstael and Kling (1988) show, the analysis of weak complementarity can be carried over to the multi-good case in a straightforward manner.

Suppose there are two market goods that are complementary to q, with Hicksian demand functions as follows:

$$h_1 = h_1\,(p_1, p_2, q, u^0)$$

and $$h_2 = h_2\,(p_2, p_1, q, u^0)$$

Using the expenditure function, the compensating surplus (CS) for a change in q is

$$CS = e\,(p_1, p_2, q', u^0) - e\,(p_1, p_2, q'', u^0) \qquad (9)$$

If the conditions for weak complementarity are satisfied, CS can be estimated from areas between the compensated demand curves for the two market goods in the following manner. First, calculate the area between the compensated demands for x_1 holding p_2 at the observed price level. Then calculate the area between the compensated demands for x_2 evaluated at the choke price (p_1^*) for x_1. Mathematically, this involves calculating the following expression:

$$CS = \int_{p_1'}^{p_1^*} h_1(p_1, p_2', q'', u^0)\, dp_1 - \int_{p_1'}^{p_1^*} h_1(p_1, p_2', q', u^0)\, dp_1$$

$$+ \int_{p_2'}^{p_2^*} h_2(p_1^*, p_2, q'', u^0)\, dp_2 - \int_{p_2'}^{p_2^*} h_2(p_1^*, p_2, q', u^0)\, dp_2 \qquad (10)$$

The first and third integrals give the value to the individual of consuming x_1 and x_2 when q is equal to q'' evaluated over a consistent path of integration. In other words, starting from (p_1', p_2'), p_1 is increased to p_1^* and then p_2 is increased to p_2^*.[7]

When each of the integrals in equation (10) is expressed as a difference in expenditures, we have:

$$
\begin{aligned}
CS = \; & e(p_1^*, p_2', q'', u^0) && - e(p_1', p_2', q'', u^0) \\
& - e(p_1^*, p_2', q', u^0) && + e(p_1', p_2', q', u^0) \\
& + e(p_1^*, p_2^*, q'', u^0) && - e(p_1^*, p_2', q'', u^0) \\
& - e(p_1^*, p_2^*, q', u^0) && + e(p_1^*, p_2', q', u^0)
\end{aligned}
$$

When like terms are canceled out, and if weak complementarity is satisfied, this reduces to equation (9).[8]

WELFARE MEASURES WITH ORDINARY DEMAND FUNCTIONS

The primary empirical requirement for utilizing weak complementarity is that we be able to obtain econometric estimates of the demand function for the private good as a function of prices, income, and q.

---

[7]Of course, the result of the calculation is independent of the order in which the prices are changed.

[8]Bockstael and Kling (1988) also consider the econometric problems that often arise when trying to include the prices of other goods in the estimation of demand functions—for example, unobservable prices of substitutes or highly correlated substitute prices. They also consider the effects of these problems on calculated welfare changes. They reach the surprising conclusion that if the price of a market good is perfectly correlated with the price of a substitute, and if the substitute price is omitted from the estimating equation, then the misspecified demand function can still yield a correct measure of welfare change.

Since the compensated demand curves are not directly observable, we must ask whether Willig's (1976) bounds on consumer surplus as an approximation to CS and ES apply in this situation. Unfortunately, since these bounds were derived for differences in areas under ordinary and compensated demand curves, they do not apply to areas between shifting ordinary and compensated demand curves.

This can best be shown graphically, as in figure 4-3.[9] The compensated demand curves for the two levels of q are denoted as $h_1(q')$ and $h_1(q'')$. CS is measured by the area $b + d$. The ordinary demand curves are denoted by $x_1(q')$ and $x_1(q'')$. Observe that at the market price of p_1' the increase in q causes the ordinary demand curve to shift out farther to the right than the compensated demand curve. This is because with the ordinary demand curve, there is no compensating reduction in income to hold utility constant. Taking areas between the ordinary demand curves would yield a consumer surplus measure of $a + b + c$. The percentage error arising when the consumer surplus measure is used to approximate the compensating surplus is:[10]

$$\% \text{ Error} = \frac{a + c - d}{b + d}$$

As can be seen by inspection, this error could be positive, negative, or, by coincidence, equal to zero. Thus the practice of using ordinary demand curves to estimate welfare changes with weak complementarity can lead to errors of unknown sign and magnitude.

Fortunately, recent analyses have shown that it is possible to use the exact welfare measurement methods described in chapter 3 to recover weakly complementary preferences from market demands. For demonstrations and discussions, see Larson (1991) and Bockstael and McConnell (1991).

WEAK COMPLEMENTARITY AND HOUSEHOLD PRODUCTION

As Bockstael and McConnell (1983) have shown, if the individual choice problem is modeled in the household production function framework, there are conditions on the production technology and preferences that yield results equivalent to the weak complementarity model. They show that if these conditions are satisfied, the welfare

---

[9]This analysis is based on Bockstael, McConnell, and Strand (1991).

[10]Similar expressions can be derived for the relationship between consumer surplus and the equivalent surplus measure for changes in q.

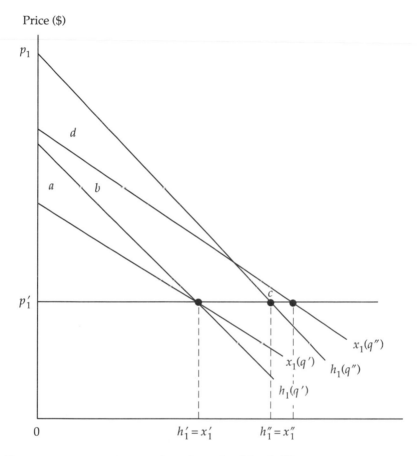

Figure 4-3. Approximating benefits using Marshallian demand curves with weak complementarity

value of a change in q can be calculated from knowledge of the market good's compensated demand function even without knowledge of the household production technology. The demand function for the market good is a derived demand. If the conditions established by Bockstael and McConnell are satisfied, the welfare value of the change in q can be calculated by the area between the two compensated demand curves for the market good at the two levels of q.

The individual's choice problem is to maximize

$$u(Z, q)$$

subject to $$Z = Z(X, q)$$

and $$P \cdot X = M.$$

The expenditure function is

$$e(P, q, u) = \min \{P \cdot X | u^0 = u[Z(X, q), q]\}$$

Suppose that z_1 is produced by the household by combining q and x_1 and that q is used only in the household production of z_1. As in the standard model, the compensated demand for x_1 is the derivative of the expenditure function with respect to p_1. But in this case it is a derived demand, so that it reflects features of both preferences and the household production technology.

An increase in q will lower the marginal cost of the household production of z_1 and therefore increase the quantity demanded of z_1. If the change in q is to increase the compensated demand for x_1, then the household production technology and preferences together must result in an increase in the demand for z_1 sufficiently large that more x_1 is required.[11] The area between the two derived compensated demand curves for x_1 is given by

$$e(p_1^*, q'', u^0) - e(p_1', q'', u^0) - e(p_1^*, q', u^0) + e(p_1', q', u^0)$$

If the first and third terms of this expression sum to zero, then this gives the welfare value of the increase in q. The sufficient conditions for these terms to sum to zero are:

(A) that x_1 be essential in the production of z_1 so that if $x_1 = 0$, $\partial z_1 / \partial q = 0$; and
(B) that q not affect utility independent of its contribution to the production of z_1, in other words, that $\partial u / \partial q = 0$ when $z_1 = 0$.

These two conditions are sufficient to make the expenditure function independent of the level of q. Thus they play a role similar to the

---

[11]In other words, restrictions on the household production technology alone are not sufficient to establish the complementarity or substitutability of the derived demand for x_1 with respect to q. A comparative statics analysis of the household maximization problem shows that with

$$\frac{\partial z_1}{\partial x_1} \cdot \frac{\partial x_1}{\partial q} > 0$$

it is possible for an increase in q to reduce both the ordinary and the compensated derived demands for x_1. See, for example, Courant and Porter (1981), pp. 325–326.

second condition of the weak complementarity model. The household production framework requires restrictions on both the household production technology and preferences for final services. But there is no restriction on the Hicksian demand analogous to the existence of a choke price.

Substitutes

Mäler (1974, pp. 116–118) has shown that the marginal willingness to pay for q can be expressed in terms of the price of any private good and the marginal utilities of that good and q. The expression is

$$w_q = - \frac{\partial e(P, q, u^0)}{\partial q}$$

$$= p_i \left(- \frac{\partial u/\partial q}{\partial u/\partial x_i} \right)$$

$$= p_i \cdot MRS_{qx_i} \tag{11}$$

This would be a useful practical result if it were possible to derive simple expressions for the marginal rates of substitution. Of course, if q could be purchased at a given price the marginal rate of substitution between q and x_1 could be inferred from the price ratio. But when the level of q is determined exogenously, the marginal rate of substitution can only be determined through some knowledge of the utility function or household production function.

Consider the case where some assumption about the separability of the utility function is used to isolate q and a good z from the other determinants of utility.[12] Suppose the utility function is weakly separable and is of the following form:

$$u = v\{u^1(X), u^2(Y), [c \cdot z^{-\alpha} + (1 - c)q^{-\alpha}]^{-1/\alpha}\}$$

Given the separability assumption, the marginal rate of substitution between z and q is independent of the quantities of X and Y. In this case, the marginal rate of substitution is

$$MRS_{zq} = \frac{c}{1 - c} \left(\frac{q}{z} \right)^{\alpha-1}$$

$$= \frac{c}{1 - c} \left(\frac{q}{z} \right)^{1/\sigma}$$

---

[12]The following discussion is based on Mäler (1974), pp. 178–183.

where σ is the elasticity of substitution, which is constant. From (11) we have

$$w_q = - p_z \cdot \left(\frac{c}{1-c}\right) \left(\frac{q}{z}\right)^{1/\sigma} \tag{12}$$

Since in general the marginal rate of substitution depends on the ratio of q/z, we need to know both the elasticity of substitution, σ, and c in order to compute w_q. This requires knowledge of the utility function.

PERFECT SUBSTITUTES

There is one special case where the expression for w_q reduces to a usable term. If z and q are perfect substitutes in consumption, the elasticity of substitution between them is infinite, and the expression for the demand price for q reduces to $p_z \cdot s$, where s is the substitution ratio between z and q $[s=c/(1-c)]$. If perfect substitutability can be assumed, s (or c) should be computable from known or observable technical consumption data or from the household production function.

The perfect substitutability assumption lies behind the defensive expenditures technique for estimating the benefits of pollution control. Defensive expenditures are made either to prevent or to counteract the adverse effect of pollution. They are also referred to as averting expenditures.[13] In effect, a defensive expenditure is spending on a good that is a substitute for higher q. An increase in q is assumed to lead to a decrease in spending on the substitute.

As equation (11) shows, the marginal change in the spending on z is the correct measure of the marginal willingness to pay for the change in q. So for marginal changes, the defensive expenditure technique is valid. But the technique is not valid for nonmarginal changes in q. The intuition behind this statement is straightforward. The benefit of a nonmarginal change in q is the reduction in the spending on z that is required to keep the individual on the original indifference curve. But in general the individual will not actually reduce spending on z by this amount. There is an income effect on z as well as a substitution effect. The increase in q means that the same level of utility can be maintained with a smaller expenditure on z. As a consequence, the individual will reallocate expenditure among all goods, including z, so as to maximize the increase in total utility. This will result in increases in the expenditures on all goods with positive income elasticities of demand. Hence,

---

[13]See, for example, Courant and Porter (1981).

the observed decrease in spending on z will be less than that necessary to hold utility constant. And the reduction in defensive spending will be an underestimate of the benefits of higher q. In fact, as we will see below, actual spending on z could increase.

A GENERAL MODEL

Since perfect substitutes represent a special case, it will be useful to explore a more general model in which q and a market good are less than perfect substitutes and in which the substitution relationship arises because q and the market good contribute to utility through the same mechanism. It seems natural to model this case in the household production framework. For example, suppose that clean air and soap are substitutes in the production of cleanliness.[14] Specifically, let

$$u = u(z, X)$$

where z is cleanliness and X is a vector of market goods. Suppose that z is produced by households by combining a market good y and air quality according to the production function

$$z = z(y, q)$$

with positive partial derivatives for both arguments. Further, assume that y contributes to utility only through its contribution to cleanliness. This is equivalent to assuming that the utility function is weakly separable, with y and q in one group.[15]

Assuming that X is normalized with a unit price, the individual chooses X and y so as to maximize:

$$u = u[X, z(y, q)]$$

subject to

$$M - X - p_y \cdot y = 0$$

---

[14]The following analysis is adapted from Courant and Porter (1981). For other analyses dealing with this type of model, see Shibata and Winrich (1983), Harford (1984), and Harrington and Portney (1987).

[15]An alternative formulation is to let z represent a bad, with y and q mitigating the adverse effects of z. In other words, u_z, z_q, and $z_y < 0$, where subscripts indicate partial derivatives. Similar conclusions are reached about willingness to pay and its relationship to changes in mitigating expenditures. This alternative model is typically referred to as an averting behavior model or a mitigating behavior model.

The first-order conditions are

$$\partial u / \partial X - \lambda = 0$$

$$\frac{\partial u}{\partial z} \cdot \frac{\partial z}{\partial y} - \lambda \cdot p_y = 0 \tag{13}$$

Substituting the expressions for the demands for X and y into the utility function yields the indirect utility function:

$$u = v(1, p_y, M, q)$$

By setting the total differential of this expression equal to zero, assuming prices do not change, and rearranging terms, we obtain the following expression for the marginal value of increasing q:

$$w_q = -\frac{dM}{dq} = \frac{\partial v / \partial q}{\lambda} = \left(\frac{\partial u}{\partial z} \cdot \frac{\partial z}{\partial q} \right) / \lambda \tag{14}$$

where λ is the marginal utility of income and the term in brackets is an expansion of $\partial v / \partial q$ making use of the chain rule. Substituting the first-order condition (13) gives the marginal willingness to pay for q as a function of the price of the private good and the marginal productivities of y and q:

$$w_q = p_y \cdot \frac{\partial z / \partial q}{\partial z / \partial y} \tag{15}$$

This is the reduction in spending on y holding z constant, that is, moving along a given isoquant in the household production function. This marginal value can be calculated if the household production function is known.[16]

---

[16]If y jointly produces two goods (or helps to avoid two bads), this result does not hold. Suppose $u = u[X, z_1(y, q), z_2(y, q)]$. Using the procedure described in the text, we obtain

$$w = p_y \left[\frac{(\partial u / \partial z_1) \cdot (\partial z_1 / \partial q) + (\partial u / \partial z_2) \cdot (\partial z_2 / \partial q)}{(\partial u / \partial z_1) \cdot (\partial z_1 / \partial y) + (\partial u / \partial z_2) \cdot (\partial z_2 / \partial y)} \right]$$

The unobservable utility terms do not cancel out.

An expression for the actual change in spending on y can be obtained by totally differentiating the household production function with respect to q and using the first-order conditions for the optimum choice of y:

$$\frac{dz}{dq} = \frac{\partial z}{\partial y} \cdot \frac{\partial y^*}{\partial q} + \frac{\partial z}{\partial q}$$

where $y^*(M, p_y, q)$ is the derived demand function. Rearranging and multiplying both sides by (13) gives

$$p_y \cdot \frac{\partial z/\partial q}{\partial z/\partial y} = w_q = \frac{\partial u/\partial z}{\lambda} \cdot \frac{dz}{dq} - p_y \cdot \frac{\partial y^*}{\partial q}$$

If spending on y decreases ($\partial y^*/\partial q < 0$), then the marginal willingness to pay for q is the sum of the observed reduction in spending and the willingness to pay for the induced increase in the final service flow z. The actual savings in spending on y is an underestimate of the marginal value, in this case. And if the lower implicit cost of household production of z leads to a sufficiently large increase in the consumption of z, then it is possible that spending on y could increase. Thus observed changes in spending on substitute goods are not reliable indicators of marginal willingness to pay for changes in q.[17]

Bartik (1988) has shown how lower and upper bounds on the welfare gains of nonmarginal changes in q can be calculated from knowledge of only the household production function. The lower bound on the compensating surplus measure of benefits of an increase in q is the reduction in the expenditures on y necessary to reach the initial level of z, other things being equal. More formally, the lower bound is given by

$$\text{Lower bound} = p_y (y' - y'')$$

This can be shown to be a lower bound on CS by defining a restricted expenditure function, $e^*(q, z', u^0)$, where z' is the initial level of z and price terms have been omitted for simplicity. The welfare gain associ-

---

[17]Similar results are obtained for the case where z requires several private goods as well as q for its production. The marginal value of a change in q can be measured by the reduction in spending on any of the market goods for movements along the original z isoquant. And observed changes in spending on market goods are not reliable measures of w.

ated with an increase in q from q' to q'' holding z at z' is

$$CS^* = e^*(q', z', u^0) - e^*(q'', z', u^0)$$

The first term on the right-hand side is $M = X' + p_y \cdot y'$. The increase in q means that less must be spent on y to produce z'. And since $u(X', z') = u^0$, the second term is $X' + p_y \cdot y''$. So

$$CS^* = X' + p_y \cdot y' - X' - p_y \cdot y''$$
$$= p_y (y' - y'')$$

This has to be less than the true CS, since relaxing the constraint holding z at z' would allow the individual to increase utility, unless the marginal utility of z were already zero. In a similar fashion, Bartik shows that an upper bound on the equivalent surplus measure of welfare gain is the decrease in spending on y that is possible while holding z constant at its new equilibrium level. And since in general $ES > CS$, the constrained changes in expenditure bound both of the true measures of welfare change.

We can also consider the case where q contributes to utility directly in addition to its contribution to z. Taking the total differential of the indirect utility function, solving for the compensating change in income, and substituting in the first-order condition for the choice of y gives us

$$w_q = \frac{dM}{dq} = \frac{\partial u / \partial q}{\lambda} + p_y \cdot \frac{\partial z / \partial q}{\partial z / \partial y}$$

Comparing this expression with equation (14), we see that the marginal willingness to pay for q now includes an unobservable marginal utility term for the direct effect of q on utility. Similarly, following the same procedures used to derive equation (15), we have

$$w_q = \frac{\partial u / \partial z}{\lambda} \cdot \frac{dz}{dq} + \frac{\partial u / \partial q}{\lambda} - p_y \cdot \frac{\partial y^*}{\partial q}$$

Again, if spending on y decreases, it is an underestimate of marginal willingness to pay because it neglects both the utility value of the induced increase in z and the direct utility value of the increase in q.

WEAK SUBSTITUTABILITY

Feenberg and Mills (1980) have proposed a special form of demand interdependence that makes it possible to measure welfare gains from nonmarginal changes in q by measuring areas between compensated demand curves for substitute goods. They refer to this form of inter-dependence as weak substitutability. The environmental good q, and a market good x_1 sold at p_1', are weak substitutes if the following two conditions hold:

(A) that for the compensated demand curve, $\partial h_1(p_1', P, M, q)/\partial q$
    < 0, indicating that the goods are substitutes; and
(B) that there be some p_1^* such that $\partial e(p_1^*, P, q, u^0)/\partial q = 0$, at $0 \le p_1^* < p_1'$.

The first condition means that at p_1', an increase in q shifts the com-pensated demand curve for x_1 to the left. The second condition means that there is some lower price for x_1 at which the individual becomes indifferent to changes in q. At this lower price p_1^*, changes in q have no effect on the quantity demanded of x_1.

The implications of weak substitutability for measuring the bene-fits of the change in q can most easily be seen graphically. In figure 4-4, the compensated demand curve for some initial level of q is $h_1(q')$, and $h_1(q'')$ is the compensated demand for x_1 at some higher level of q. If the p_1^* of condition (B) is equal to zero, the two demand curves will intersect at point C. The measurement of the benefits of the increase in q, given p_1', proceeds as follows:

(1) Given the initial level of q and $h_1(q')$, assume that p_1 falls to zero. The gain to the individual is equal to the area $0p_1'BC$.
(2) Given $p_1 = p_1^* = 0$, an increase in q has no effect on the welfare of the individual. However, the demand curve above and to the left of point C shifts to the left to $h_1(q'')$.
(3) Now assume an increase in price back to p_1'. The welfare loss associated with this price increase is equal to the area $0p_1'AC$. The net gain in welfare for these three steps combined is equal to the triangle ABC.

In summary, given the conditions of weak substitutability, the benefits of an improvement in q can be measured by the area between the initial and final compensated demand curves for x_1 bounded above by p_1' and below by p_1^*. Consistent with the earlier analysis of the general case of substitutes, any decrease in actual spending on x_1 will be an underestimate of the true welfare gain. Recall that observed changes in spending are generated by shifts in the ordinary demand curve. If the ordinary demand curve shifts to the left, it will be by a smaller amount than the shift in the compensated demand curve. And the ordinary demand curve could shift to the right.

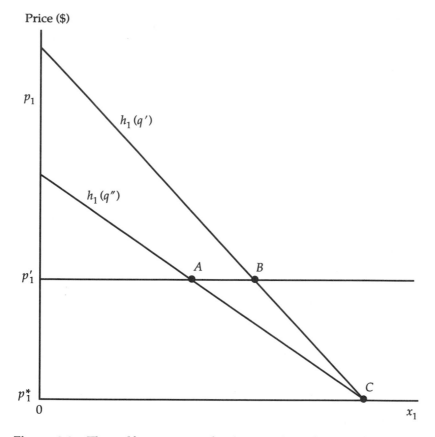

Figure 4-4.   The welfare measure of an increase in q when q and x_1 are weak substitutes

It is tempting to consider whether in the substitute case there is an analog to the weak complementarity measure to be found in areas between shifting demand curves for the market good. If q increases, the compensated demand curve for the substitute x_1 shifts to the left. What is the welfare significance of the change in the area under the demand curve for x_1 but above p_1'?

If p_1^* is now the choke price, the change in the area under the compensated demand curve for x_1 is given by

$$\Delta A = \left[e\left(q'', p_1^*, u^0\right) - e\left(q'', p_1', u^0\right) \right]$$
$$- \left[e\left(q', p_1^*, u^0\right) - e\left(q', p_1', u^0\right) \right] \quad < 0$$

The CS for the increase in q is given by

$$CS = e(q', p_1', u^0) - e(q'', p_1', u^0)$$

If the following condition holds,

$$e(q'', p_1^*, u^0) = e(q', p_1^*, u^0)$$

then $CS = \Delta A$. In other words, the reduction in the area under the demand curve for x_1 measures the benefit of an increase in q only if the individual is indifferent about the level of q when that person's consumption of x_1 has been driven to zero by its high price. It is difficult to imagine circumstances in which this would be the case when x_1 and q are substitutes. If the individual gains from an increase in q when x_1 is zero, then the area between the two demand curves for x_1 will be an underestimate of the welfare gain from the increase in q. And since the degree of underestimation could be substantial, this does not appear to be a useful result.

Bounds on Willingness to Pay

In this section I approach the relationship between changes in spending on market goods and changes q in a more general way. The result is a set of expressions that place bounds on and provide an exact measure of marginal willingness to pay and which may prove to be useful in some cases. The limitation of this line of analysis is that the bounds may prove to be too wide to be useful for policy purposes; in fact, the lower bound estimate of the welfare improvement can even be negative. Also, the exact measure requires a good deal of information on preferences for unrelated goods, which may be difficult to obtain in practice.

First, let us suppose that each of the goods an individual consumes can be classified as being either a complement to, independent of, or a substitute for q, where the definitions and notation are as follows:

The set of complements is $\quad X_C: \quad \partial h_{Ci}/\partial b > 0 \, (i = 1, ..., m)$

The set of independent goods is $\quad X_I: \quad \partial h_{Ij}/\partial b = 0 \, (j = 1, ..., n)$

The set of substitutes is $\quad X_S: \quad \partial h_{Sk}/\partial b < 0 \, (k = 1, ..., p)$

where b is the virtual price of q (see chapter 3). Neill (1988) proves that given these definitions, the following relationships between the compensated quantities and q hold:

$$\partial h_{Ci}/\partial q > 0 \qquad (i = 1, ..., m),$$
$$\partial h_{Ij}/\partial q = 0 \qquad (j = 1, ..., n), \text{ and}$$
$$\partial h_{Sk}/\partial q < 0 \qquad (k = 1, ..., p)$$

The compensated demands and the ordinary demands for each these goods are related according to the following lemma:

LEMMA: $\partial h_{Ji}/\partial q = \partial x_{Ji}/\partial q - w_q \cdot \partial x_{Ji}/\partial M$ where $J = C, I, S, i$ subscripts are goods, x are the ordinary demands, and w_q is the marginal value of a change in q.

Proof: In equilibrium, $h_{Ji} = x_{Ji}$ for all J and i. Since M is equal to the expenditure required to attain the initial level of utility, we can write

$$h_{Ji} = x_{Ji}[P, q, e(P, q, u^0)]$$

Differentiating yields

$$\partial h_{Ji}/\partial q = \partial x_{Ji}/\partial q + (\partial x_{Ji}/\partial M)(\partial e/\partial q)$$

for all J and i. Since $-\partial e/\partial q = w_q$, the lemma is proved. This means that we can write w_q as a function of the compensated and uncompensated changes in demand for any good:

$$w_q = \frac{\partial x_{Ji}/\partial q}{\partial x_{Ji}/\partial M} - \frac{\partial h_{Ji}/\partial q}{\partial x_{Ji}/\partial M}$$

or as a similar function of changes in the expenditure on the good:

$$w_q = \frac{\partial E_{Ji}/\partial q}{\partial E_{Ji}/\partial M} - \frac{\partial H_{Ji}/\partial q}{\partial E_{Ji}/\partial M} \tag{16}$$

where $\qquad\qquad E_{Ji} = p_i \cdot x_{Ji}$

and $\qquad\qquad H_{Ji} = p_i \cdot h_i$

Given knowledge of how the expenditure on a good changes with changes in income and q, the last term in (16) represents the error involved in using that data as an estimate of w_q. For independent goods, this error term is zero. Thus an exact measure of marginal willingness to pay can be obtained from observable changes in the demand for and expenditure on any independent good.[18] The trick in applying this measure is to know enough about the structure of preferences to identify those goods that are independent by our definition and to know how the ordinary demand functions shift with changes in q.

If equation (16) is applied to data on a complement, the result is an overestimate of w_q since the error term in (16) is negative, assuming that the complements are normal goods. The first term represents an upper bound on the true w_q. Theory does not shed any light on how big this error is likely to be. One approach would be to compute equation (16) for each of the private goods known to be a complement to q and select as the "best" upper bound measure the good producing the lowest computed value. In the case of substitute goods, the first term on the right-hand side of (16) will often be negative for individual substitutes and will always be negative for the set of substitutes taken as a whole. Although this expression is a lower bound, this is not a very useful result concerning a number that is presumed to be positive.

A Hopeless Case

As these analyses have shown, there are conditions under which the benefits of a change in q can be estimated from information on the demand functions for market goods. What is necessary is that the specification of these market good demand functions reflect the interaction between q and the market goods in the utility function or the household production function. But there is one form of utility function in which there are no interactions between q and market goods and with which it is impossible to estimate the value of q from market data.

Suppose that the utility function is strongly separable, with q as the single argument in one of the subsets. In other words,

$$u = v[u^1(X) + u^2(Y) + u^3(q)]$$

where X and Y are subsets of marketable goods. Strong separability means that the marginal rates of substitution between any x_i and y_i are independent of q. Since changes in q have no effect on marginal rates of substitution of any of the marketable goods, q can be excluded

---

[18]Actually, Neill's expression for w_q is based on the sum of the expenditure changes for the whole set of independent goods.

as an argument in all of the market demand functions. Although changes in q affect utility, they leave no record of this impact in the data on market transactions. In principle, it is not possible to estimate the demand for q from observable market data on transactions in x_i or y_i when the utility function is strongly separable in q.

Strong separability is a property of two commonly used functional forms for utility functions—the Cobb-Douglas

$$u = a \cdot \prod_i \cdot x_i^{\alpha_i} \cdot q^\beta$$

and the constant elasticity of substitution (CES) function

$$u = a\left(\sum_i b_i \cdot x_i^{-\alpha} + c \cdot q^{-\alpha}\right)^{-1/\alpha}$$

This can be seen by writing them in their log transformations. Separability may be a characteristic of an important class of benefits. For example, those amenities of the urban environment that are not directly associated with private goods consumption may be separable. Also, as discussed in chapter 5, if use of the environment is defined to mean purchase of a marketed good such as travel to a resource site, then nonuse values flow from a separable component of the utility function and cannot be recovered from data on market demands.

Hedonic Prices and the Value of q

The techniques described so far have been developed for the case where the level of the resource service or environmental quality is fixed and is the same for all individuals. Although this represents the textbook version of the public goods problem, it is not descriptive of all possible cases involving public goods or environmental quality. In some circumstances, the level of q can be considered to be a qualitative characteristic of a differentiated market good. In these cases individuals have some freedom to choose their effective consumption of the public good or environmental quality through their selection of a private goods consumption bundle. In effect, the market for the differentiated private good functions also as a market for the public good or environmental quality. For example, people can choose the level of consumption of local public goods through their choice of a jurisdiction to reside in; and the housing market functions also as a market for the purchase of local public goods.

Where these choices are possible, information on public good demand is embedded in the prices and consumption levels for private goods. For example, if air quality varies across space in an urban area, individuals may choose their exposure to air pollution through their decisions about residential location. Residential housing prices may include premiums for locations in clean areas and discounts for locations in dirty areas. If they do, it may be possible to estimate the demand for public goods such as clean air from the price differentials revealed in private markets. These price differentials are implicit prices for different levels of the public good.

A job can also be considered as a differentiated good. Different jobs have different bundles of characteristics and different wage rates. Wage differentials can be interpreted as the implicit prices of job characteristics. These wage and price differentials and the implicit prices they reflect are the subject matter of hedonic price theory.

The hedonic price technique is a method for estimating the implicit prices of the characteristics that differentiate closely related products in a product class.[19] In principle, if the product class contains enough models with different combinations of characteristics, it should be possible to estimate an implicit price relationship that gives the price of any model as a function of the quantities of its various characteristics. This relationship is called the hedonic price function. The partial derivative of the hedonic price function with respect to any characteristic gives its marginal implicit price, that is, the additional expenditure required to purchase a unit of the product with a marginally larger quantity of that characteristic.

More formally, let Y represent a product class. Any model of Y can be completely described by a vector of its characteristics. Let $Q = q_1, ..., q_j, ..., q_n$ represent the vector of characteristics of Y. Then any model of Y, say y_i, can be described by its characteristics, that is $y_i = y_i(q_{i1}, ..., q_{ij}, ..., q_{in})$, where q_{ij} is the quantity of the jth characteristic provided by model i of good Y. The hedonic price function for Y gives the price of any model, as a function of its characteristics. Specifically, for y_i,

$$p_{y_i} = p_y(q_{i1}, ..., q_{ij}, ..., q_{in}) \qquad (17)$$

If $p_y(\cdot)$ can be estimated from observations of the prices and charac-

---

[19]The hedonic price technique was developed by Griliches and others initially for the purpose of estimating the value of quality change in consumer goods. See Griliches (1971). Rosen (1974) has used the hedonic price concept to analyze the supply and demand of the characteristics that differentiate products in competitive markets. For an early discussion of the application of the concept to measuring the demand for environmental quality characteristics, see Freeman (1974).

teristics of different models, the price of any model can be calculated from knowledge of its characteristics.

Before turning to the interpretation of the hedonic price function, it will be useful to describe briefly how the hedonic price function is generated in a competitive market for a differentiated product. Following the analysis of Rosen (1974), assume that each individual purchases only one unit of Y in the relevant time period. An individual's utility depends upon that person's consumption of the numeraire, X, and the vector of characteristics provided by the unit of Y purchased:

$$u = u(X, Q) \qquad (18)$$

or $$u = u(M - p_{y_i}, q_{i1}, \ldots, q_{ij}, \ldots, q_{in}) \qquad (19)$$

if the individual purchases model i at p_{y_i}. In order to maximize (18) subject to the budget constraint $M - p_y - X = 0$, the individual must choose levels of each characteristic to satisfy

$$\frac{\partial u/\partial q_j}{\partial u/\partial X} = \partial p_y/\partial q_j \qquad (20)$$

Alternatively, the marginal willingness to pay for q_j must just equal the marginal cost of purchasing more of q_j, other things being equal.

Inverting (19) and holding all but characteristic j constant, we obtain an indifference curve or bid curve that gives the maximum amount the individual would pay to obtain a model as a function of q_j, holding other things constant. It takes to form

$$B_j = B_j(q_j, Q^*, u^*)$$

where u^* is the solution to the constrained utility maximization problem and Q^* represents the optimally chosen quantities of the other characteristics.

Because of differences in preferences and/or incomes, individuals can have different bid functions. Two such functions for individuals a and b are shown in figure 4-5. They both show diminishing willingness to pay for q_j or a diminishing marginal rate of substitution between q_j and X. Given the hedonic price function, these two individuals choose q_j^a and q_j^b respectively.

Turning to the supply side of the market, firms' costs of production depend upon the levels of the characteristics of the models they

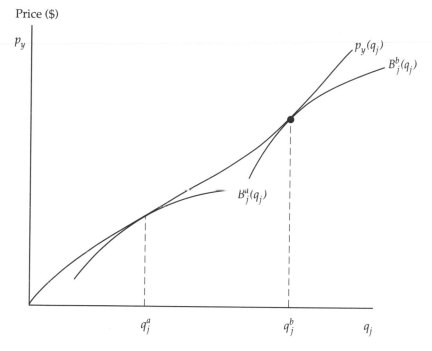

Figure 4-5.   Bid curves of buyers in a hedonic market

produce. Assume that firms are heterogeneous so that their cost functions are different. Inverting a firm's profit function yields an offer curve for the characteristic of the form

$$C_j = C_j\,(q_j\,,\,Q^*,\,\pi^*)$$

where π^* is the maximum attainable profit. Offer curves and the optimal quantities of q_j supplied for firms α and β are shown in figure 4-6 for a given hedonic price function.

For all firms and individuals to be in equilibrium, all of the bid and offer curves for characteristics, for each participant in the market, must be tangent to the hedonic price function. Thus the hedonic price function is a double envelope of the two families of bid curves and offer curves (Rosen, 1974). As a double envelope, the hedonic price function depends on the determinants of both the supply side and the demand side of the characteristics markets. It is a locus of equilibria between bids for and offers of characteristics.

The marginal implicit price of a characteristic can be found by differentiating the hedonic price function with respect to that

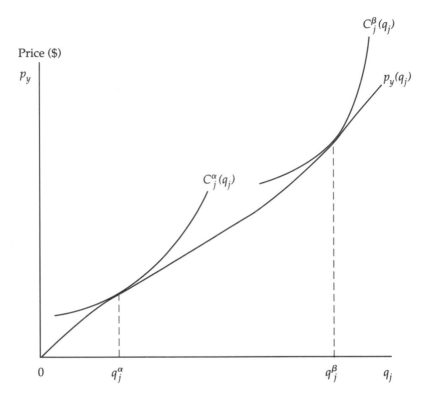

Figure 4-6.   Offer curves of sellers in a hedonic market

characteristic. That is,

$$\partial p_y / \partial q_j = p_{q_j} (q_{i1} , \ldots, q_{ij} , \ldots, q_{in}) \tag{21}$$

This gives the increase in expenditure on Y that is required to obtain a model with one more unit of q_j, other things being equal. If equation (17) is linear in the characteristics, then the implicit prices are constants for individuals. But if (17) is nonlinear, then the implicit price of an additional unit of a characteristic depends on the quantity of the characteristic being purchased.

Equation (17) need not be linear. Linearity will occur only if consumers can "arbitrage" attributes by untying and repackaging bundles of attributes (Rosen, 1974, pp. 37–38). For example, if individuals are indifferent between owning two two-door cars and one four-door car, other things being equal, they can create equivalents of four-door cars by repackaging smaller units. If both sizes exist on the market,

the larger size must sell at twice the price of the smaller one, and the
hedonic price of a car will be a linear function of the number of doors.
The example suggests that nonlinearity will be a common feature of
hedonic price functions.

Another way of looking at the market equilibrium for a characteristic helps to make clearer the welfare implications of the hedonic
price model. First, individual i's choice problem can be solved to
obtain that person's uncompensated inverse demand function for q_j:

$$b_{ij}^* = b_{ij}^* (q_j, P, M - p_y)$$

where i indexes the individual. This is shown in figure 4-7, along with
the marginal implicit price function for q_j, $p_{q_j} (\cdot)$. For each individual,
the quantity of q_j purchased is known by observation and its implicit
price is known from equation (21). This point (q_j', p_{q_j}') can be interpreted as a utility-maximizing equilibrium for this individual

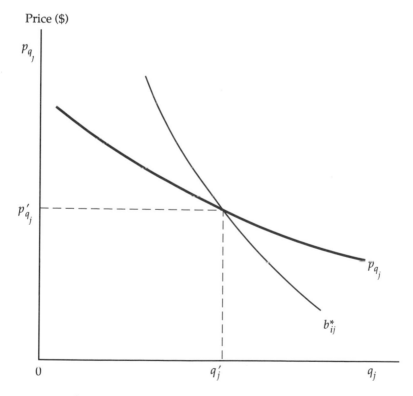

Figure 4-7.   The marginal implicit price and inverse demand curves for q

resulting from the intersection of the individual's inverse demand curve and the locus of opportunities to purchase q_j, as defined by the marginal implicit price function. Thus from the first-order conditions of equation (20), p'_{q_j} can be taken as a measure of the individual's equilibrium marginal willingness to pay for q_j.

Individuals can be viewed as moving out along p_{q_j} as long as their willingness to pay, as reflected in their inverse demand curves, exceeds the marginal implicit price. Thus p_{q_j} is a locus of individuals' equilibrium marginal willingness to pay.

It is also possible to specify a compensated marginal willingness-to-pay function for q_j,

$$b_{ij} = b_{ij}(P, q_j, u^*) \tag{22}$$

This function gives the individual's marginal willingness to pay for q_j, holding utility constant. The value of a nonmarginal change in q_j can be easily calculated if the compensated marginal willingness-to-pay function (22) is known. It is given by the integral of equation (22) over the range of the change in q_j.

This leads us naturally to the question of whether individuals' marginal willingness-to-pay and/or their inverse demand functions can be identified from observations of marginal implicit prices and quantities. The answer depends on the circumstances of the case. If the implicit price function is linear in q_j, then it is not possible to identify a demand curve for q_j. The price observation is the same for all individuals. However, p_{q_j} can be interpreted as the marginal willingness to pay or marginal benefit for small changes in q_j for each individual.

If the hedonic price function is nonlinear, different individuals selecting different bundles of characteristics will have different marginal implicit prices for q_j. There is one situation where the inverse demand function can be immediately identified—that is if all individuals have identical incomes and utility functions. Then the marginal implicit price function is itself the inverse demand function. Recall that the marginal implicit price curve is a locus of equilibrium points on individuals' inverse demand curves. With identical incomes and preferences, all individuals have the same inverse demand curve. Since all the equilibrium points fall on the same inverse demand curve, they fully identify it.

In the case where differences in incomes, preferences, or other variables result in individuals having different inverse demand functions, Rosen (1974) argued that implicit price and quantity data from a single market could be used to estimate this inverse demand func-

tion, provided that the standard identification problem of econometrics could be solved. It is now clear that this analysis is incorrect. The
problem is that the data from a single hedonic market are insufficient
to identify how the same individuals would respond to different implicit prices and incomes.

There are at least two ways in which estimates of inverse demand
functions for q_j can be obtained from hedonic analysis. The first is to
increase the quantity of information obtained from marginal implicit
prices by estimating hedonic price functions for several separate markets, and then pooling the cross-sectional data on the assumption
that the underlying structure of demand is the same in all markets
(Freeman, 1974; Brown and Rosen, 1982; and Palmquist, 1984). The
second approach is to impose additional structure on the problem by
invoking a priori assumptions about the form of the underlying utility
function. The nature of the identification problem in hedonic price
models and alternative approaches to dealing with it are major topics
of chapter 11.

The Household Production Model Again

As we have seen, some of the existing models for interpreting demands for market goods (for example, the averting behavior/
defensive expenditures model and the weak complementarity model)
can be interpreted as simpler versions of a household production
function model. In fact, the household production framework may be
a useful way of thinking about whether a specific market good is
likely to be a substitute or a complement for environmental quality.[20]
It is now time to ask whether a more formal and comprehensive
application of the household production framework will be useful.

Hori (1975) has examined this question. He has shown that under fairly general conditions, knowledge of the household production
function, of the demands for market goods, and of the level of q is
sufficient to determine the ratios of the marginal costs of production—
that is, the marginal rate of transformation (MRT)—$\text{MRT}_{z_j z_j'}$ for any
pair of final service flows. If the individual is in equilibrium, knowledge of the z_j values and the marginal costs gives the individual's
marginal rates of substitution, since $\text{MRS}_{z_j z_j'} = \text{MRT}_{z_j z_j'}$. Hori argued
that this establishes the marginal rate of substitution as a function of
the arguments of the marginal rate of transformation, that is, market
goods (or their prices) and the vector Q. If this is so, integration to the
utility function is straightforward.

---

[20]See Smith (1991) for a discussion of this point of view.

Hori's main objective was to show that under four different specifications of the household production technology it is generally possible to compute the values of the $z_j s$ (which are not directly observable) and their marginal costs. In effect, the Hori model transforms the conventional problem of public goods preference revelation into one of household production function revelation. The usefulness of this approach depends on the ability to determine the household production function. Whether Hori's approach represents a practical advance in technique over models such as those discussed above may be open to question. As Smith (1979) has noted, Hori's model imposes more restrictive assumptions even in those cases where other models can be used.

To see Smith's line of reasoning, consider the simplest case of nonjoint production where each market good enters the production of only one final service flow:

$$
\begin{aligned}
z_1 &= z_1(x_1, \ldots, x_s, q_1), \\
z_2 &= z_2(x_{s+1}, \ldots, x_t, q_2),
\end{aligned}
\tag{23}
$$

$$
\vdots
$$

$$
z_m = z_m(x_v, \ldots, x_n, q_m),
$$

and
$$
u = u(z_1, \ldots, z_j, \ldots, z_m)
\tag{24}
$$

Note that the q_j do not yield utility directly. Hori shows that with knowledge of the $q_j s$, of the demand functions for the $x_i s$, and of the household production function, it is possible to compute the z_j values and their marginal costs for any given set of market goods prices. This case can also be represented in the conventional framework as a weakly separable utility function by substituting equation (23) into the utility function equation (24):

$$
u = [z_1(x_1, \ldots, x_s, q_1); z_2(x_{s+1}, \ldots, x_t, q_2), \ldots]
$$

Weak separability means that the marginal rate of substitution between any two goods in one partition—say, between x_1 and q_1—is independent of the quantities in any other element. As discussed above, the marginal willingness to pay for q_1 can be expressed as a simple function of this MRS. Alternatively, it is the marginal rate of technical substitution between x_1 and q_1 in the household production of z_1. There is no essential difference between the nonjoint production framework and the separable utility framework. Any inference that can be drawn about technical production relationships between public and private goods can also be expressed in the conventional framework in terms of a specification of the MRS_{xq}. Under Hori's approach, knowledge of the whole household production technology is required to solve for the utility function, while under the conventional approaches the analysis of the demand for one public good can proceed with a more limited set of data. And as Pollack and Wachter (1975) and Bockstael and McConnell (1983) have argued, there are formidable data and econometric problems involved in estimating the complete household production technology.

DISCRETE CHOICE MODELS AND MEASURES OF VALUE

The models described in this chapter so far have all in one way or another exploited the marginal equalities revealed when individuals optimize over choice variables that are continuously variable. For example, in the averting behavior model it was assumed that the individual could choose any level of the activity that was consistent with the budget constraint. A marginal rate of substitution could be inferred from knowledge of the slope of the budget constraint. This is not always a realistic way to model the individual choice problem. Some problems are better viewed as involving the choice of one option from a range of discrete alternatives. For example, the choice might be whether or not to take a once-in-a-lifetime cruise around the world, or whether to travel to work by private auto, bus, or on foot. The solutions of discrete choice problems of this sort are essentially corner solutions. As a consequence, there are no tangencies from which a marginal rate of substitution can be inferred. Discrete choice models have been developed both to predict individuals' behavior in these choice contexts and to draw inferences about welfare change on the basis of observed choices.

In this section, I describe a simple discrete choice model and show how measures of welfare change and value can be derived from it. In subsequent chapters I present more detailed discussions of applications of the model to several specific problems, including voting yes or no on a referendum question or accepting or rejecting a hypothetical offer for an environmental commodity (see chapter 6), the choice of which of several alternative houses to live in (see chapter 11), and the choices of whether or not to undertake a specific recreation activity or to visit a specific recreation site (see chapter 13).

Consider now an individual's decision regarding which one of several alternative goods to purchase. The set of alternatives is Y and

P_y denotes the vector of prices attached to the elements of Y. Let y_i $(i = 1, \ldots, n)$ indicate the alternative chosen and p_i be its price, where p_i is an element in P_y. Let the utility of the individual be a function of y_i and the consumption of a numeraire good X purchased at a normalized price of 1. Assume that only one unit of y_i is purchased, so that $X = M - p_i$. There is a set of conditional utility functions

$$u_i = u(M - p_i, y_i, C) \text{ if } y_i \text{ is purchased,}$$

where C is a vector of individual characteristics that affect preferences. If the individual purchases y_i, that person reveals that $u_i > u_j$ for all $j \neq i$.

Now assume that we as researchers cannot fully observe utility. Rather, we can measure an observable component of utility, $v(M - p_i, y_i, C)$. Thus the utilities of all of the possible outcomes are given by:

$$u_i = v(M - p_i, y_i, C) + \epsilon_i \tag{25}$$

where ϵ_i is the random unobservable component of utility. This stochastic specification of the utility function means that the probability that the individual chooses to purchase y_i can be expressed as the probability that the utility associated with i is greater than the utilities associated with all the other alternatives:

$$Pr(y_i) = Pr[v(M - p_i, y_i, C) + \epsilon_i > v(M - p_j, y_j, C) + \epsilon_j] \text{ for all } j \neq i$$

The logit model of McFadden (1973) shows that if the error terms are independently and identically distributed with a Type I Extreme Value distribution, this probability can be written as a function of the utilities, as follows:[21]

$$Pr(y_i) = \frac{e^{v(M - p_i, y_i, C)}}{e^{v(M - p_i, y_i, C)} + \sum_{j \neq i}^{n} e^{v(M - p_j, y_j, C)}}$$

$$= \left(1 + \sum_{j \neq 1}^{n} e^{-\Delta v_{ij}}\right)^{-1} \tag{26}$$

where $\Delta v_{ij} = v_i - v_j$.

---

[21]For more recent expositions of the specification, estimation, and interpretation of discrete choice models, see Ben-Akiva and Lerman (1985) and Train (1986).

The logit model of choice implies certain restrictions on individuals' choices and preferences. The most notable restriction is that choices must have the property of the Independence of Irrelevant Alternatives (IIA). Consider two alternatives from the choice set, i and k. From equation (26) we can see that the ratio of the probabilities of choosing y_i and y_k is independent of the characteristics of any other alternative, that is,

$$\frac{Pr(y_i)}{Pr(y_k)} = \frac{e^{v\,(M-p_i,\ y_i,\ C)}}{e^{v\,(M-p_k,\ y_k,\ C)}}$$

One implication of the property of the IIA is that the addition or deletion of options from the choice set does not affect the ratios of the probabilities associated with any other pair of alternatives. Yet it is easy to think of examples of choice problems in which the property of IIA is implausible. For example, suppose that the individual is choosing from among three sites for a day of fishing. The sites are lake A, lake B, and river C. Suppose that the two lakes are quite similar but that river fishing is a much different experience. Finally, suppose that the estimated probabilities are: $Pr(A) = .25$; $Pr(B) = .25$; $Pr(C) = .5$. It seems plausible that if lake A were eliminated from the choice set, the probability of visiting lake B would rise relative to the probability of visiting the river. But this would violate IIA. The logit model predicts that $Pr(B) = .33$ and $Pr(C) = .67$.

One approach to dealing with the property of IIA is to redefine the choice set so that two or more very close substitutes are modeled as one alternative. In the example of recreational choice, the problem disappears if individuals are modeled as choosing between the alternatives of river fishing and lake fishing.

Now, suppose that we wish to estimate the willingness to pay associated with the choice of y_i over some model, say y_j, for a group of similar individuals. If all individuals face the same price set P_y, and we observe that different individuals make different choices, it must be because of differences in income and/or differences in the vector of characteristics C. If we make some assumptions about the functional form of the observable component of utility in equation (25), the parameters of $v(\cdot)$ can be estimated from equation (26) using maximum likelihood methods.[22] Once the relevant parameters of the util-

---

[22]Some studies have specified the utility difference as being a linear function of the parameters. But Sellar, Chavas, and Stoll (1986) showed that a nonlinear specification may yield results more consistent with demand theory.

ity function are known, it is possible to calculate welfare changes directly.[23] For example, for the individual who chooses y_i when it is available but chooses y_j otherwise, the welfare value of y_i compared to y_j is given implicitly by

$$u(M - p_i, y_i, C) = u(M - p_j + CS_i, y_j, C)$$

or $\quad v(M - p_i, y_i, C) + \epsilon_i - \epsilon_j = V(M - p_j + CS_i, y_j, C) \quad\quad (27)$

In other words, if y_i is removed from the choice set, CV_i compensates for this restriction in choice.[24] The inclusion of the error terms makes CS_i a random variable with a known probability distribution. It can be calculated if the parameters of the utility function are known.[25]

Hanemann (1984) describes three alternative approaches to obtaining a welfare measure from estimates of the parameters of $v(\cdot)$.[26] The first is simply the researcher's expected value of CS_i derived from (27). The second is based on the expectation of $v(\cdot)$ and gives the sum of money that equates the researcher's expectation of the individual's utility with and without the good. The third is that payment which if imposed as a price would make the individual indifferent as to the two alternatives (that is, equally likely to accept or reject the good); it is equivalent to using the median of the distribution of CS_i. Hanemann argues for using the median-based measure, since it will not be as sensitive to outliers and extreme responses in the data.

---

[23]For more rigorous treatments of the basic theory of welfare measurement with discrete choice models, see Small and Rosen (1981) and Hau (1985).

[24]Alternatively, the willingness to pay in order to have y_i added to the available choices is

$$u(M - p_i - CS_i^*, y_i, C) - u(M - p_y, y_j, C)$$

Similar equivalent variation measures can also be defined.

[25]Kaoru and Smith (1990) have shown that although aggregation of alternatives can preserve the IIA property, the full set of alternatives is required for welfare measurement.

[26]See also Johansson, Kriström, and Mäler (1989) and Hanemann (1989). Cameron has developed an alternative specification of the discrete choice model for deriving welfare measures from binary choice data. See Cameron and James (1987), Cameron (1988), and McConnell (1990) for more details. This alternative specification is described in chapter 5 in the context of referendum contingent valuation questions.

SUMMARY

We have reviewed the conceptual basis for empirical techniques that attempt to take advantage of market information. Where q is an input into the production of marketed goods and services, our techniques focus on the effects of changes in q on output and factor markets, not on the utility of q per se. We have shown that there is a valid theoretical basis for examining changes in factor incomes such as land rents, costs savings in production, and changes in consumer surplus associated with the private good outputs.

When q is a consumption good and enters directly into the utility function, the problems are more severe. One of the more promising techniques is based on the concept of weak complementarity. Benefits are measured in terms of shifts in the demand curve for the private complementary good. The approach may be applicable to estimating recreation benefits due to the water quality changes.

Where the public good is either implicitly or explicitly an input, along with one or more market goods, in the household production of a final service flow, benefit estimation is straightforward if the substitution relationship or marginal rate of technical substitution is known and the problem has a sufficiently simple structure so that unobservable utility terms cancel out. Defensive expenditures and measures of additional costs (for example, for household cleaning, or for medical care and drugs in the case of health) are examples of estimates that are based on approximations of this approach.

Another promising case is where q varies across space, as in air pollution, or as a characteristic embodied in some private good. Then individuals can choose different quantities of q by varying residential locations or by choosing different private good models. The hedonic price approach can be used to measure the implicit price of q; and under some circumstances the demand curve for q can be identified. Finally, where individuals make choices from among discrete alternatives, the fact that the actual choice is revealed to yield higher utility than the alternative can be exploited to draw inferences about the utility function and value.

In most of the models described in this chapter, it makes intuitive sense to speak of the individual as *using* the environmental or resource service in question. For example, in the models of substitution and averting behavior, the individual can be interpreted as using clean air to produce a clean household or good health. Thus the values estimated with the models described here are sometimes referred to as use values. It is also possible for individuals to value environmental services independently of any use they might make of those services. The concept of so-called nonuse values is the topic of the next chapter. As we will see there, when nonuse values are defined and modeled in a plausible way, the result is that transactions in market goods reveal nothing about these nonuse values. So the indirect methods described in this chapter can shed no light on the possible magnitude of these values; instead, we must rely on hypothetical direct methods. These are the topic of chapter 6.

REFERENCES

Bartik, Timothy J. 1988. "Evaluating the Benefits of Non-marginal Reductions in Pollution Using Information on Defensive Expenditures," *Journal of Environmental Economics and Management* vol. 15, no. 1, pp. 111–127.

Ben-Akiva, Moshe, and Steven B. Lerman. 1985. *Discrete Choice Analysis: Theory and Application to Travel Demand.* Cambridge, Mass.: MIT Press.

Bockstael, Nancy E., and Catherine L. Kling. 1988. "Valuing Environmental Quality: Weak Complementarity with Sets of Goods," *American Journal of Agricultural Economics* vol. 70, no. 3, pp. 654–662.

Bockstael, Nancy E., and Kenneth E. McConnell. 1983. "Welfare Measurement in the Household Production Framework," *American Economic Review* vol. 73, no. 4, pp. 806–814.

Bockstael, Nancy E., and Kenneth E. McConnell. 1991. "Public Goods as Characteristics of Non-Market Commodities," unpublished paper, University of Maryland, College Park, Md.

Bockstael, Nancy E., Kenneth E. McConnell, and Ivar Strand. 1991. "Recreation," in John B. Braden and Charles D. Kolstad, eds., *Measuring the Demand for Environmental Quality.* Amsterdam, The Netherlands: North-Holland.

Bradford, David F., and Gregory G. Hildebrandt. 1977. "Observable Preferences for Public Goods," *Journal of Public Economics* vol. 8, no. 2, pp. 111–121.

Brown, James N., and Harvey S. Rosen. 1982. "On the Estimation of Structural Hedonic Price Models," *Econometrica* vol. 50, no. 3, pp. 765–768.

Cameron, Trudy A. 1988. "A New Paradigm for Valuing Non-Market Goods Using Referendum Data: Maximum Likelihood Estimation by Censored Logistic Regression," *Journal of Environmental Economics and Management* vol. 15, no. 3, pp. 355–379.

Cameron, Trudy A., and Michelle D. James. 1987. "Efficient Estimation Methods for Contingent Valuation Surveys," *Review of Economics and Statistics* vol. 69, no. 2, pp. 269–276.

Courant, Paul N., and Richard Porter. 1981. "Averting Expenditure and the Cost of Pollution," *Journal of Environmental Economics and Management* vol. 8, no. 4, pp. 321–329.

Deaton, Angus, and John Muellbauer. 1980. *Economics and Consumer Behavior.* Cambridge: Cambridge University Press.

Feenberg, Daniel, and Edwin S. Mills. 1980. *Measuring the Benefits of Water Pollution Abatement.* New York: Academic Press.

Freeman, A. Myrick, III. 1974. "On Estimating Air Pollution Control Benefits from Land Value Studies," *Journal of Environmental Economics and Management* vol. 1, no. 1, pp. 74–83.

Freeman, A. Myrick, III. 1982. *Air and Water Pollution Control: A Benefit-Cost Assessment.* New York: John Wiley.

Goldman, S. M., and Hirofumi Uzawa. 1964. "A Note on Separability in Demand Analysis," *Econometrica* vol. 32, no. 3, pp. 387–398.

Griliches, Zvi, ed. 1971. *Price Indexes and Quality Change.* Cambridge, Mass.: Harvard University Press.

Hanemann, W. Michael. 1984. "Welfare Evaluations in Contingent Valuation Experiments with Discrete Responses," *American Journal of Agricultural Economics* vol. 66, no. 3, pp. 332–341.

Hanemann, W. Michael. 1989. "Welfare Evaluations in Contingent Valuation Experiments with Discrete Response Data: Reply," *American Journal of Agricultural Economics* vol. 71, no. 4, pp. 1057–1061.

Harford, Jon D. 1984. "Averting Behavior and the Benefits of Reduced Soiling," *Journal of Environmental Economics and Management* vol. 11, no. 3, pp. 296–302.

Harrington, Winston, and Paul R. Portney. 1987. "Valuing the Benefits of Health and Safety Regulations," *Journal of Urban Economics* vol. 22, no. 1, pp. 101–112.

Hau, Timothy Doe-Kwong. 1985. "A Hicksian Approach to Cost-Benefit Analysis with Discrete Choice Models," *Economica* vol. 52, no. 209, pp. 479–490.

Hori, H. 1975. "Revealed Preference for Public Goods," *American Economic Review* vol. 65, no. 5, pp. 197–991.

Johansson, Per-Olov, Bengt Kriström, and Karl-Göran Mäler. 1989. "Welfare Evaluations in Contingent Valuation Experiments with Discrete Response Data: Comment," *American Journal of Agricultural Economics* vol. 71, no. 4, pp. 1054–1056.

Kaoru, Yoshiaki, and V. Kerry Smith. 1990. " 'Black Mayonnaise' and Marine Recreation: Methodological Issues in Valuing a Cleanup," Resources for the Future Discussion Paper QE91–02. Washington, D.C.: Resources for the Future.

Katzner, Donald W. 1970. *Static Demand Theory.* New York: Macmillan.

Larson, Douglas M. 1991. "Recovering Weakly Separable Preferences," *Journal of Environmental Economics and Management* vol. 21, no. 2, pp. 97–108.

Mäler, Karl-Göran. 1974. *Environmental Economics: A Theoretical Inquiry.* Baltimore, Md.: The Johns Hopkins University Press for Resources for the Future.

Mathtech, Inc. 1982. *Benefits Analysis of Alternative Secondary National Ambient Air Quality Standards for Sulfur Dioxide and Total Suspended Particulates.* Princeton, N.J.: Mathtech, Inc.

McConnell, Kenneth E. 1990. "Models for Referendum Data: The Structure of Discrete Choice Models for Contingent Valuation," *Journal of Environmental Economics and Management* vol. 18, no. 1, pp. 19–34.

McFadden, Daniel. 1973. "Conditional Logit Analysis of Discrete Choice Behavior," in Paul Zarembka, ed., *Frontiers of Econometrics*. New York: Academic Press.

Mjelde, James W., Richard M. Adams, Bruce L. Dixon, and Phillip Garcia. 1984. "Using Farmers' Actions to Measure Crop Loss Due to Air Pollution," *Journal of the Air Pollution Control Association* vol. 34, no. 4, pp. 360–364.

Neill, Jon. R. 1988. "Another Theorem on Using Market Demands to Determine Willingness to Pay for Non-traded Goods," *Journal of Environmental Economics and Management* vol. 15, no. 2, pp. 224–232.

Palmquist, Raymond B. 1984. "Estimating the Demand for Characteristics of Housing," *Review of Economics and Statistics* vol. 64, no. 3, pp. 394–404.

Pollack, Robert A., and Michael Wachter. 1975. "The Relevance of the Household Production Function and Its Implications for the Allocation of Time," *Journal of Political Economy* vol. 83, no. 2, pp. 255–277.

Rosen, Sherwin. 1974. "Hedonic Prices and Implicit Markets: Product Differentiation in Perfect Competition," *Journal of Political Economy* vol. 82, no. 1, pp. 34–55.

Sellar, Christine, Jean-Paul Chavas, and John R. Stoll. 1986. "Specification of the Logit Model: The Case of Valuation of Nonmarket Goods," *Journal of Environmental Economics and Management* vol. 13, no. 4, pp. 382–390.

Shibata, H., and J. S. Winrich. 1983. "Control of Pollution when the Offended Defend Themselves," *Economica* vol. 50, no. 200, pp. 425–438.

Small, Kenneth A., and Harvey S. Rosen. 1981. "Applied Welfare Economics with Discrete Choice Models," *Econometrica* vol. 49, no. 1, pp. 105–130.

Smith, V. Kerry. 1979. "Indirect Revelation of the Demand for Public Goods: An Overview and Critique," *Scottish Journal of Political Economy* vol. 26, no. 2, pp. 183–189.

Smith, V. Kerry. 1991. "Household Production Functions and Environmental Benefit Estimation," in John B. Braden and Charles D. Kolstad, eds., *Measuring the Demand for Environmental Quality*. Springfield, Ill.: Illinois Department of Energy and Natural Resources.

Train, Kenneth. 1986. *Qualitative Choice Analysis: Theory, Econometrics and an Application to Automobile Demand*. Cambridge, Mass.: MIT Press.

Willig, Robert D. 1976. "Consumers' Surplus Without Apology," *American Economic Review* vol. 66, no. 4, pp. 589–597.

5

Nonuse Values

The hypothesis that underlies this chapter is that people place monetary values on natural resources that are independent of their present use of those resources.[1] For example, people may gain utility from their knowledge that the Grand Canyon is preserved even though they expect never to visit the Canyon. People may be willing to pay to assure the survival of whales, eagles, and other endangered species even though they never expect to see one of them. Lying behind this hypothesis is the assumption that there is a meaningful way to define use so that values that arise from use can be distinguished from those that are independent of use. This assumption is discussed below.

In the economics literature, natural resource values that are independent of people's present use of the resource have been variously termed "existence," "intrinsic," "nonuser," and "nonuse" values. These values are said to arise from a variety of motives, including a desire to bequeath certain environmental resources to one's heirs or future generations, a sense of stewardship or responsibility for preserving certain features of natural resources, and a desire to preserve options for future use.

I think that a majority of the economists working in the field of environmental and resource economics accept the hypothesis of nonuse values, at least in principle, and that many economists believe that nonuse values can be large in the aggregate, at least in some circumstances.[2] If nonuse values are large, ignoring them in natural resource policymaking could lead to serious errors and resource misallocations. Beyond this, however, there is very little

---

[1]Much of the material in this chapter is drawn from Freeman (1993).

[2]For recent reviews of the evidence bearing on the hypothesis of nonuse values, see Randall (1991) and Freeman (1993). An earlier review is in Fisher and Raucher (1984).

agreement among economists as to terminology, definitions, what motivates people to hold nonuse values, and how to measure them empirically.

In this chapter I argue that the appropriate starting place for theoretical analysis is the total value of a change in the quality or quantity of a resource. Given an agreed-upon definition of "use," nonuse values are all of those components of total value that arise independent of use. I then develop a comprehensive framework for defining and measuring this inclusive concept of nonuse values in a manner that is consistent with standard welfare theory. I also propose a nomenclature that is consistent with this framework and that reflects the variety of ways in which welfare theoretic economic values can arise independent of individuals' use of the environment.

In the next section of this chapter, I review some of the early literature in which various forms of nonuse values are discussed. The most restrictive definition links use to the purchase of some complementary good, usually visits to the site of a resource. This definition has been convenient in that it provides a way of measuring use value—for example, through weak complementarity or the hedonic price model. But there are other ways to define use. And we must consider the question of whether the distinction between use and nonuse values is itself useful.

In the third section of this chapter I develop a consistent theoretical framework for defining various forms of nonuse value—a framework based on defining use in terms of the purchase of a related market good. I show that a meaningful distinction can be made between what I call "pure existence values" and other forms of nonuse value. Pure existence values reflect what would be lost if a resource ceased to exist, while the other forms of nonuse value represent what would be lost if a resource were degraded in quality or quantity but continued to exist. Existence values arise from preventing the extinction of a species or preventing the complete destruction of a resource (say, the flooding of the Grand Canyon behind a dam). Other nonuse values arise from such things as preventing the premature death of individual members of highly valued species (such as sea otters and bald eagles) and preventing the pollution of a special environment (such as the pollution caused by the Exxon *Valdez* oil spill in Prince William Sound).

In the fourth section of this chapter I discuss empirical techniques for the measurement of whatever we choose to call nonuse values. I consider the possibility of deriving measures of nonuse values from observed behavior of individuals. But this does not appear to be a fruitful approach; thus we are left with the hypothetical methods that are the focus of chapter 6. I briefly discuss some of the issues involved in applying hypothetical valuation techniques to the measurement of nonuse values.

BACKGROUND: MOTIVATION AND DEFINITIONS

John V. Krutilla can be credited with first introducing the concept of existence or nonuse values into the mainstream economics literature (Krutilla, 1967). In his classic article "Conservation Reconsidered," he argued that individuals do not have to be active consumers of a resource, whose willingness to pay can be captured by a price-discriminating monopoly owner, in order to derive value from the continuing existence of unique, irreplaceable environmental resources. He wrote that "when the existence of a grand scenic wonder or a unique and fragile ecosystem is involved, its preservation and continued availability are a significant part of the real income of many individuals" (Krutilla, 1967, p. 779). In an accompanying footnote he added that "these [individuals] would be the spiritual descendants of John Muir, the present members of the Sierra Club, the Wilderness Society, the National Wildlife Federation, the Audubon Society and others to whom the loss of a species or the disfigurement of a scenic area causes acute distress and a sense of genuine relative improvishment" (Krutilla, 1967, p. 779, n).

Krutilla went on to suggest at least two reasons why people might hold values unrelated to their current use of a resource. These reasons were related to preserving options for future use and bequeathing natural resources to one's heirs.[3] "An option demand may exist," Krutilla wrote, " ... not only among persons currently and prospectively active in the market for the object of the demand, but among others who place a value on the mere existence of biological and/or geomorphological variety and its widespread distribution" (Krutilla, 1967, p. 781). In a footnote he also suggested that the "phenomenon discussed may have an exclusive sentimental basis, but if we consider the bequest motivation in economic behavior ... it may be explained by an interest in preserving an option for one's heirs to view or use the object in question" (Krutilla, 1967, p. 781n). Later,

---

[3]More recent literature has recognized that the value of preserving options arises out of uncertainty about future demand and/or supply. For a brief discussion of this issue and references to the literature, see chapter 8 of this book.

Krutilla and Fisher wrote:

> In the case of existence value, we conceived of individuals valuing an environment regardless of the fact that they feel certain they will never demand *in situ* the services it provides ... however, if we acknowledge that a bequest motivation operates in individual utility-maximizing behavior ... the existence value may be simply the value of preserving a peculiarly remarkable environment for benefit of heirs. (Krutilla and Fisher, 1975, p. 124)

While Krutilla and Fisher offer a bequest motivation as one of several possible explanations for a pure existence value, McConnell (1983) takes a different point of view:

> The notion that a good is valued only for its existence, that it provides no *in situ* services, is far fetched. In most cases, resources are valued for their use. Existence value occurs only insofar as bequest or altruistic notions prevail. We want resources there because they are valued by others of our own generation or by our heirs. Thus use value is the ultimate goal of preferences that yield existence demand, though the existence and use may be experienced by different individuals. (McConnell, 1983, p. 258)

Thus, in McConnell's view, an altruistic attitude toward other people's use of a resource can be the source of existence value. Although the source of existence value is related to someone's use, it is independent of any use made of the resource by the person holding the existence value.

Some have argued that people can have what are essentially existence values out of an ethical or altruistic concern for the status of nonhuman species or proper rules of human conduct (Kopp, 1992). While philosophers of ethics are not in agreement as to the validity and proper form of such concerns,[4] it is possible that some people hold such values and are willing to commit resources on the basis of those values. Such values could be entirely independent of any *human* use of the environment.

As noted above, there is also disagreement about definitions and the classification of various nonuse values. A typical approach in the literature is to define use value as the economic value associated with the in situ use of a resource—as, for example, through visiting a

---

[4]For discussion of these issues, see, for example, Norton (1982), Sagoff (1980), Rescher (1980), and Van De Veer and Pierce (1986).

recreation site or observing a natural wildlife population (see Smith, 1987a). Total value is viewed as an individual's willingness to pay to preserve or maintain a resource in its present state. If total value exceeds use value, the difference is a nonuse value, or as some have called it, an "existence value," an "intrinsic value," or a "preservation value." Several authors have chosen one of these terms to represent the total difference and then identified various possible components of this total. For example, Fisher and Raucher (1984) use the term "intrinsic value" to refer to the aggregate, and state that the total intrinsic value is the sum of option value, aesthetic value, existence value, and bequest value. Sutherland and Walsh (1985) use "preservation value" to refer to the aggregate and state that it is the sum of option, existence, and bequest values.

I question the amount of attention given to questions of classification and motivation in the literature. Motivations do not play an important role in the empirical analysis of the demands for market goods. There is little talk of "prestige value" and "speed value" in the literature on the demand for automobiles. So why should motivations be important in the case of nonuse values? Arguments about motivations seem to be offered primarily to persuade the reader of the plausibility of the hypothesis that nonuse values are positive. But the real test of this hypothesis will come from the data. Rather than further debating definitions and possible motivations, it would be more useful to proceed with a test of the hypothesis that nonuse values (defined in a way that makes testing of the hypothesis feasible) are positive. If the evidence supports this hypothesis, then further research might be devoted to testing hypotheses about the determinants of the size of nonuse values in different cases. At that point, investigation of motivations might be useful in formulating hypotheses for testing. But then the choice of terms and explanatory variables would be governed by what are empirically meaningful distinctions.

One of the major issues in the literature on nonuse values is how to define the use that lies behind use value. The most common approach is to identify some market good or service that is a complement to the resource, as regards consumption, and to define and measure use in terms of the purchased quantity of this complementary good. If the resource is a park, the complementary good is the purchase of travel services to the park. Use of the resource is measured by the number of trips purchased. However, this approach to defining use is clearly a simplification of a more complex reality. The physical proximity that one normally thinks of as being an essential part of use can occur independently of the purchase of a complementary good such as travel. For example, people who live within the natural range of an endangered species such as the bald eagle or peregrine falcon may be able to view one of these birds and experience the utility and value associated with the sightings as an incidental part of their daily routine; there is no connection between that kind of use and any market good.

Some have argued that use does not require the physical proximity of the user and the resource. In their view, use can be defined to include the purchase of a complementary market good that embodies some visual or literary representation of the resource. Randall and Stoll (1983), for example, argue that there can be off-site uses, which they label as "vicarious consumption." "Thus," they explain, "we consider the values generated by reading about Q [the resource] in a book or magazine, looking at it in photographic representations, for example, to be use values. Clearly our definition of use includes vicarious consumption" (Randall and Stoll, 1983, p. 267). This is essentially what Boyle and Bishop (1987) choose to call indirect use value.

Defining use in this way creates some problems for the measurement of value. One problem with such vicarious or indirect forms of use is that the observable market transaction—say the purchase of a nature magazine—often entails the simultaneous use of many environmental resources, so that allocation of the market transaction to specific resources is not possible. Furthermore, where vicarious uses involve information conveyed by photographs, films, and the like, the public-good dimension of information seems likely to virtually destroy any meaningful relationship between observed market behavior and underlying values. And vicarious use has the odd feature that it can occur even though the resource itself no longer exists (as in the viewing of films and photographs).

Another unresolved question in the literature is whether the relevant distinction is between use and nonuse or whether it is more meaningful to distinguish between values to users and nonusers. I prefer the use-versus-nonuse distinction because it focuses on the presence or absence of activities involving the resource directly, rather than on the characteristics of the individuals holding the values. There is no logical reason why a user of a resource could not also hold values that are independent of that use and related to preservation, existence, or bequest motivations. Nonusers of a resource by definition can only hold nonuse values (if any); but users may hold both use and nonuse values for a resource.

This discussion of the possible definitions of and motivations for various types of use and nonuse values is inconclusive. Definitions can be considered in part a matter of taste. A set of definitions can be considered useful if it furthers research objectives and leads to useful answers to meaningful questions, and if the definitions are based on operationally meaningful distinctions. If use values are limited by definition to those associated with in situ use as measured by the purchase of a complementary good, then the definition has the virtue of distinguishing between cases where use of a resource generates observable market data and cases where no meaningful data can be obtained by observing market transactions. However, this definition leaves out vicarious or indirect uses as well as what I would call incidental uses. Perhaps it would be more useful to drop the distinction between use and nonuse values and instead to distinguish between values associated with the in situ consumption of a complementary market good on the one hand and all other values on the other hand. As we will see in the next section, such a classification leads in turn directly into the classification of available empirical techniques for measuring resource values.

A THEORETICAL FRAMEWORK

The theoretical framework for defining nonuse value presented here is based on defining use in terms of the purchase of a market good.

Total Value

Assume that an individual has a preference ordering over a vector of market goods X and some nonmarketed resource q. The individual has no control over the level of q, but takes it as given. Here q is taken to be a scalar measure of some characteristic of the environment, such as the population of some species or the value of some parameter of water quality. In the abstract, q can represent a measure of either a quantity or a quality. The choice of a unit for measuring q has important implications for measurement in practice, but that question is not addressed here. The assumption that the environmental resource can be described by a single attribute is clearly a simplification. A more realistic model would allow for simultaneous changes in two or more quantitative and/or qualitative characteristics of the resource.

Following the analysis in the section on welfare measures for quantity changes in chapter 3, and assuming that r (the price of q) is zero, the compensating surplus (CS) measure of welfare change is

$$CS = e(P, q^0, u^0) - e(P, q^1, u^0) \tag{1}$$

where u^0 is the solution to

$$\text{Max: } u(X, q^0) \text{ subject to } P \cdot X \leq M$$

The derivative of the expenditure function with respect to q gives the compensated inverse demand function for q. If the inverse demand functions are known, the compensating and equivalent surpluses can be measured by integrating over the range of change for q. These are measures of total value and do not distinguish between use and non-use values. In order to distinguish between use and nonuse value, it is necessary to add more structure to the model through assumptions about either preferences or the structure of household production. We turn now to this task.

Weak Complementarity and Use Values

One way to give additional structure to the model is to assume that use can be represented by the in situ purchase of some market good, say x_1, and to define use to be zero when the individual's purchase of x_1 is zero. (For examples of this modeling strategy, see Smith [1987a and 1987b], and McConnell [1983].) Market good x_1 could be travel to a site for recreation, rental of a boat for fishing, or the use of the services of a guide for hunting or fishing. It is further assumed that q is an argument in the demand function for x_1 such that an increase in q increases the demand for x_1, other things being equal. Although q might be weakly complementary with several goods (see chapter 4), for simplicity q is assumed here not to be an argument in the demand function for any other goods.

   If the conditions for weak complementarity are satisfied, then the total value of an increase in q is reflected in the compensated demand function for x_1, and the value of an increase in q that shifts the individual's Hicks-compensated demand curve out to the right can be measured by the geometric areas between the two demand curves. The resource value measured in this way is a use value because it is associated with the use of the resource as measured by the consumption of x_1. The conditions of weak complementarity mean that the individual derives no utility from an increase in q when x_1 is zero. If the second condition of weak complementarity does not hold, the area between the two demand curves is still a measure of use value. But because the individual derives additional utility from an increase in q when x_1 (use) is zero, there is an additional nonuse value that is not reflected in the demand curves for x_1. If the first condition of weak complementarity does not hold (that x_1 be nonessential), there is no finite price that reduces use to zero, and there is no economically meaningful distinction between use and nonuse values.

Separable Preferences and Use Versus Nonuse Values

An alternative structure for preferences gives rise to a case where changes in q have no effect on purchases of market goods. Suppose that the utility function is strongly separable, with q being the single argument in one of the separable subsets. This is the "hopeless case" described in chapter 4. Changes in q have no effect on the demand functions for any market goods. If use is defined as consumption of a complementary market good, then the value of a change in q must be deemed a nonuse value. The utility associated with q is independent of the level of use of any market good. Use values as typically defined are therefore zero.

Alternatively, if use is measured by physical contact or proximity, use is possible without the purchase of any complementary market good. An example of this would be a person who sees members of an endangered species as an incidental part of his daily activities. Given this conception of use, there may be use values. But nothing can be inferred about these values from the demand curves for market goods. Thus estimation of this type of use value by indirect methods based on market data will not be possible.

The Household Production Framework and Use Values

The theory of household production provides a useful framework for thinking about how resources affect utility and welfare and what it means to assume that x_1 is a measure of the use of a resource. Recall that in the household production framework, the demand functions for market goods are derived demands. Assume that q is an input in the production of only one final commodity z_1. If x_1 is also an input in the production of z_1, then (as shown in chapter 4) q will be an argument in the ordinary and compensated demand functions for x_1. If x_1 and q are the only inputs in z_1, then the value of q in producing z_1 will be reflected in the derived demand for x_1 (Bockstael and McConnell, 1983). If x_1 and q are both essential inputs, then the marginal product of each is zero if the other input is absent. A positive level of input for x_1 is required to utilize q in household production. This is equivalent to saying that the use of q and hence its use value is reflected in the level of x_1.

Another advantage of thinking about resource values in the context of the household production framework is the light that it sheds on the nature of what have been called indirect or vicarious use values. It appears that there are two categories of indirect or vicarious values—those that derive from household production and those that derive from market production. If the vicarious use involves the viewing of home-produced films and photos, then the theory suggests that the place to look for measures of indirect use values is in the influence of q on the demands for other inputs in the production of films and photos. If a market good is an essential input to this indirect use, then in principle the use value of changes in q can be measured by areas between the demand curves for the input in home production.

On the other hand, if indirect or vicarious use is the viewing of commercially produced films and television programs or the reading of books and magazines, the source of value for q is in its influence either on the production function or the quality-differentiated demand functions for these marketed goods. The place to look for measures of this type of value is either in changes in costs and prices or in the hedonic price functions for these goods. Thus the role of q is analytically no different from the role of, say, air quality in the production of agricultural crops. There is a well-developed methodology for measuring values in this case, as shown in chapter 9. But it seems doubtful that a reduction in the availability of a specific resource would increase the cost of producing the editorial content of films, books, and magazines, so the empirical significance of this approach to measuring values of changes is in question.

It seems reasonable to assume that the demand for a specific magazine or television program will depend in part on the quality and the information content of the material presented. But the link between changes in q and changes in the demand for information about a specific resource is complex and may not meet the requirements of the standard models for estimating welfare values from changes in demand. For example, a major pollution event that damages a unique resource system may increase the demand for information about that resource; and producers are likely to respond with more articles and programs about that resource. This would be an increase in vicarious use, but it could hardly be called a benefit resulting from damage to the resource.

Defining Nonuse and Existence Values

In this section I maintain the assumption that use of a resource is measured by the consumption of a market good x_1 and that x_1 and q

are weak complements. I show that within this framework use and nonuse values can be defined in a consistent manner that satisfies an adding-up condition. I also show that if there is some threshold for the resource, it is possible to define an existence value that is distinct from the nonuse value associated with increasing the quality or quantity of the resource above that threshold. My purpose in defining this threshold is to show the distinction between the loss of nonuse values associated with the degradation of a resource that continues to exist and the loss associated with the destruction of a resource (the extinction of a species, the flooding of the Grand Canyon, or the clear-cutting of an old-growth redwood forest).

In this section I develop a model of preferences and demand, and use the model to consider five specific cases that are distinguished by the way in which changes in q and p_1 affect the level of use as measured by x_1. I begin by defining a range of values for q, where the relationship between q and use as measured by x_1 at its current price p_1 has certain specific characteristics. Let q^{min} represent the minimum level of q at which it can be said that the resource exists and is available for use; q^{min} represents a threshold or minimum viable level of the resource.

Assume that the choke price for use, p_1^*, is an increasing function of q. Also assume that the compensated demand for the complementary good representing use has the following characteristics:

I. For $0 \leq q < q^{min}$,

$$x_1 = x_1(p_1, P, q, u) = 0$$

for all $p_1 \geq 0$

II. For $q \geq q^{min}$

$$x_1 = x_1(p_1, P, q, u) > 0$$

for $0 \leq p_1 < p_1^*$, and

$$x_1 = x_1(p_1, P, q, u) = 0$$

for $p_1 \geq p_1^*$

Characteristic I says that if the level of q is below the threshold, no use would occur even if use were free. This can be given a reasonable interpretation in the context of household production. Let utility be a function of final services Z, produced according to a household production function, where inputs include market goods X. Let z_1 (x_1, q) represent the household production function for one of these commodities. Also assume that q and x_1 are inputs in the production of only z_1. Characteristic I is equivalent to saying that q is an essential input in the production of z_1 and that q^{min} is the minimum level of q at which positive production of z_1 can be sustained. Thus if q is less than q^{min}, z_1 must be zero and the derived demand for x_1 is zero at all nonnegative prices.

Characteristic II means that q^{min} is sufficient to sustain production. Thus at $q \geq q^{min}$, there are nonnegative prices for x_1 at which its demand is positive. The choke price can be interpreted as the price that raises the marginal cost of household production of z_1 to the point where the quantity demanded of z_1 falls to zero.

Since most discussion of nonuse values occurs in the context of proposals for irreversible development of a natural resource or the destruction or degradation of some natural amenity, the model presented here is framed in terms of the payment required to compensate for a reduction in q. In other words, welfare measures are defined for decreases in q using the compensating surplus definition of welfare change. With appropriate modification, similar conclusions can be reached for increases in q and/or equivalent surplus (ES) measures of welfare change.

Specifically, let q^0 and q^1 represent the initial and final level of the resource, with $q^0 > q^1$. We can distinguish between use value and nonuse value and among three forms of the latter, which I will call pure nonuse value, user's nonuse value, and pure existence value. The definitions of each form of value and their interrelationships are examined in the five cases that follow.

Case I. This case defines pure nonuse value. Let $p_1' > p_1^*$ be the market price for x_1 so that this individual's demand for x_1 is equal to zero both at q^0 and q^1. Let q^0 and q^1 both be greater than q^{min}. In other words, this individual is a nonuser of the resource both before and after the decrease in q. The value for the change in q is given by the difference in the expenditures that are necessary to obtain the initial utility level before and after the change in q, or[5]

$$CS_N \equiv e(p_1', q^0, u^0) - e(p_1', q^1, u^0) < 0$$

This corresponds to the measure of total value (CS) of the resource change defined by equation (1). Note that since use would still be zero if p_1 fell to p_1^*, this can be rewritten as

$$CS_N = e(p_1^*, q^0, u^0) - e(p_1^*, q^1, u^0) \qquad (2)$$

---

[5]For notational simplicity, I suppress the term for all other prices.

Case II. If the market price for x_1 were below the choke price, there would also be use value. Again assume $q^0 > q^1 > q^{min}$. Suppose now that $p_1' < p_1^*$. The use value of the resource is measured by the increase in expenditure necessary to compensate for an increase in the price of x_1 that reduces the use of x_1 to zero. This value provides a dollar measure of the welfare change associated with the use that takes place at the existing price p_1'.

The lost use value of a decrease from q^0 to q^1 is the difference in the CS associated with moving to the choke price at the two levels of q, that is,

$$CS_U \equiv e(p_1', q^0, u^0) - e(p_1^*, q^0, u^0)$$
$$- e(p_1', q^1, u^0) + e(p_1^*, q^1, u^0) < 0 \qquad (3)$$

Notice that CS_U can be defined only if there is a price that chokes off demand. CS_U is the area between the compensated demand curves for x_1 at the two levels of q.

Now define user's nonuse value as that change in expenditure that holds total utility constant, given that the price of visits has increased so as to eliminate use of the site:

$$CS_{UN} \equiv e(p_1^*, q^0, u^0) - e(p_1^*, q^1, u^0) < 0 \qquad (4)$$

This expression defines pure nonuse value for a user.

A comparison of equations (1), (3), and (4) shows that these values are related in the following way:

$$CS = CS_U + CS_{UN}$$

or

$$CS = e(p_1', q^0, u^0) - e(p_1^*, q^0, u^0)$$
$$- e(p_1', q^1, u^0) + e(p_1^*, q^1, u^0)$$
$$+ e(p_1^*, q^0, u^0) - e(p_1^*, q^1, u^0)$$
$$= e(p_1', q^0, u^0) - e(p_1', q^1, u^0)$$

If the initial level of q and the price of x are such that use is positive, then use, user's nonuse, and total values can be defined in a consistent manner that satisfies an adding-up condition. Also, if the conditions for weak complementarity are met, CS_{UN} equals zero and CS_U equals total value CS.

Case III. Now let q^1 be less than q^{min} and q^0 be equal to q^{min}. Also, assume that at $q^0 = q^{min}$, the choke price is equal to the market price $(p_1' = p_1^*)$ so that

$$x_1(p_1', q^0, u^0) = x_1(p_1^*, q^0, u^0) = 0$$

From equation (1), the total loss due to a decrease from q^{min} to q^1 is

$$CS = e(p_1', q^{min}, u^0) - e(p_1', q^1, u^0) < 0$$

Since $p_1' = p_1^*$, there is no use; so this is a loss of nonuse value. Call this a pure existence value, since it is related to falling below an existence threshold. In order to adequately capture the idea of the threshold and the impact of the loss of existence of the resource when q crosses the threshold, we must assume that the expenditure function is discontinuous in q at q^{min} and that the derivative of the expenditure function with respect to q is zero for $q < q^{min}$.

Case IV. Suppose that, at q^{min}, $p_1' < p_1^*$. Now the change from $q^0(q^{min})$ to q^1 can cause losses of both pure existence and use values. The loss of value associated with losing the use made possible at the threshold is

$$CS_U = e(p_1', q^{min}, u^0) - e(p_1^*, q^{min}, u^0)$$

The loss of existence value associated with falling below the threshold can be found by subtracting the lost use value from the total loss:

$$
\begin{aligned}
CS_E &= CS - CS_U \\
&= e(p_1', q^0, u^0) - e(p_1', q^1, u^0) \\
&\quad - e(p_1', q^{min}, u^0) + e(p_1^*, q^{min}, u^0) \\
&= e(p_1^*, q^{min}, u^0) - e(p_1', q^1, u^0)
\end{aligned}
\tag{5}
$$

or, since $e(p_1', q^1, u^0) = e(p_1^*, q^1, u^0)$ when use is precluded by nonexistence,

$$CS_E = e(p_1^*, q^{min}, u^0) - e(p_1^*, q^1, u^0)$$

This shows that it is possible to define existence and use values in a consistent manner such that total value satisfies the adding-up condition. Also, if weak complementarity holds, CS_E is zero and total value equals CS_U.

Case V. Finally, consider the general case where $q^1 < q^{min} < q^0$. The loss has three components. The first is a lost use value associated with the reduction in q to below the threshold. It is given by

$$CS_U = e(p_1', q^0, u^0) - e(p_1^*, q^0, u^0)$$

The second is a loss of user's nonuse value given by a revised version of equation (4) ($q^{\min}$ replaces q^1). The third is a lost existence value as given by equation (5). These components of value satisfy the adding-up condition, as can be seen by substituting these expressions as follows and comparing the result with equation (1):

$$CS = CS_U + CS_{UN} + CS_E$$
$$= e(p_1', q^0, u^0) - e(p_1^*, q^0, u^0)$$
$$+ e(p_1^*, q^0, u^0) - e(p_1^*, q^{\min}, u^0)$$
$$+ e(p_1^*, q^{\min}, u^0) - e(p_1', q^1, u^0)$$
$$= e(p_1', q^0, u^0) - e(p_1', q^1, u^0)$$

This analysis has shown that in the general case where there is a threshold value for q, the total value of a decrease in q that crosses that threshold is the sum of three components:

- a loss of use value because of crossing the threshold that makes use no longer possible;
- the loss of user's nonuse value associated with degradation of q down to the threshold; and
- the loss of pure existence value associated with falling below the threshold of existence.

In a similar manner, it can be shown that the individual whose choke price is below the current price experiences a loss of both pure nonuse value and pure existence value when q falls below the threshold of existence.

These components of total value can also be given an interpretation in the context of the household production framework, as follows.

- The first is a use component that is related to the marginal productivity of q in household production when q is greater than $q^{\min}$ and x_1 is greater than zero. This *use value* can be measured by the difference in consumer surpluses associated with use at the two levels of q.
- The second is a nonuse component that arises because x_1 is nonessential in the household production of z_1 and the marginal productivity of q is greater than zero when x_1 is equal to zero (provided that q is greater than $q^{\min}$). This is a user's *nonuse value*.
- The third relates to crossing the threshold for q and arises because q is an essential input in the household production of commodity z_1 such that the marginal productivity of x_1 in household production is zero when q is less than $q^{\min}$. This is an *existence value*.

The Temporal Dimension of Nonuse Values

The focus of this section so far has been on the structure of preferences in a static, timeless framework. I now take time explicitly into account by looking at the implications of different temporal patterns of changes in q.

First consider a permanent reduction in q from q^0 to q^1 such that $q_t^1 < q_t^0$ for all future t, where t indexes future time. An example would be the irreversible destruction of a unique natural asset. As developed more fully in chapter 7, the proper measure of the welfare loss is the compensating increase in wealth (CW) required to maintain lifetime utility at its original level. This is approximated by the present value of the CSs for all future years. This measure could, in principle, reflect either or both use and nonuse values. The classification and definitions established in this section for the current-period case could easily be modified to be expressed in terms of lifetime utility and welfare.[6]

Now consider a temporary reduction in q, that is, assume that q returns to its original level at, say, t^* ($q_{t^*}^1 = q_t^0$ for some $1 < t^* \leq \infty$). This could be the result of natural recovery or of investment of resources in cleanup and restoration. The temporary reduction in q would likely result in lost use values. But would there also be lost nonuse values? There is nothing in the logic of the structure we have imposed on preferences up to this point that would preclude there being lost nonuse values. But the early literature on nonuse values emphasized irreversible change and the irreplaceability of natural assets as justifications for the presence of nonuse values (see Krutilla, 1967, and Krutilla and Fisher, 1975). It seems to me that whether irreversibility or the irreplaceable character of a natural asset is a prerequisite for nonuse values is an open question that can only be settled on the basis of empirical evidence. This will probably prove to be an important issue in some natural resource damage cases involving long-term but perhaps not irreversible damage to special natural areas, such as Prince William Sound in Alaska.

Conclusions

In this section I have presented some results derived from one way of defining and modeling the use of a resource. I have shown that where there is a complementary market good that can be chosen to repre-

---

[6]Alternatively, the lost values for each period could be measured on the basis of each period's current utility function. But as is shown in chapter 7, the present value of the current-period lost values will be greater in absolute value than CW.

sent use, something about the value of the resource is revealed in the observable market demand for the complementary good. If the assumption of weak complementarity holds, then market demand information can reveal everything there is to know about the value of the resource; that is, value measures derived from market data reflect total value. If the conditions of weak complementarity do not hold, then there are other components of value that are not reflected in market demands. There is also a consistent way of distinguishing between pure nonuse and existence values within this framework. But whether preferences actually take the form that leads to this distinction is an empirical question.

If we adopt a broader concept of use based on physical proximity of the individual to the resource, we must confront the question of how to model those uses that do not require purchase of a complementary good. One possibility is that the requisite proximity is a characteristic of some of the houses in a residential housing market. If that is the case, then hedonic property value models can be used to learn something about that type of use value. Similarly, if some jobs provide opportunities for incidental viewing of peregrine falcons out of office windows, hedonic wage models might be used to estimate this type of use value. Many (but not all) of these incidental uses require time. In principle, it would be possible to infer the "time value" of incidental use from data on individuals' allocation of time across activities. But it is possible that some types of incidental use do not leave any traces in market behavior or in the allocation of time.[7] If that is the case, the distinction between use and nonuse values may well be irrelevant as far as empirical work is concerned, since something other than indirect market-based methods must then be utilized to measure all values.

TOWARD MEASUREMENT

In this section I first ask whether there are any methods based on observed behavior that might lead to measurement of nonuse values. Since my conclusions on this question are pessimistic, I then consider how hypothetical methods might be used as part of a research strategy for measuring nonuse values. I conclude with a discussion of

---

[7]A personal observation: I value the bald eagle population in Maine in part because of the thrill I experience when I see an eagle from my boat. But the number of trips I take, and their duration, are independent of the likelihood that I will see an eagle.

other issues that must be resolved in any effort to measure nonuse values.

Because preservation of a resource for nonusers has the properties of nonexcludability and nondepletability, we would expect markets to fail to provide these preservation services, or at least to provide them in suboptimal quantities. Unless there is a market for preservation of a resource or for enhancing its quality for nonusers, there will be no market transactions to reflect the nonuse values of individuals, and only direct methods of estimating nonuse values will be feasible.

We do observe environmental organizations undertaking a variety of activities to protect and preserve natural environments, and people support these activities through voluntary contributions of time and money.[8] The question is whether this revealed behavior provides an adequate basis for measuring willingness to pay for nonuse values. For several reasons, I do not think that economic data on either the activities of these organizations or individuals' contributions to them can be relied upon as measures of nonuse values for policymaking purposes. Here I discuss a range of these activities and why I think that data on these activities are not appropriate for the task of valuation.

The activities of environmental organizations can be placed in one of two categories: direct provision of preservation through acquisition, and advocacy in an effort to influence public-sector provision of preservation. Organizations such as The Nature Conservancy accept private donations and use the funds to purchase lands with special ecological, geological, or scenic characteristics for the purpose of protection and preservation. Individuals' donations and dues paid to such organizations are manifestations of willingness to pay for preservation, that is, nonuse values. But if "free-rider" behavior is significant, these donations would be only a lower bound of true aggregate willingness to pay for preservation. Furthermore, at least in some instances the lands acquired by such organizations are accessible to individual use. So individual donations could reflect a combination of use as well as nonuse values.

Many environmental organizations devote a substantial portion of their budgets to advocacy activities on behalf of environmental preservation in general and for policy actions to protect specific natural resources. Again, individuals' membership dues and donations in support of these activities reflect individuals' willingness to pay for

---

[8]See Mitchell (1979) for an insightful discussion of the activities of environmental lobby groups.

nonuse values. But because of free-rider behavior, aggregate donations are·likely to represent less than total willingness to pay for preservation. Also, the organizations undertaking these activities frequently have multiple purposes; they provide such services to members as magazines, other publications, and field trips. This means that only that portion of dues and donations supporting the incremental cost of advocacy to the organization is relevant for estimating nonuse values.

In the policymaking arena, another factor weakens the relationship between individual donations to support advocacy and nonuse values. This factor is the uncertainty concerning the outcome of the policy process and the contribution of advocacy activities to the desired outcome. A rational organization with limited resources would estimate the probabilities of successful advocacy as a function of the resource commitment for each specific issue, and would allocate resources so as to maximize the expected value of the outcome. The observed allocation of advocacy resources across specific issues would reflect the interaction of the probabilities of a successful advocacy, the marginal productivity of advocacy in increasing the probability of success, and the value of success to the organization (Freeman, 1969). It would probably prove difficult, if not impossible, to model this complex policy process so as to identify individuals' willingness to pay on the basis of observed contributions to advocacy on specific issues.

It is reasonable to conclude that individuals' donations to environmental organizations involved in acquisition and advocacy reflect willingness to pay for preservation and nonuse values. These activities provide evidence in support of the hypothesis of significant nonuse values. But for the reasons outlined here, aggregate nonuse values are likely to be larger, and perhaps very much larger, than observed acquisition and advocacy expenditures.

Also for the reasons discussed above, the hypothetical direct valuation methods treated in the next chapter are likely to offer the only feasible approaches to estimating nonuse values for policy purposes. Broadly speaking, there are two approaches to estimating nonuse values. The first is to use a hypothetical question to obtain an estimate of total value and to deduct from this a separately obtained estimate of use value. The latter could be obtained, for example, by the travel cost method for estimating the demand for visits to a recreation site. As should be apparent, a problem with this approach to imputing nonuse values is that the imputation is based on the difference between two other values, both of which are measured with some unknown error. So without some understanding of the error properties of the other measures, one cannot know whether the imputed value is simply the result of measurement error or is a true nonuse value.

The second approach to estimating nonuse values is to ask people explicitly about their nonuse values, either by questioning people who are known to be nonusers or by asking people to assume that their own use is zero.[9] Any value revealed by a nonuser is, by definition, a nonuse value. But if present users are asked to assume that they are not using the resource, their responses to hypothetical questions may not be valid indicators of nonuse value. This is because the situation they are being asked to see themselves in is unfamiliar to them, and they might misunderstand it or reject it outright.

Four additional issues concerning research study design deserve at least brief mention. The first, which applies only to studies designed to distinguish between use and nonuse values, is how use is to be defined and measured. This has been discussed above.

The second issue concerns the definition and description of the resource to be valued. Resources have both quantitative and qualitative dimensions that can be affected by policy decisions or damaged by pollution events. Yet in the theoretical discussion above, I assumed that a resource was measured in a single dimension. If the purpose of the empirical research is to estimate total values, then respondents must be given a clear description of the changes in all relevant dimensions of resource quantity and quality. The determination of what is relevant must come in part from the judgment of experienced researchers in this field and in part from research specifically designed to determine what characteristics of the resource are important to people.

The third issue concerns the relevant population for sampling when nonuse values are involved. If the resource to be valued is in California, for example, should the sample include East Coast residents? Or should it be limited only to westerners or to California residents? Casting the sampling net too wide wastes scarce research resources; but important values may be missed if the geographic scope of the sample is too narrow. Even small per capita values can loom large when aggregated over a large population. Again, experience and the results of research designed specifically to shed light on this set of issues can help to guide research study design. It is unclear what more can be said about this issue until we know a lot more about what characteristics of resources are likely to give rise to significant nonuse values. Some re-

---

[9]For further discussion of approaches to measurement that have been used in other studies, see Freeman (1993)

sources such as the Amazonian rain forest and African elephants may have worldwide significance, implying that the relevant population is the world population. And one cannot rule out the possibility that where there are regionally significant resources, important nonuse values are held only by people within that region. It can be said that this is an important research question.

The fourth issue concerns the choice of the temporal dimension of the value measure being elicited. In the preceding section, I showed that the correct welfare measure was the compensating change in wealth, that is, a "one-shot" payment. This amount would be elicited by asking individuals a lump-sum question. Some studies, on the other hand, have asked respondents about their willingness to pay in equal annual payments over some period of time. For this form of question, it is necessary to specify the time period involved. Individuals are likely to find this form of question difficult to answer because they are being asked implicitly to carry out an annuitization calculation. This seems likely to reduce the reliability of responses.

SUMMARY

As this discussion of the concept of nonuse values and approaches to their measurement has shown, several questions remain unresolved. One of the most fundamental is how to distinguish between use and nonuse values. The present practice is a consequence of how the field of natural resource valuation has evolved over time. Economists have devoted their energies to developing techniques for measuring particular pieces of the total value puzzle. The success of such a piece-meal strategy depends on how we choose to break up the pieces. The easy pieces have been those associated with identifiable uses such as recreation. As more specific uses have been identified, modeled, and measured, the use-versus-nonuse distinction has seemed to take on some validity. And some attention has been devoted to the separate development of a theory of nonuse values.

But as I have tried to show in this chapter, the question of how to model nonuse values cannot be separated from that of modeling use values. The theory of economic welfare gives unambiguous guidance only on defining total values as compensating income changes for changes in a resource. The question of whether nonuse values, however defined, are positive takes on meaning only after some decision has been made about what use values measure, since nonuse value is simply total value minus whatever has been called use value.

Most theorizing on nonuse value starts by defining use in terms of observable market-related behavior, that is, the purchase of a complementary market good. This has been a useful strategy because of the availability of indirect methods for estimating values from observed market data. But it is not totally unfair to characterize this strategy as one of shaping the definitions to fit the available empirical methods and data. This problem cannot be avoided by declaring that the relevant distinction is between users and nonusers, since users cannot be identified until we have decided what activity it is that characterizes users. And I have shown that there is no a priori reason why users cannot also hold nonuse values.

I urge that we not accept uncritically the present conventions for distinguishing between use and nonuse values. Ultimately we want to be able to measure total value. Any distinction between use and nonuse values is itself useful only if it helps in the task of measuring total values. To be fair, I think that the present convention has been useful because it makes possible the measurement of certain kinds of use values by invoking the weak complementarity model. But this is not the only way to look at the problem.

Another important question is, when are nonuse values likely to be important? The long literature on nonuse values emphasizes the uniqueness or specialness of the resource in question and the irreversibility of the loss or injury. For example, economists have suggested that there are important nonuse values in preserving the Grand Canyon in its natural state and in preventing the global or local extinction of species and the destruction of unique ecological communities. In contrast, resources such as ordinary streams and lakes or a subpopulation of a widely dispersed wildlife species are not likely to generate significant nonuse values because of the availability of close substitutes. Moreover, the literature does not suggest that nonuse values are likely to be important where recovery from an injury is quick and complete, either through natural processes or restoration. But there are problems in giving operational meaning to the idea of uniqueness. In economic terms, uniqueness would be reflected in the absence of substitutes and a low price elasticity of demand. This is an idea that comes from the literature on industrial economics. But there is no threshold on price elasticity that distinguishes between the presence or absence of close substitutes. Similarly, long-term injury with slow recovery could give rise to nonuse values that are of the same order of magnitude as those associated with irreversible injury.

It would be desirable to give empirical researchers some guidance about how to determine when nonuse values are likely to be significant. However, at the present time there is no general method for determining whether a resource is sufficiently unique or a resource change is of sufficient duration to generate important nonuse values. This important question should be the subject of continued research.

REFERENCES

Bockstael, Nancy E., and Kenneth E. McConnell. 1983. "Welfare Measurement in the Household Production Framework," *American Economic Review* vol. 73, no. 4, pp. 806–814

Boyle, Kevin J., and Richard C. Bishop. 1987. "Valuing Wildlife in Benefit-Cost Analyses: A Case Study Involving Endangered Species," *Water Resources Research* vol. 23, no. 5, pp. 943–950.

Fisher, Ann, and Robert Raucher. 1984. "Intrinsic Benefits of Improved Water Quality: Conceptual and Empirical Perspectives." In V. Kerry Smith and Ann Dryden Witte, eds., *Advances in Applied Microeconomics.* Greenwich, Conn.: JAI Press.

Freeman, A. Myrick, III. 1969. "Advocacy and Resource Allocation Decisions in the Public Sector," *Natural Resources Journal* vol. 9, no. 2, pp. 166–175.

Freeman, A. Myrick, III. 1993. "Nonuse Values in Natural Damage Assessment." In Raymond J. Kopp and V. Kerry Smith, eds., *Valuing Natural Assets: The Economics of Natural Resource Damage Assessment.* Washington, D.C.: Resources for the Future.

Kopp, Raymond J. 1992. "Ethical Motivations and Nonuse Value," RFF Discussion Paper QE92-10. Washington, D.C.: Resources for the Future.

Krutilla, John V. 1967. "Conservation Reconsidered," *American Economic Review* vol. 57, no. 4, pp. 777–786.

Krutilla, John V., and Anthony C. Fisher. 1975. *The Economics of Natural Environments: Studies in the Valuation of Commodity and Amenity Resources.* Baltimore, Md.: The Johns Hopkins University Press.

McConnell, Kenneth E. 1983. "Existence and Bequest Value." In Robert D. Rowe and Lauraine G. Chestnut, eds., *Managing Air Quality and Scenic Resources at National Parks and Wilderness Areas.* Boulder, Colo.: Westview Press.

Mitchell, Robert Cameron. 1979. "National Environmental Lobbies and the Apparent Illogic of Collective Action." In Clifford S. Russell, ed., *Collective Action: Applications from Public Choice Theory.* Baltimore, Md.: The Johns Hopkins University Press for Resources for the Future.

Norton, Bryan G. 1982. "Environmental Ethics and Nonhuman Rights," *Environmental Ethics* vol. 4, pp. 17–36.

Randall, Alan. 1991. "Total and Nonuse Values." In John B. Braden and Charles D. Kolstad, eds., *Measuring the Demand for Environmental Quality.* Amsterdam, The Netherlands: North-Holland.

Randall, Alan, and John R. Stoll. 1983. "Existence Value in a Total Valuation Framework." In Robert D. Rowe and Lauraine G. Chestnut, eds., *Managing Air Quality and Scenic Resources at National Parks and Wilderness Areas.* Boulder, Colo.: Westview Press.

Rescher, Nicholas. 1980. *Unpopular Essays on Technical Progress.* Pittsburgh, Pa.: University of Pittsburgh Press.

Sagoff, Mark. 1980. "On the Preservation of Species," *Columbia Journal of Environmental Law* vol. 7, pp. 33–67.

Smith, V. Kerry. 1987a. "Non-Use Values in Benefit-Cost Analysis," *Southern Economic Journal* vol. 54, no. 1, pp. 19–26.

Smith, V. Kerry. 1987b. "Uncertainty, Benefit-Cost Analysis, and the Treatment of Option Value," *Journal of Environmental Economics and Management* vol. 14, no. 3, pp. 283–292.

Sutherland, Ronald J., and Richard G. Walsh. 1985. "Effect of Distance on the Preservation Value of Water Quality," *Land Economics* vol. 61, no. 3, pp. 281–291.

Van De Veer, Donald, and Christine Pierce. 1986. *People, Penguins, and Plastic Trees: Basic Issues in Environmental Ethics.* Belmont, Calif.: Wadsworth.

6

Hypothetical Methods for Direct and Indirect Valuation

In chapter 4 I have described several methods for inferring individuals' values for environmental amenities from their observed choices of quantities of market goods when they are assumed to make optimal choices subject to income and other constraints. The resulting data on individual choices and behavior can be combined with some model of how the environmental amenity affects utility to infer a willingness to pay. But discussions in chapters 4 and 5 also showed that there are many circumstances under which value measures cannot be derived from observations of individuals' choices. In this chapter I describe a variety of methods that have in common their source of data for analysis: individuals' responses to hypothetical questions that take the form, What would you do if ...? or Would you pay ...?

The most commonly used hypothetical questions simply ask people what value they place on a specified change in an environmental amenity or the maximum amount they would be willing to pay to have it occur. The responses, if truthful, are direct expressions of value and would be interpreted as measures of compensating surplus (CS). The term contingent valuation method (CVM) is conventionally used to refer to approaches based on this form of question.

A second major type of hypothetical question asks for a yes or no answer to the question, Would you be willing to pay $X ...? Each individual's response reveals only an upper bound (for a no) or a lower bound (for a yes) on the relevant welfare measure. Questions of this sort are often referred to as referendum questions because of the analogy with voting on such things as bond issues. Discrete choice methods can be used to estimate willingness to pay functions or indirect utility functions for data on responses and on the characteristics of the people surveyed.

The third and fourth major types of hypothetical questions do not reveal monetary measures directly. Rather, they require some form of analytical model to derive welfare measures from responses to questions. The third approach to questioning is known as contingent ranking. Respondents are given a set of hypothetical alternatives, each depicting a different situation with respect to some environmental amenity and other characteristics that are presumed to be arguments in the respondent's preference function. Respondents are asked to rank the alternatives in order of preference. These rankings can then be analyzed to determine, in effect, the marginal rate of substitution between any other characteristic and the level of the environmental amenity. If one of the other characteristics has a monetary price, then it is possible to compute the respondent's willingness to pay for the good on the basis of the ranking of alternatives.

In the fourth type of hypothetical question—known as contingent activity questions—individuals can be asked how they would change the level of some activity in response to a change in an environmental amenity. If the activity can be interpreted in the context of some behavioral model, such as an averting behavior model or a recreation travel cost demand model, the appropriate indirect valuation method can be used to obtain a measure of willingness to pay.

Some issues and problems in hypothetical methods are specific to the particular form of the question being asked. For example, when people are asked how much they would be willing to pay for something, they might say "zero" because they reject the idea of having to pay for something they consider to be rightfully theirs. Other problems are generic to all methods based on hypothetical questions—for example, problems in scenario specification, sampling, and item nonresponse. In this chapter I focus more on the generic issues, but since the choice of a hypothetical method for dealing with a specific problem must necessarily depend upon the relative strengths and weaknesses of each type of hypothetical question, I also discuss the more specific problems.

The major questions regarding all hypothetical methods concern the validity and reliability of the data, that is, whether the hypothetical nature of the questions asked inevitably leads to some kind of bias or results in so much "noise" that the data are not useful for drawing inferences. In the first section below I describe a framework for defining and assessing validity and reliability. In the second section I describe each of the four forms of questions for obtaining data from hypothetical questions in more detail, and I briefly describe some of the problems that are specific to each form. In the third section I take up the question of the validity of hypothetical responses. Because the answers to some forms of hypothetical questions can be interpreted directly as value measures but other responses may require some kind of analytical model from which values can be inferred, in the fourth section I describe the estimation of welfare measures through the use of data from direct questions and the two forms of discrete choice question: referenda and contingent ranking.

My treatment of the issues surrounding hypothetical methods is selective. I will not discuss a number of design and implementation problems, such as the form of contact with respondents (mail, telephone, personal interview) and sampling design. And I will not review the now extensive literature reporting empirical results of hypothetical surveys. For a thorough and comprehensive treatment of these issues, readers should consult Mitchell and Carson (1989) and the overview by Carson (1991). For an earlier review of some of the issues and an assessment of the state of the art at that time, see Cummings, Brookshire, and Schulze (1986).

A FRAMEWORK FOR ANALYZING VALIDITY AND RELIABILITY

It is tempting to think of hypothetical methods as being cheap and easy substitutes for the data-hungry indirect methods, with their sophisticated analytical and econometric models. But the hypothetical methods have their own difficulties, subtleties, and pitfalls. They require more than just going out and asking people questions. In a commentary on the assessment of the state of the art of contingent valuation, Kenneth Arrow has said, "If you ask somebody a question, you will get an answer" (Arrow, 1986, p. 181). But will the answer you get convey the desired information on the respondent's preferences and values?

This question in one form or another has been at the heart of the debate on the merits of CVM and related techniques from the very beginning. And it has provided a major motivation for the recent development and testing of hypothetical methods generally. In my earlier book (Freeman, 1979) I identified two types of concern with hypothetical questions: the incentive for respondents to behave strategically—that is, to provide biased responses in order to influence public policy—and the absence of an incentive for respondents to provide accurate responses when asked about purely hypothetical situations. There are, in fact, many other sources of potential bias in addition to strategic behavior. The twin problems of bias from all sources and accuracy were the primary focus in the assessment of the state of the art of contingent valuation conducted by Cummings, Brookshire, and Schulze (1986).

Until recently, the literature has lacked a formal framework for classifying types of error in hypothetical responses, and in particular for defining bias. What follows is my adaptation of the framework recently presented by Mitchell and Carson (1989, pp. 120–125) for defining and classifying types of errors. Although the following discussion is presented in terms of a directly revealed value, with appropriate modification the framework can be applied to any form of hypothetical response data.

Let Wt_i be the ith individual's true willingness to pay and Wr_i be his or her revealed willingness to pay (WTP). Wt_i depends on the size of the environmental change being valued (Δq), income (M_i), and a vector of other socioeconomic variables (S_i):

$$Wt_i = Wt_i(\Delta q, M_i, S_i)$$

Wr_i can differ from Wt_i for three reasons. The first is a random error process affecting Wr_i with a mean of zero. It is described by $f_1(X, a)$, where X is a vector of variables and α is a vector of parameters describing this process. The second is a systematic error process affecting Wr_i which may include those errors arising from the way the scenario is specified. It is described by $f_2(Wt_i, Y, \beta)$, where Y and β represent another set of variables and parameters describing this process. The third reason is related to the probability that Wr_i will actually be observed. It is described by $f_3(Wt_i, Z, \gamma)$ where Z and γ are the variables and parameters describing this process. Combining these three processes, we have

$$Wr_i = g\{Wt_i(\cdot), f_1(\cdot), f_2(\cdot), f_3(\cdot)\}$$

where $g\{\cdot\}$ aggregates the three sources of divergence between true and revealed values.

The process captured by $f_1(\cdot)$ reflects what I had termed the problem of accuracy in my earlier book (Freeman, 1979, pp. 97–99) and what some others have called hypothetical "bias."[1] In the framework presented here this would not be called a bias, since by assumption the random inaccuracies in responses have a zero mean. The process captured by $f_2(\cdot)$ includes such things as strategic be-

---

[1]See, for example, Rowe, d'Arge, and Brookshire (1980) and Schulze, d'Arge, and Brookshire (1981). Cummings, Brookshire, and Schulze (1986) provide a more critical discussion of the term.

havior, starting point bias, anchoring, and implied value cues, all of which come under the heading of validity and which are discussed in a later section of this chapter. The $f_3(\cdot)$ process represents such things as sampling bias and item nonresponse bias.

If the probability that Wr_i as observed is independent of Wt_i and of other individual characteristics, then the $f_3(\cdot)$ mechanism is not operative.[2] If there is no systematic error in the individual's revealed willingness to pay, then

$$E[Wr_i - Wt_i] = 0 \qquad \text{for all } i.$$

Under these conditions there is no bias in the measurement of individual willingness to pay, and the survey instrument is valid. Then the only source of divergence between sample means of Wr_i and Wt_i is the random error process. And the impact of this error process can be reduced (and the reliability of the mean as a measure increased) by increasing the sample size and improving the sample design (Mitchell and Carson, 1989, p. 122). The reliability-sample size issue and the problem of nonresponse bias are reasonably well understood. It is the question of the validity of the measurement instrument that deserves further attention. I will return to this question in the second section below.

FORMS OF HYPOTHETICAL QUESTIONS

The types of hypothetical questions described in this section can be administered by trained interviewers in person or by telephone, by a written survey instrument administered through the mail, or, more recently, by an interactive computer program (see, for an example, Magat, Viscusi, and Huber, 1988). Studies often incorporate experimental designs with different treatments administered to different subsets of the survey sample. The purpose of the different treatments is to test hypotheses about how responses are influenced by certain features of the study, such as the format of the question. It is important that all individuals in one treatment group receive the same information and are asked the same questions, so in-person and telephone interviews will normally follow a script. The survey instrument normally consists of three components:[3]

---

[2] If this source of bias is present, there are standard statistical methods for coping with it. See Mitchell and Carson (1989, especially chapter 12) for an extensive treatment.

[3] The following discussion is based on Mitchell and Carson (1989, pp. 2–4).

1. *A description of the choice setting in which the respondent is to imagine himself or herself.* This should include a description of the amenity or resource to be valued or ranked or the conditions under which some activity is being undertaken.

2. *The choice questions from which values will be inferred.* These can take any of the forms described below, that is, they may be direct expression, referendum, contingent ranking, or contingent activity questions.

3. *Questions about the respondents.* Information on respondents' socioeconomic characteristics, such as income, age, gender, and education, may make it possible to estimate bid functions, value functions, or indirect utility functions that control for these variables. Questions about attitudes and beliefs may make it possible to test hypotheses about the influence of (for example) pro-environmental attitudes on revealed values for environmental amenities.

Following the terminology of Mitchell and Carson (1989), I will refer to these three components together as a scenario. The remainder of this section focuses on the choice questions; choice setting and other aspects of the scenario are discussed in the following section.

Direct Expressions of Values

Data gained from hypothetical direct expression of value questions (hypothetical data) are the simplest to interpret. Each respondent is asked in one way or another to state his or her maximum willingness to pay for an environmental improvement (compensating surplus) or to avoid a loss (equivalent surplus). Alternatively, the survey could attempt to determine each individual's maximum required compensation to accept a loss (compensating surplus) or forgo a gain (equivalent surplus).[4] There are several ways to elicit this number. Many of the earliest studies used an iterative technique, which has come to be

---

[4]There is a substantial body of evidence of large differences between willingness-to-accept (WAC) and willingness-to-pay (WTP) values for the same environmental change. See, for example, Cummings, Brookshire, and Schulze (1986, pp. 35–36). The general view is that WTP values are more reliable indicators of welfare change than are the WAC results (Cummings, Brookshire, and Schulze, 1986; Mitchell and Carson, 1989). For this reason, the following discussion is in terms of willingness to pay questions rather than compensation questions. The differences between WTP and WAC responses are discussed in a later section.

called the bidding game.[5] In the bidding game, individuals are first asked whether they would be willing to pay $X. If the answer is yes, the question is repeated with a higher "price." The procedure is repeated until the individual answers no. The highest price with a yes response is interpreted as the maximum willingness to pay. If the original response is no, the iteration proceeds downward until a yes response is received.

When researchers divided their samples and tested for the effects of different starting points for the bidding game, they often found that the starting point did influence the outcomes. This has been called the problem of starting point bias. Mitchell and Carson have argued that this is just one of several ways in which the format and content of the survey instrument can provide what they call implied value cues to respondents (Mitchell and Carson, 1989, pp. 240–246). I will return to this issue in the section on the validity of responses.

An alternative elicitation technique is simply to ask open-ended questions of the form, How much would you be willing to pay? The major problem with this approach is that it confronts people with an unfamiliar problem. In most real market settings, individuals are faced with choices among sets of goods with listed prices. Seldom are they asked to offer a one-shot bid that may be either accepted or rejected by the seller. People appear to have difficulty dealing with the open-ended form of direct question. As a consequence, surveys using this form of elicitation method typically result in high rates of nonresponse to the valuation question and/or high proportions of implausibly high or low stated values.[6]

A variation on the open-ended approach is to show respondents a card with a range of alternative payment values on it and ask them either to pick a number from the card or to state their own value if that is not to be found on the card. Mitchell and Carson and others have also experimented with payment cards that indicate the amounts that typical respondents are paying in the form of taxes for such public programs as police protection, health care, and national defense. There is evidence that the use of payment cards and the

---

[5]Mitchell and Carson (1989, appendix A) present an annotated summary of more than one hundred contingent valuation surveys conducted between 1963 and 1987. The annotations include information on the elicitation method, sample size, means of contact (mail, telephone, personal interview), and other characteristics of the surveys, analyses, and results.

[6]The problems of failure to respond to some questions (known as item nonresponse) and implausible responses (outliers and protest zeros) are discussed below in the section on bidding data.

provision of information on what is being paid for other public goods reduces the variance in individuals' responses. But it may also introduce other forms of bias.[7]

Another feature of scenario design that may be important is the specification of a payment vehicle for collecting the respondents' willingness to pay. In order to make contingent market scenarios plausible to respondents, it is necessary to inform them how their stated value will be collected. Examples of payment vehicles used in past studies include an increase in the prices of all goods resulting in a higher expenditure of $X, an increase in taxes of $X, and a surcharge on utility bills of $X.[8]

There is evidence that the specification of a payment vehicle can exert an independent influence on bids, a phenomenon that has been termed "vehicle bias" in the literature. But judging whether this influence is a bias in the usual sense of the word is a more difficult matter, since the respondent's "true" value is probably not known. The influence of the payment vehicle on bids is one example of the importance of what Fischhoff and Furby (1988) have called the social context of the hypothetical transaction. They argue that in both designing and interpreting surveys that elicit hypothetical values, researchers must be sensitive to how individuals understand and interpret the environmental good being valued, the payment they are being asked to consider, and the social context in which the exchange takes place.

Another problem with the direct expression of values is a potential for strategic bias. Strategic bias results from conscious attempts by individuals to influence either their payment obligation or the level of provision of the environmental good through their stated valuations. I return to this issue below in the section on incentives to misrepresent values.

---

[7]For further discussion, see Mitchell and Carson (1989), especially, pp. 100–101 and 241–243.

[8]If the vehicle involves a higher price for a market good, the question must be clear about whether it refers to a change in expenditure or a change in price. If the question is about a change in price, then the welfare calculation must take account of the reduction in quantity that a price increase will cause. In some studies where respondents were asked for a maximum increase in a price per unit, total willingness to pay was calculated by multiplying the increase in price by the observed quantity. But assuming that respondents accurately stated the maximum price increase they would accept, the correct welfare measure is the area to the left to the demand curve between the two prices. This will be less than the original quantity times the increase in price. For examples see Randall, Ives, and Eastman (1974) and Brookshire, Ives, and Schulze (1976).

Responses to bidding games and open-ended willingness-to-pay questions require no analysis to obtain a measure of individual welfare change. Responses from a sample of individuals can be analyzed econometrically to test hypotheses about the influence of income, taste variables, and so forth. And econometric analyses can also be used to test for the effects of differences in treatments, such as differences in starting point and payment vehicle.

Referendum Questions

In its simplest form, the referendum question format entails asking each respondent whether he or she would be willing to pay a specified amount of money to obtain the environmental change in question. If a respondent answers yes, that person has indicated a willingness to pay that is greater than or equal to the specified sum. If the response is no, then that sum of money can be taken as an upper bound on true willingness to pay. Respondents should be assigned randomly to different subsamples, with each subsample being asked to respond to a different dollar amount. It is then possible to test the hypothesis that the proportion of yes responses decreases with an implied increase in the price of the environmental good. These data can then be analyzed with a model of discrete choice to obtain estimates of indirect utility functions or bid functions. How this can be done is described below in the section on welfare analysis of referendum data.

While the referendum question makes relatively inefficient use of a sample, the information content from a given sample size can be increased by asking each respondent a follow-up question. If the initial answer is yes, the respondent can be asked a second referendum question on a higher amount chosen randomly. And if the initial response is no, the respondent can be asked a second question about a lower amount.

The referendum format has several advantages relative to the bidding game and willingness-to-pay formats. First, it places people in a relatively familiar social context. Many private market transactions involve goods offered on a take-it-or-leave-it basis in which the individual decides whether or not to purchase the good at the offered price. And if the payment vehicle is a tax, the referendum question simulates a true referendum of the sort found everywhere from small New England town meetings to statewide votes on highway bond issues.

The second advantage is that since only a yes or no answer is required, the referendum question format poses a relatively simple decision problem for individuals. This may result in lower levels of item nonresponse and fewer refusals to participate in the survey.

A third advantage is that there is no reason to expect starting point bias in responses. At least this is true if the referendum question is framed in such a way that respondents do not pick up value cues. For example, respondents could be told that "there is no right answer to the following question—some people will say yes (no) and others will say no (yes); we want to know what you think."[9]

Finally, if respondents believe that any resource allocation decisions made on the basis of the survey results will be based on a plurality voting rule, then the referendum question format is incentive-compatible, at least if the question asks for only one payment value (see Hoehn and Randall, 1987, p. 237; Carson, 1991, p. 142). If the respondent's willingness to pay exceeds the stated dollar amount, the respondent has no reason to say anything other than yes. And if the respondent's willingness to pay is less than the dollar payment, nothing is gained by saying yes. So there is no reason to expect strategic behavior in properly framed referendum questions.

Contingent Ranking

In chapter 4, I have described the use of the discrete choice model to estimate the parameters of a utility function from individuals' choices of one alternative from a set of n. The referendum format is a simplified version of this approach in that individuals make their selection of one alternative from a set of two (the referendum alternatives of yes or no). The contingent ranking method of obtaining hypothetical data asks individuals to provide more information about their preferences by giving them more alternatives than the referendum approach and by asking them to rank these alternatives in order of preference. Each alternative could have several different attributes. Normally one attribute would have a monetary dimension—for example, a price—in order facilitate calculation of monetary values.

The analytical model used to extract information about preferences from the ranking is a straightforward extension of the discrete choice model. It is an ordered logit model. This model provides that set of parameter weights on the attributes which maximizes the likelihood of realizing that rank ordering. Provided one of the attributes is a money measure, parameter weights can then be used to calculate

---

[9]If the order of yes and no phrases in this statement is randomized across the sample, then responses can be tested to see if there is an order effect. If there is none, this suggests the absence of what Mitchell and Carson have called yea-saying and nay-saying (Mitchell and Carson, 1989, pp. 101, 120–241).

the marginal willingness to pay for an attribute (the marginal rate of substitution between income and the attribute) or the willingness to pay for a nonmarginal change in its level.

One of the earliest applications of the ordered logit model to contingent ranking data was that of Beggs, Cardell, and Hausman (1981). They used the model to estimate the values of characteristics of alternative models of cars, including electric-powered vehicles. Their respondents were asked to rank sixteen alternative vehicle designs; each design had nine attributes, including purchase price and fuel costs per mile. The method has subsequently been used to value rural visibility (Rae, 1983), the avoidance of diesel odor (Lareau and Rae, 1989), and water quality (Smith and Desvousges, 1986). The Smith and Desvousges study was perhaps the simplest for respondents: they were asked to rank only four alternatives, each of which involved only two attributes, a description of water quality and an annual fee.

In comparing contingent ranking with other methods of obtaining hypothetical data, one question is whether individuals find it easier to rank alternatives ordinally than to establish a monetary (cardinal) value for a commodity. Intuitively, it would seem that ordinal ranking is easier than establishing a value under the open-ended and bidding game formats. On the other hand, ranking several alternatives, each with several attributes, seems likely to be more difficult and certainly more time-consuming than ranking only two alternatives, each with only two attributes, as in the case of the referendum format.[10]

Contingent Activity

This term refers to the use of hypothetical questions about activities to obtain data for use in one of the behavioral models of the sort described in chapter 4. For example, suppose one had used travel cost data to estimate a demand function for visits to a recreation site, but one wanted to know the value of a change in one of the environmental attributes of that site. The relevant behavioral model would be weak complementarity. But in the absence of an observed change in the environmental attribute, it might not be possible to predict the shift in the demand curve for visits to the site. Visitors could be asked how their visitation behavior would change if the environmental at-

---

[10]Mitchell and Carson offer a skeptical view about contingent ranking, at least in part on these grounds (1989, pp. 81–86). But see Lareau and Rae (1989) and Hausman and Ruud (1987) for evidence that respondents can provide consistent and stable rankings at least for some attributes and with a relatively small number of alternatives.

tribute were to change in a specified way. This contingent change in visitation rates could be used to estimate the shift in the demand curve for visits. McConnell (1986) used this approach to estimate the benefits to visitors of local beaches if the pollution of New Bedford Harbor, Massachusetts, by polychlorinated biphenyls (PCBs) could be eliminated. In another application, Thayer (1981) asked recreationists how their choices of sites to visit would be altered by construction of a geothermal plant in the region they currently used for recreation.

THE VALIDITY OF HYPOTHETICAL RESPONSES

To paraphrase Mitchell and Carson (1989, p. 190), the validity of any piece of hypothetical data is the degree to which it measures the theoretical construct of interest. The theoretical construct of interest is the individual's true value, true probability of accepting an offer, true ranking of alternatives, or true change in the level of an activity. Ideally, one would like to assess the validity of a hypothetical value by comparing it with the true value. But the true value is usually not known, so this option is not available. There are two strategies available for assessing validity, and they are not mutually exclusive. One involves the careful assessment of the survey instrument and scenario to see if all known sources of bias have been removed or avoided. The other strategy involves the empirical analysis of responses to determine whether they are consistent with economic theory, or to compare them with estimates of the same value derived from other methods, or both. Economists have a comparative advantage over other social scientists in pursuing the second strategy, but cognitive psychologists and sociologists may be better equipped than economists to pursue the first strategy.

Examining the Instrument

Practitioners in the field of hypothetical valuation methods have identified a number of problem areas and issues that can affect the validity of responses. There is general agreement, for example, that the form of the valuation question is important. Referendum questions work better than the bidding game format. Willingness-to-pay (WTP) questions work better than willingness-to-accept compensation (WAC) questions. And a number of sources of potential bias have been identified. Thus much can be learned from a systematic examination of the survey instrument itself.

One important characteristic of a scenario is the nature of the property rights implied by the valuation question. WTP and WAC questions imply different structures of property rights in the environmental change being valued. So I first take up the issue of property rights and the selection of WTP or WAC questions.

WTP OR WAC

Direct value and closed-end questions can be asked either in WTP or WAC formats. As discussed at the end of chapter 3, each form of question implies something different about property rights and the reference level of utility. For example, if the change being valued is a gain, a WTP question will yield an estimate of the compensating surplus, and it implies that the individual has no property right in the increase in the thing being valued. Alternatively, if the individual is asked to state the compensation he or she requires to forgo the improvement, it is implied that the individual has a property right in the improvement. The relationships are symmetrical in the case of a loss. An individual's WTP to avoid the loss is an equivalent surplus measure and is consistent with having property right in the status quo. A WAC question gives the compensating surplus measure of welfare loss and implies that the individual does have a property right in the status quo.

Different policy questions and welfare measures imply different property right structures and question formats. For example, if an environmental improvement is being valued and the Kaldor compensation test is being employed, the proper form of question is WTP and the reference utility level is the status quo. But if the question is the value lost under a legal requirement to assess damages or to compensate losers, the appropriate form of question is WAC. There is a problem with the WAC form, however. A large body of evidence suggests that WTP and WAC formats for hypothetical values lead to quite different results. As noted earlier, WAC responses are typically several times larger (and, in at least one case, two orders of magnitude larger) than WTP responses for the same change.[11]

These large differences are troubling. The theoretical analyses of Willig (1976) and Randall and Stoll (1980) suggest that differences between WTP and WAC should be small.[12] This is because the differ-

---

[11]See Cummings, Brookshire, and Schulze (1986, especially, p. 35) for a review of earlier studies comparing WTP and WAC. A more recent study—Brookshire and Coursey (1987)—shows even larger differences.

[12]Their analyses are summarized in chapter 3 of this book. See pp. 61–66 and pp. 81–85.

ences arise from an income effect associated with the change in the thing being valued, and the implied change in income has been very small in most of the cases studied.

Efforts at explaining the observed disparity between WTP and WAC have generally taken one of two paths (see, for example, Gregory, 1986). On the first path, theoretical arguments for small differences between true WTP (*WTPt*) and true WAC (*WACt*) are accepted as correct and explanations for the disparity between revealed WTP (*WTPr*) and revealed WAC (*WACr*) are sought in problems having to do with the hypothetical question format. Analysts have generally argued that *WACr* is greater than *WACt*. For example, Hoehn and Randall (1987) modeled the strategy of a rational individual who was risk-averse and uncertain about the true value of the postulated environmental change. They showed that in the absence of incentives to misrepresent values, *on average* the individual's *WACr* would be greater than *WACt*, and *WTPr* would be less than *WTPt*. Thus the difference between *WACr* and *WTPr* would be greater than the true difference, on average. Carson (1991) has gone further in suggesting that at least when questions take the form of direct expressions of value, WAC questions are not incentive-compatible. For example, they may encourage respondents to state the highest amounts that they think could be extracted in compensation rather than the minimum compensation they would require.

The second path involves examining the theory of preferences and value more closely to see if theory and observation can be reconciled in the sense that theory predicts the observed disparity between WAC and WTP. One example of this is in the work of Hanemann (1991). As explained in chapter 3, he has shown that standard welfare theory predicts large differences between the two measures if the good being valued is unique in the sense that there is a low elasticity of substitution between it and market goods. However, Hanemann's analysis does not explain the persistent differences between the two measures in simulated markets involving commonplace goods such as lottery tickets, coffee mugs, and pens (see Knetsch and Sinden, 1984; and Kahneman, Knetsch, and Thaler, 1990). In these studies, the disparity cannot be attributed to the hypothetical nature of the questions, since it is based on observations from simulated markets and experimental settings.[13]

---

[13]But see also Harless (1989). The experiments cited in the text involved comparisons of mean or median responses between groups of subjects. Harless faced each individual in his experiment with both WAC and WTP forms

Thaler (1980) has proposed that the reconciliation of theory with observation can be brought about by postulating an "endowment effect" on individuals' valuation functions and a kink in this function at the status quo point.[14] The idea of the endowment effect and the differential valuation of gains and losses can be shown with the aid of figure 6-1. The horizontal axis shows the quantity of an environmental good q. The vertical axis shows the compensating welfare measure for changes in q. This measure is positive (WTP) for increases and negative (WAC) for decreases from some status quo point. Suppose that the status quo is q^0. The associated valuation function w_0 shows the monetary payment (compensation) that holds utility constant for a given increase (decrease) in q from q^0. This function is kinked at the status quo point of q^0, showing that the marginal valuation of increases in q is substantially lower than the marginal valuation of losses from q^0. A change in the endowment of q from q^0 to q^1 shifts the valuation function. And as figure 6-1 shows, the willingness to pay for an increase from q^0 to q^1 is substantially less than the required compensation for the decrease from q^1 to q^0.

In conclusion, the issues surrounding WTP and WAC formats are not settled. There are reasons to believe that hypothetical WAC questions may result in responses that are not valid. But the observed large differences between WAC and WTP do not prove lack of validity, since large differences are not inconsistent with some forms of value and preference theory. Thus I think that Cummings, Brookshire, and Schulze (1986, pp. 104, 107) go too far in saying that WTP questions are always to be preferred. If a WTP question is asked when people believe they have a right to the thing being valued (implying a WAC format), there may be a high rate of scenario rejection and loss of validity in the responses.

I would make three points. First, many valuation problems fall naturally into a WTP format. Examples include valuation of an increase in the provision of some public good and of an increase in the level of an environmental amenity. There are no concerns about using WTP questions in such a setting. Second, as Mitchell and Carson have argued (1989, pp. 37–41), there is a class of problems for which the WAC format seems appropriate but which can be framed in WTP

---

of questions. In his within-subject comparisons, the disparity between WTP and WAC was substantially reduced.

[14]Thaler suggests that this is a reasonable extension and generalization of the prospect theory of Kahneman and Tversky (1979) to choices not involving uncertainty.

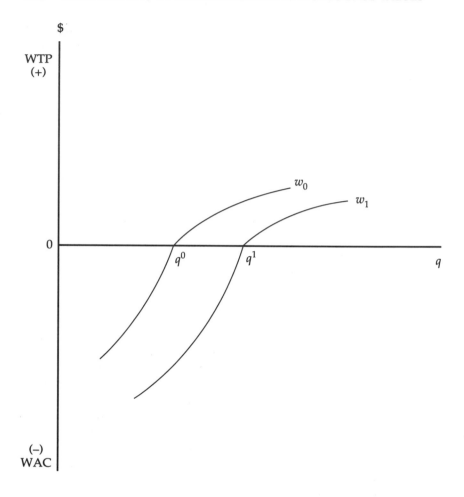

Figure 6-1.   The value function and the endowment effect

terms in a plausible way. For example, suppose that the question concerns the value of an improvement in air quality but that it is widely believed that people have a "right" to clean air. This implies that the "correct" question would be a WAC to forgo the improvement. But in this situation a plausible WTP question could be framed in the following way:

> You have a right to clean air. But if you enforce that right, it will impose costs on producers to control their air pollution. And these costs will necessarily be passed on to consumers

in the form of higher prices for a wide range of products. Recognizing the cost to you of enforcing your right, what is the maximum you would be willing to pay in the form of these higher prices in order to achieve the stated level of air pollution control?

Third, where the WAC framework is appropriate and unavoidable, it may be possible to design a closed-ended WAC referendum question to elicit required compensation (Mitchell and Carson, 1989, p. 37). An example of such a case might be determining the compensation required for local residents who are being asked to accept a locally unwanted land use (LULU).

SOURCES OF ERROR

As noted earlier, Mitchell and Carson (1989, especially chapter 11) have provided a classification of ways in which scenarios can affect individuals' responses so as to introduce systematic error or bias into the value measures that are derived from these responses. Systematic errors arise from three sources. The first is scenario misspecification. Were the question and the information provided with it to establish its context understood by the respondent in the way intended by the investigator? The second source is implied value cues. To what extent did the information provided to the respondent predispose him or her toward a particular response? The third is incentives for misrepresentation. Did the question and its context provide any incentive to the respondent to misrepresent his or her values in some way? Part of the art of asking hypothetical questions involves ensuring that these questions can be answered in the negative. I now take up each of these sources of error in turn.

Scenario Misspecification. The various types of scenario misspecification discussed by Mitchell and Carson (1989, pp. 246–258) all have the effect of creating a divergence between what the respondent understands about the choice situation and what the investigator intends the respondent to understand. The investigator can increase the likelihood that the questions will be understood as intended by developing an awareness of how people think about the issues at hand and what language and terminology they use in talking about it. This can be accomplished through the use of focus groups, pilot surveys, pretests, and the like. Focus groups involve structured interviews with small groups of individuals from the population to be surveyed. Focus groups can be used to investigate perceptions of environmental problems, to determine which terminologies have particular salience, and to try out various descriptions and question formats.[15] Pretests and pilot surveys may be useful in testing different scenarios and question formats, especially if respondents are queried afterwards to identify questions and problems they may have with specific wording.

Implied Value Cues. If a respondent has well-formed preferences regarding the hypothetical choice problem posed to him, he should find the task of formulating a response to the question fairly easy. But if the choice problem posed to the individual is unfamiliar and the individual is not clear about his preferences, he might, quite unconsciously, seek clues regarding the "correct" choice or value from the information supplied as part of the scenario specification. If such clues are present, they may systematically bias individuals' choices and values. Such clues are termed implied value cues. The most well-studied implied value cue is the starting point in bidding game versions of direct expression of value questions. Other forms of direct value question may avoid starting point biases. As noted earlier, this problem is not present in the referendum form of valuation question. However, Mitchell and Carson (1989, pp. 240–241) suggest that there may be a form of yea-saying bias even in referendum questions, when respondents take the offered price as a cue to a value they should be willing to accept, even if it is somewhat above their true willingness to pay.

Implied value cues can be found outside of the scenario itself. For example, respondents who are asked about the value of access to a recreation site or to a fishing experience may anchor their responses to the implied value provided by the normal admission fee or fishing license fee. The task for the investigator is to find ways to thwart this tendency, perhaps by specifying a payment vehicle that focuses attention away from the customary admission fee or license fee. For example, recreationists might be asked about their willingness to pay higher travel costs to reach a site.

Contingent ranking and contingent activity questions appear to be less susceptible to implied value cues. However, there could be a tendency for the results of a contingent ranking question to depend upon the order in which options are initially presented for ranking. This kind of order effect can be prevented by randomizing the order of presentation of alternatives across respondents.

Incentives to Misrepresent Values. Biased responses can occur if the scenario specification and the framework for eliciting values are not

---

[15]See Desvousges and coauthors (1984) for a description of the use of focus groups in developing a contingent valuation survey instrument.

incentive-compatible. In an incentive-compatible framework there is no reason for an individual to respond to any valuation question other than in terms of that person's true preferences. If an individual has any reason to believe that his or her response will have any effect on either the provision of public goods or environmental quality or on the individual's taxes or other payment responsibilities, then the respondent may try to use the response to influence the outcome of the public policy process.

One form of strategic bias arises from the efforts of respondents to "free ride," that is, to reduce their repayment obligation by stating low values. If a respondent believes that the level of provision of the public good is independent of his response, but that his tax liability does depend on the response, that person has an incentive to understate his true value for the good. Alternatively, if a respondent believes that her response can have some effect on the level of provision of the good but will not affect her payment obligation, then she has an incentive to bid high if she prefers more of the public good or bid low (or zero) if she prefers less of the good.

The opportunity for strategic behavior can arise only when the valuation questions are asked in a setting in which it at least appears that the actual outcome will be affected by individuals' responses, that is, in other than the purely hypothetical contingent market setting. In addition to the incentive, there must also be an opportunity to influence the outcome through manipulation of responses. Incentive-compatible mechanisms such as the direct referendum question format are preferable because the best strategy for respondents is to provide truthful answers (Mitchell and Carson, 1989, pp. 148–149).

Strategic bias has been discussed extensively in the literature. See, for example, Cummings, Brookshire, and Schulze (1986) and Mitchell and Carson (1989). The general view is that strategic bias has not been a serious problem in most hypothetical valuation studies, at least in part because the experimental nature of these studies provided no incentive for strategic behavior. But if hypothetical valuation studies come to be used on a regular basis to determine values for making decisions on resource allocation or for purposes of litigation, then strategic bias might become a problem. And the incentive-compatibility, or lack thereof, of a survey framework may become an important consideration.

Empirical Analysis of Responses

A number of possible strategies for experimental design and data analysis can contribute to an assessment of the validity of hypothetical responses. These strategies are not necessarily mutually exclusive.

In this section I briefly describe a number of the strategies that have appeared in the literature and provide references to examples where appropriate.

Many studies of direct expressions of value include regression analyses of revealed values as a function of the characteristics of the good being valued and characteristics of the respondent. One test of validity is whether expressed values vary systematically and in the expected manner with the explanatory variables that theory suggests should be important. For example, values should be an increasing function of income, other things being equal, and the estimated income elasticity or price flexibility of income should be "reasonable."

When an experimental design involves facing different individuals with different quantities and/or implied prices, the choices should be consistent with the axioms of revealed preference. There are nonparametric tests for consistency with these axioms. For an example of an application of this strategy, see Adamowicz and Graham-Tomasi (1991).

Another possible strategy is to explore the stability of responses over time by using a test-retest design. In such a design, the same survey instrument is administered to the sample a second time after a suitable interval of time has elapsed. For examples, see Loomis (1989) and Reiling and coauthors (1990).

Still another strategy is to develop different scenarios to test hypotheses about the effects of certain features of a scenario on the mean value of the sample. These tests can involve either comparing mean responses across sample groups given different scenarios (between-subject comparisons) or comparing responses of the same sample group to different scenarios (within-subject comparisons). The hypotheses can take two forms. One form is the hypothesis that individual or sample mean values vary in a consistent fashion with relevant and meaningful variations in the scenario. For example, an increase in the quantity of the good being valued should produce an increase in bids. For examples of applications of tests of this form, see Smith and Desvousges (1987) and Samples, Dixon, and Gowen (1986).[16] The other form of hypothesis is that bids are invariant to irrelevant changes in the scenario. Experiments in settings different from contingent valuation have produced a large body of evidence that responses are not independent of irrelevant changes in descriptions or contexts. For some examples, see Tversky, Slovic, and Kahneman (1990). But at least one

---

[16]See also the recent exchange between Kahneman and Knetsch (1992a and 1992b) and Smith (1992) on embedding.

study (Boyle, 1989) has shown that contingent valuation responses can be invariant to irrelevant changes in the scenario.

A further strategy is to compare hypothetical values with measures derived from some indirect method. This strategy is as old as the contingent valuation method itself (see Knetsch and Davis, 1966). More recently, Brookshire and coauthors (1982) compared contingent valuation measures of the value of improved air quality with values derived from a hedonic property value model. Bishop and Heberlein (1979) compared hypothetical values of goose hunting permits with travel cost measures of value. These and other comparisons are discussed in some detail in Cummings, Brookshire, and Schulze (1986) and in Mitchell and Carson (1989). The problem with this strategy, of course, is that if a discrepancy between the hypothetical value and the alternative is discovered, it could be because of the lack of validity of the hypothetical measure, the lack of validity of the alternative measure derived from observation, or both.

Finally, experiments can be designed to make it possible to compare responses to hypothetical questions with responses in similarly designed experimental or simulated markets. In simulated or experimental markets transactions actually take place and respondents live with the consequences of their choices. One of the pioneering studies of this type was conducted by Bishop and Heberlein (1979),[17] who made hypothetical closed-ended offers to purchase special goose hunting permits from a sample of hunters. At the same time, they made actual cash offers for a limited number of permits from a different sample, so they could compare the values revealed. More recent experiments of this type include those by Kealy, Dovidio, and Rockel (1988) and Kealy, Montgomery, and Dovidio (1990).

WELFARE ANALYSIS OF HYPOTHETICAL RESPONSE DATA

In this section, I describe how measures of welfare change can be derived from responses to three of the forms of hypothetical questions. Responses to direct questions about willingness to pay and to bidding games can be interpreted as estimates of each individual's compensating surplus (CS) or equivalent surplus (ES), depending upon the format of the question. But there remain the problems of how to treat extreme responses and protest zero bids. These problems are addressed in the following subsection on bidding data. I then describe the models for deriving welfare measures from direct

---

[17]Earlier work by Peter Bohm (1971, 1977) was in the same vein.

referendum and contingent ranking questions. Responses to contingent activity questions must be analyzed with one of the indirect measurement methods described in chapter 4, so responses to this form of question will not be considered in this section.

Bidding Data

A direct valuation question produces a set of welfare measures W_i ($i = 1, ..., i, ..., n$) for the n respondents in the sample. As noted above, W_i is either a CS or ES measure, depending on the format of the question. An estimate of the total value of the welfare change for the population from which the sample is drawn can be obtained by calculating the sample mean W^* and multiplying by the total population. Alternatively, the responses can be regressed on income and other socioeconomic characteristics to obtain a bid or inverse demand function for a given change in q:

$$b_i^* = b^* (M_i, S_i)$$

where S is a vector of socioeconomic characteristics that affect individuals' valuations. Then data on the characteristics of the relevant population can be used to calculate b_i^* for every member of the population. An alternative approach is to include variation in the size of the change in q across the sample as part of the survey design. Then the bid function

$$b_i^* = b^* (\Delta q_i, M_i, S_i)$$

can be estimated and used to calculate values for alternative scenarios of environmental or resource change.

Mean values, and therefore aggregate values for the population, are sensitive to the presence of large individual bids. Many surveys obtain at least a few bids that are so large relative to the sample mean as to be obviously invalid for valuation purposes. It would be desirable to have an unambiguous rule for identifying such outliers so they could be deleted from the sample before aggregate values or valuation functions are estimated. Several procedures for identifying outliers have been used in the literature, but each has a degree of arbitrariness associated with it.

The simplest and most commonly used procedure is to adopt a rule of thumb concerning the relationship between the stated bid and the respondent's income. For example, the rule could be to delete all responses representing more than $X\%$ of reported income.

The chosen value for X could depend on the analyst's assessment of the likely importance to people of the environmental or resource change being valued. As an alternative to this procedure, Mitchell and Carson (1989, pp. 226–227) advocate the calculation of the α-trimmed mean, where the analyst chooses the value of α. The procedure is to delete a proportion of responses equal to the selected value of α from the top and bottom of the rank-ordered responses. For $\alpha = 0$, the α-trimmed mean is the true mean of the sample. For $\alpha = .5$, the α-trimmed mean is equal to the true median. Regression analysis of the data would be done after the α-trimming was carried out. A third alternative is to use a set of regression diagnostic procedures (those of Belsley, Kuh, and Welsch, 1980, for example) to delete observations having unduly large influences on regression coefficients.[18] It should be noted that the deletion of extreme values from the sample by any procedure may be creating a different sort of bias problem—that due to item nonresponse. Item nonresponse bias and strategies for dealing with it are discussed extensively in Mitchell and Carson (1989, pp. 267–278).

In addition to the problem of invalid large responses, a sample could contain invalid zero responses, so-called protest zeros. Protest zeros occur when respondents reject some aspect of the constructed market scenario by reporting a zero value even though they place a positive value on the amenity or resource being valued. The process of α-trimming the sample is not a solution to the problem of protest zeros, since α could be greater or less than the true proportion of protest zeros. Rather, some means must be found to identify protest zeros for deletion before applying the procedure for deleting outliers. One approach is to ask every respondent who gives a zero value to indicate a reason for doing so. For example, respondents could be asked:

Which statement best expresses your reason for giving a zero response?
    I. I can't afford to pay for the good.
    II. The good is not important to me.
    III. I don't think that I should have to pay for the good.

Responses of those choosing the third statement would be classified as protest zeros and deleted from the sample, while responses of those choosing the first and second would be considered valid zeros. The deletion of protest zeros can also lead to item nonresponse bias if

---

[18]For an example of this approach to identifying outliers, see Smith and Desvousges (1986).

those who protest are systematically different in some respect from those who give proper responses.

Referendum Data

In order to convert data on yes or no responses to a referendum question into a monetary measure, it is necessary to employ some explicit utility-theoretic model of choice. The discrete choice model introduced in chapter 4 is well suited to the task. The hypothetical nature of referendum surveys makes it possible to gain more information than would be available by observing individual choices in most real-world markets or discrete choice settings. In real-world choice settings it is typically the case that all individuals face the same set of prices for the set of alternatives from which they must choose. For example, in choosing a mode of transportation, all individuals face the same fares. Differences in responses must then be due to differences in individual characteristics. But in a hypothetical referendum it is possible to present respondents in different randomly chosen sub-samples with different referendum prices. As Cameron (Cameron, 1988; and Cameron and James, 1987) has pointed out, this feature of referendum data makes it possible to explain individuals' choices in terms of a willingness-to-pay function rather than in terms of differences in indirect utility.

McConnell (1990) has compared the Cameron model with the model based on utility functions outlined by Hanemann (1984) and by Sellar, Chavas, and Stoll (1986). He has shown that the two models can be derived from the same underlying utility-theoretic framework. In a deterministic formulation, the two models yield the same predictions about behavior and choice. Since the stochastic forms of the models introduce random components in different ways, the two models are not in general equivalent, although there are special cases in which the two models are dual to each other. Moreover, as Cameron (1988) shows, the willingness-to-pay function model permits the straightforward calculation of marginal values for all of the arguments in the willingness-to- pay function, while this is not possible with the utility function model.

In what follows, I first model the individual's response in the random utility framework and show how welfare measures can be derived. I then show the parallel development of the willingness-to-pay function model.

Consider an individual who must decide whether to answer yes or no to the following question: Would you vote for a program to permanently increase environmental quality from q^0 to q^1 if it would increase your taxes by T for this year? Let the indirect utility function be $u(M, q, C)$, where C is a vector of individual characteristics and the vector of market prices, P, is omitted since prices are assumed to be constant. The individual responds yes if

$$u(M - T, q^1, C) - u(M, q^0, C) \geq 0 \tag{1}$$

and no otherwise.

Letting $v(\cdot)$ represent the observable component of utility, the probability of a yes response is given by

$$Pr(Y) = Pr[v(M - T, q^1, C) + \epsilon_1 > v(M, q^0, C) + \epsilon_0]$$

where the ϵ_i ($i = 0, 1$) are the random, unobserved components of utility. As discussed in chapter 4, if the random terms are independently and identically distributed with a Type I Extreme Value distribution, then this probability can be expressed as

$$Pr(Y) = (1 + e^{-\Delta v})^{-1} \tag{2}$$

where $\Delta v = v^1 - v^0$. Also, reversing the sign on the probability difference gives the expression for the probability of rejecting the offer:

$$Pr(N) = (1 + e^{\Delta v})^{-1} \tag{3}$$

The willingness to pay for q^1 (CS) is defined implicitly by

$$u(M - CS, q^1) = u(M, q^0)$$

and
$$v(M - CS, q_1) + \epsilon_1 - \epsilon_0 = v(M, q^0)$$

In terms of the observable utility function, CS is thus a random variable, because of the term $\epsilon_1 - \epsilon_0$. The probability of accepting the offer is also, then, the probability that $CS \geq T$, and the probability of rejecting the offer is the probability that $CS < T$. This is a cumulative density function (c.d.f.), denoted here as $F(T)$, and shown in figure 6–2. As Hanemann (1984) has pointed out, the expected value of the random variable CS can be found from the c.d.f. as follows:

$$E[CS] = \int_0^\infty [1 - F(T)]dT \tag{4}$$

Graphically, in figure 6–2, the expected value of CS is the shaded area above the c.d.f. and below $Pr(N) = 1$. For example, for any given T, an individual with a lower willingness to pay for the change in q would have a higher probability of rejecting the offer, and the shaded area would be smaller. Alternatively, an offer of $q^2 > q^1$ at any T would decrease the probability of an individual rejecting the offer, so $F(T)$ would be shifted down and $E[CS]$ would be larger.

Specifying a functional form for the observable component of utility makes it possible to estimate the parameters of the utility difference in equations (2) and (3). For example (following Hanemann, 1984), if $v = a + b \cdot \ln M + c \cdot \ln q$ (ignoring the C vector for simplicity), then

$$\Delta v = (a_1 - a_0) + b \cdot \ln (1 - T/M) + c \cdot \ln (q^1/q^0) \qquad (5)$$

Then with the parameters from (5), equations (2) (or (3)) and (4) can be used to calculate $E[CS]$.[19]

As equation (4) shows, in principle $F(T)$ is integrated over the range to infinity. In order to avoid the implausibly high estimates of $E[CS]$ that can sometimes result, some researchers have truncated the integration of $F(T)$ at some "reasonable" finite value for T. But estimates of $E[CS]$ can be highly sensitive to the value of T_{max} chosen for truncation, at least in some cases.[20]

There is an alternative approach to dealing with the sensitivity of estimates of CS to outliers in the distribution. As Hanemann (1984) has suggested, the median willingness to pay for the sample can be found by setting $Pr(Y)$ in equation (2) equal to 0.5 and solving for T. This gives the value for T that makes the representative individual indifferent as to accepting or rejecting the offer. And generalizing to the sample, one would expect half the respondents to accept (and half

---

[19]See, for example, Hanemann (1984); Seller, Chavas, and Stoll (1986).

[20]See Bishop and Heberlein (1979); Hanemann (1984); and Mitchell and Carson (1989, pp. 103, 196–197). Alternatively, one could say that estimates of $E[CS]$ are likely to be sensitive to the goodness of fit of the estimated function at the upper tail of $F(T)$. Bishop and Heberlein (1990, pp. 90–91) have suggested an approach to selecting the offer amounts at the survey design stage to improve the goodness of fit at the tails. Briefly, they suggest using responses to an open-ended pretest question to obtain a preliminary estimate of $F(T)$. Then for a survey sample design of N, $N/2$ probabilities are randomly chosen from the interval 0, 1. For each probability, $F^{-1}(T)$ and $1 - F^{-1}(T)$ are solved for a T value to be offered to a member of the sample. They say that estimates of $E[CS]$ are not affected by truncation at values for T that imply $F(T) = .99$.

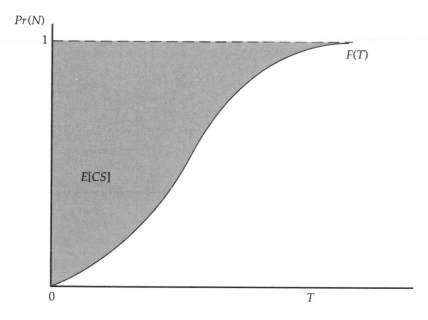

Figure 6-2. The cumulative density function for rejecting the offer of q^1 at T and the expected value of CS

to reject) the offer, other things (including income and C) being held constant. This procedure is less sensitive to extreme values in the data.

The alternative model for analyzing direct referendum data is based on the bid or willingness-to-pay function.[21] This function can be derived from the expenditure function as follows

$$B(q^0, q^1, u^0, C) = e(q^0, u^0, C) - e(q^1, u^0, C)$$

The individual will respond yes if

$$B(q^0, q^1, u^0, C) \geq T$$

and no otherwise. The probability of accepting the offer of q^1 at T can be expressed in terms of the function

---

[21]Cameron (1988) calls this a valuation function; McConnell (1990) calls it a variation function, because it can be defined for either compensating or equivalent welfare measures. The formulation in the text here gives the compensating surplus measure of value for an improvement. Other versions can easily be specified for losses and for ES measures. See Carson (1991, especially pp. 143–144) for further discussion of this model.

$$Pr(Y) = Pr[B^*(q^0, q^1, u^0, C) - T > \eta]$$

where B^* is the observable component of the bid function and η is the unobserved random component of willingness to pay. In other words,

$$B(q^0, q^1, u^0, C) = B^*(q^0, q^1, u^0, C) + \eta$$

The next step is to make some assumptions about the distribution of the random component of the bid. Cameron and James (1987) assume that η is normally distributed with zero mean and standard deviation of σ. The result is a form of probit model in which

$$Pr(Y) = 1 - \Phi[T - X'\beta/\sigma]$$

where $X'\beta$ is the set of explanatory variables and coefficients and Φ is the standard normal cumulative density. The bid is given by

$$B^* = X'\beta + \sigma \tag{6}$$

Once the parameters of (6) have been estimated, calculations of total and marginal bids for individuals and in the aggregate are straightforward.

Alternatively, if η is assumed to be distributed as in the logistic model, then (following Cameron, 1988)

$$Pr(Y) = [1 + e^{-(B^* - T)}]^{-1}$$

Again, estimation of the parameters of the bid function makes it possible to calculate individual total and marginal willingnesses to pay as well as the probabilities of accepting an offer.

Just as in the case of direct value questions, respondents might express rejection of the referendum market scenario by saying no to all offers even though their willingness to pay exceeds the lowest offered price. Respondents who refuse the lowest offer should be queried about their reasons, as discussed above. And respondents who signal that their no vote is a protest should be deleted from the sample.

Contingent Ranking Data

Welfare measures from contingent ranking data are also based on a form of random utility model. Suppose that an individual is asked to rank a set of alternatives Y $(Y = y_1, \ldots y_i, \ldots y_n)$ and that each element in Y can be described by its price p_i and a vector of nonprice attributes Q $(Q = q_1, \ldots q_j, \ldots q_m)$. The individual's indirect utility associated with alternative i consists of two components. The observable component is $v(M, Q_i, p_i, C)$ where C is a vector of individual taste parameters. The unobservable component is ϵ_i, which is assumed to be random. From chapter 4, we know that if the ϵ_i are independently and identically distributed with a Type I Extreme Value distribution, the probability of the individual ranking alternative 1 first is

$$Pr[u_1 > u_j; \text{ for all } j \neq 1] = \frac{e^{v_1}}{\sum\limits_{i=1}^{n} e^{v_i}}$$

Now suppose that the individual values the alternatives in the following order from most preferred: $y_1, y_2, \ldots, y_n$. The probability of this ranking is the probability of ranking y_1 above the remaining $n - 1$ alternatives times the probability of ranking y_2 above the remaining $n - 2$ alternatives, and so forth. Or

$$Pr[u_1 > u_2 > \ldots u_n] = \frac{e^{v_1}}{\sum\limits_{j-1}^{n} e^{v_j}} \cdot \frac{e^{v_2}}{\sum\limits_{j-2}^{n-1} e^{v_j}} \cdots$$

$$= \prod_{i=1}^{n} \left[\frac{e^{v_r}}{\sum\limits_{j=r}^{n+1-r} e^{v_j}} \right]$$

where r is the rank ordering of the alternatives.

As in the other applications of the discrete choice model, once a functional form for the observable component of the indirect utility function has been specified, its parameters can be estimated from data on the rankings of a sample of individuals, using maximum likelihood methods. Welfare measures can be then derived by computing the marginal rate of substitution between income and any of the elements of Q.

SUMMARY

Economists and other social scientists have made a great deal of progress in the design and implementation of hypothetical methods since the earliest contingent valuation studies of the mid 1960s and early 1970s. One major advance is the recognition that the original bidding game format is just one of a family of hypothetical elicitation methods. Others include the referendum question, contingent ranking, and contingent activity methods. Each of these methods has its strengths and limitations relative to the others; but together they provide a more powerful array of tools for dealing with a variety of valuation problems.

A second advance has been the development of a much greater awareness of and sensitivity to the ways in which scenario specification can influence responses and thereby affect the validity of the value measures. Practitioners in economics owe a debt to researchers from other fields, especially those in cognitive psychology, for their contributions to our understanding of how people respond to questions and of the importance of how questions are asked.

A third advance has been the development of more sophisticated approaches to the analysis of data and the calculation of welfare measures, especially with respect to discrete choice models. And finally, there has developed a much greater sensitivity to a variety of sampling issues and related problems such as the need to deal with item nonresponse and to identify and treat outliers and protest zeros. All of these developments have helped to make the set of hypothetical methods a more powerful tool for dealing with amenity and resource valuation problems.

REFERENCES

Adamowicz, Viktor L., and Theodore Graham-Tomasi. 1991. "Revealed Preference Tests of Nonmarket Goods Valuation Methods," *Journal of Environmental Economics and Management* vol. 20, no. 1, pp. 29–45.

Arrow, Kenneth J. 1986. "Comments," in Ronald G. Cummings, David S. Brookshire, and William D. Schulze, eds., *Valuing Environmental Goods: An Assessment of the Contingent Valuation Method*. Totowa, N.J.: Rowman and Allanheld.

Beggs, S., S. Cardell, and J. Hausman. 1981. "Assessing the Potential Demand for Electric Cars," *Journal of Econometrics* vol. 16, no. 1, pp. 1–19.

Belsley, David A., Edwin Kuh, and Roy E. Welsch. 1980. *Regression Diagnostics: Identifying Influential Data and Sources of Collinearity*. New York: John Wiley.

Bishop, Richard C., and Thomas A. Heberlein. 1979. "Measuring Values of Extra Market Goods: Are Indirect Measures Bias?" *American Journal of Agricultural Economics* vol. 61, no. 5, pp. 926–930.

Bishop, Richard C., and Thomas A. Heberlein. 1990. "The Contingent Valuation Approach," in Rebecca L. Johnson and Gary V. Johnson, eds., *Economic Valuation of Natural Resources: Issues, Theory, and Applications.* Boulder, Colo.: Westview Press.

Bohm, Peter. 1971. "An Approach to the Problem of Estimating Demand for Public Goods," *Swedish Journal of Economics* vol. 73, no. 1, pp. 94–105.

Bohm, Peter. 1977. "Estimating Access Values," in Lowdon Wingo and Alan Evans, eds., *Public Economics and the Quality of Life.* Baltimore, Md.: The Johns Hopkins University Press for Resources for the Future and the Centre for Environmental Studies.

Boyle, Kevin J. 1989. "Commodity Specification and the Framing of Contingent-Valuation Questions," *Land Economics* vol. 65, no. 1, pp. 57–63.

Brookshire, David S., and Don L. Coursey. 1987. "Measuring the Value of a Public Good: An Empirical Comparison of Elicitation Procedures," *American Economic Review* vol. 77, no. 4, pp. 554–566.

Brookshire, David S., Berry C. Ives, and William D. Schulze. 1976. "The Valuation of Aesthetic Preferences," *Journal of Environmental Economics and Management* vol. 3, no. 4, pp. 325–346.

Brookshire, David S., Mark Thayer, William D. Schulze, and Ralph C. d'Arge. 1982. "The Valuation of Public Goods: A Comparison of Survey and Hedonic Approaches," *American Economic Review* vol. 72, no. 1, pp. 165–177.

Cameron, Trudy A. 1988. "A New Paradigm for Valuing Non-Market Goods Using Referendum Data: Maximum Likelihood Estimation by Censored Logistic Regression," *Journal of Environmental Economics and Management* vol. 15, no. 3, pp. 355–379.

Cameron, Trudy A., and Michelle D. James. 1987. "Efficient Estimation Methods for Contingent Valuation Surveys," *Review of Economics and Statistics* vol. 69, no. 2, pp. 269–276.

Carson, Richard T. 1991. "Constructed Markets," in John Braden and Charles Kolstad, eds., *Measuring the Demand for Environmental Quality.* Amsterdam, The Netherlands: Elsevier.

Cummings, Ronald G., David S. Brookshire, and William D. Schulze. 1986. *Valuing Environmental Goods: An Assessment of the Contingent Valuation Method.* Totowa, N.J.: Rowman and Allanheld.

Desvousges, William H., V. Kerry Smith, Diane H. Brown, and D. Kirk Pate. 1984. "The Role of Focus Groups in Designing a Contingent Valuation Survey to Measure the Benefits of Hazardous Waste Management Regulations," draft report for the U.S. Environmental Protection Agency. Research Triangle Park, N.C.: Research Triangle Institute.

Fischhoff, Baruch, and Lita Furby. 1988. "Measuring Values: A Conceptual Framework for Interpreting Transactions with Special Reference to Contingent Valuation of Visibility," *Journal of Risk and Uncertainty* vol. 1, no. 2, pp. 147–184.

Freeman, A. Myrick, III. 1979. *The Benefits of Environmental Improvement: Theory and Practice*. Baltimore, Md.: The Johns Hopkins University Press for Resources for the Future.

Gregory, Robin. 1986. "Interpreting Measures of Economic Loss: Evidence from Contingent Valuation and Experimental Studies," *Journal of Environmental Economics and Management* vol. 13, no. 4, pp. 325–337.

Hanemann, W. Michael. 1984. "Welfare Evaluations in Contingent Valuation Experiments with Discrete Responses," *American Journal of Agricultural Economics* vol. 66, no. 3, pp. 332–341.

Hanemann, W. Michael. 1991. "Willingness to Pay and Willingness to Accept: How Much Can They Differ?" *American Economic Review* vol. 81, no. 3, pp. 635–647.

Harless, David W. 1989. "More Laboratory Evidence on the Disparity Between Willingness to Pay and Compensation Demanded," *Journal of Economic Behavior and Organization* vol. 11, no. 3, pp. 359–379.

Hausman, Jerry A., and Paul A. Ruud. 1987. "Specifying and Testing Econometric Models for Rank-Ordered Data," *Journal of Econometrics* vol. 34, no. 1/2, pp. 83–104.

Hoehn, John P., and Alan Randall. 1987. "A Satisfactory Benefit Cost Indicator from Contingent Valuation," *Journal of Environmental Economics and Management* vol. 14, no. 3, pp. 226–247.

Kahneman, Daniel, and Jack L. Knetsch. 1992a. "Contingent Valuation and the Value of Public Goods: Reply," *Journal of Environmental Economics and Management* vol. 22, no. 1, pp. 90–94.

Kahneman, Daniel, and Jack L. Knetsch. 1992b. "Valuing Public Goods: The Purchase of Moral Satisfaction," *Journal of Environmental Economics and Management* vol. 22, no. 1, pp. 57–70.

Kahneman, Daniel, Jack L. Knetsch, and Richard H. Thaler. 1990. "Experimental Tests of the Endowment Effect and the Coase Theorem," *Journal of Political Economy* vol. 98, no. 6, pp. 1325–1347.

Kahneman, Daniel, and Amos Tversky. 1979. "Prospect Theory: An Analysis of Decisions Under Risk," *Econometrica* vol. 47, no. 1, pp. 263–291.

Kealy, Mary Jo, John F. Dovidio, and Mark L. Rockel. 1988. "Accuracy in Valuation Is a Matter of Degree," *Land Economics* vol. 64, no. 2, pp. 158–171.

Kealy, Mary Jo, Mark Montgomery, and John F. Dovidio. 1990. "Reliability and Predicted Validity of Contingent Values: Does the Nature of the Good Matter?" *Journal of Environmental Economics and Management* vol. 19, no. 3, pp. 244–263.

Knetsch, Jack L., and Robert K. Davis. 1966. "Comparisons of Methods for Recreation Evaluation," in Allen V. Kneese and Stephen C. Smith, eds., *Water Research*. Baltimore, Md.: The Johns Hopkins University Press for Resources for the Future.

Knetsch, Jack L., and John A. Sinden. 1984. "Willingness to Pay and Compensation Demanded: Experimental Evidence of an Unexpected Disparity in Measures of Values," *Quarterly Journal of Economics* vol. 99, no. 3, pp. 507–521.

Lareau, Thomas J., and Douglas A. Rae. 1989. "Valuing WTP for Diesel Odor Reductions: An Application of Contingent Ranking Technique," *Southern Economic Journal* vol. 55, no. 3, pp. 728–742.

Loomis, John B. 1989. "Test-Retest Reliability of the Contingent Valuation Method: A Comparison of General Population and Visitor Responses," *American Journal of Agricultural Economics* vol. 71, no. 1, pp. 76–84.

Magat, Wesley A., W. Kip Viscusi, and Joel Huber. 1988. "Paired Comparison and Contingent Valuation Approaches to Morbidity Risk Valuation," *Journal of Environmental Economics and Management* vol. 15, no. 4, pp. 395–411.

McConnell, Kenneth E. 1986. *The Damages to Recreational Activities for PCBs in New Bedford Harbor*. Cambridge, Mass.: Industrial Economics.

McConnell, Kenneth E. 1990. "Models for Referendum Data: The Structure of Discrete Choice Models for Contingent Valuation," *Journal of Environmental Economics and Management* vol. 18, no. 1, pp. 19–34.

Mitchell, Robert Cameron, and Richard T. Carson. 1989. *Using Surveys to Value Public Goods: The Contingent Valuation Method*. Washington, D.C.: Resources for the Future.

Rae, Douglas A. 1983. "The Value to Visitors of Improving Visibility at Mesa Verde and Great Smoky Mountain National Parks," in Robert D. Rowe and Lauraine G. Chestnut, eds., *Managing Air Quality and Scenic Resources and National Parks and Wilderness Areas*. Boulder, Colo.: Westview Press.

Randall, Alan, Berry Ives, and Clyde Eastman. 1974. "Bidding Games for Evaluation of Aesthetic Environmental Improvement," *Journal of Environmental Economics and Management* vol. 1, no. 2, pp. 132–149.

Randall, Alan, and John R. Stoll. 1980. "Consumer's Surplus in Commodity Space," *American Economic Review* vol. 70, no. 3, pp. 449–455.

Reiling, Steven D., Kevin J. Boyle, Marcia L. Phillips, and Mark W. Anderson. 1990. "Temporal Reliability of Contingent Values," *Land Economics* vol. 66, no. 2, pp. 128–134.

Rowe, Robert D., Ralph C. d'Arge, and David S. Brookshire. 1980. "An Experiment on the Economic Value of Visibility," *Journal of Environmental Economics and Management* vol. 7, no. 1, pp. 1–19.

Samples, Karl C., John A. Dixon, and Marsha M. Gowen. 1986. "Information Disclosure and Endangered Species Valuation," *Land Economics* vol. 62, no. 3, pp. 306–312.

Schulze, William D., Ralph C. d'Arge, and David S. Brookshire. 1981. "Valuing Environmental Commodities: Some Recent Experiments," *Land Economics* vol. 57, no. 2, pp. 151–169.

Sellar, Christine, Jean-Paul Chavas, and John R. Stoll. 1986. "Specification of the Logit Model: The Case of Valuation of Nonmarket Goods," *Journal of Environmental Economics and Management* vol. 13, no. 4, pp. 382–390.

Smith, V. Kerry. 1992. "Arbitrary Values, Good Causes, and Premature Verdicts," *Journal of Environmental Economics and Management* vol. 22, no. 1, pp. 71–89.

Smith, V. Kerry, and William H. Desvousges. 1986. *Measuring Water Quality Benefits*. Norwell, Mass.: Kluwer-Nijhoff.

Smith, V. Kerry, and William H. Desvousges. 1987. "An Empirical Analysis of the Economic Value of Risk Changes," *Journal of Political Economy* vol. 95, no. 1, pp. 89–114.

Thaler, Richard. 1980. "Toward a Positive Theory of Consumer Choice," *Journal of Economic Behavior and Organization* vol. 1, no. 1, pp. 39–60.

Thayer, Mark A. 1981. "Contingent Valuation Techniques for Assessing Environmental Impacts; Further Evidence," *Journal of Environmental Economics and Management* vol. 8, no. 1, pp. 27–44.

Tversky, Amos, Paul Slovic, and Daniel Kahneman. 1990. "The Causes of Preference Reversal," *American Economic Review* vol. 80, no. 1, pp. 204–217.

Willig, Robert D. 1976. "Consumers' Surplus Without Apology," *American Economic Review* vol. 66, no. 4, pp. 589–597.

7

Aggregation of Values across Time

For purposes of economic planning, management and policymaking, it may often be useful to view environmental and natural resources as assets that yield flows of services over time. We have now developed the theoretical framework necessary for defining and measuring the values of these service flows to the individuals who receive them. However, this theory of value has been developed without explicit consideration of the temporal dimension of these service flows. Specifically, the theory has considered the values of services only at a single point in time or over some relatively short period of time (a day, a week, or a year) during which it is reasonable to treat other things such as income, prices, and the consumption of other goods and services as fixed. And the theory has defined values only at that moment in time at which the service flow is received. Thus we are left with the question of how to aggregate values that are realized at different moments in time and that might vary from one moment to another.

In this chapter I deal explicitly with the temporal dimension of value and welfare theory. In the first section, I review the standard theory of individual preferences and choice in an intertemporal setting. In the absence of taxes and other capital market imperfections, utility maximizing individuals borrow or lend so as to equate their marginal rates of substitution between present and future consumption with the market rate of interest. Thus this rate can be taken as a revealed preference indicator of individuals' intertemporal marginal rates of substitution.

In the second section, I show that an individual's marginal rate of substitution can be used to convert a marginal welfare measure for one moment in time into its equivalent at any other point in time. Measures for several different points in time can be made commensurate by converting them all to one period—for example, the present. Then these measures can be added to obtain an intertemporal aggregate welfare measure.

After reviewing the theory of intertemporal choice and using this theory to develop measures of intertemporal welfare change, I turn in the third section to the question of choosing an interest rate in a world with many different market interest rates and other intertemporal prices. The multitude of market interest rates arises because of inflation, taxation, and risk, among other things. As a consequence, there is no single number that can be identified as the ''correct'' interest rate for intertemporal welfare measurement; but it is possible to suggest a reasonable range that will be appropriate in many circumstances.

One consequence of the multiplicity of interest rates is that environmental policies that divert resources from other investments with high rates of return may have intertemporal opportunity costs at the margin, which are different from individuals' intertemporal marginal rates of substitution. In the fourth section, I briefly describe one approach to calculating these opportunity costs and making them commensurate with the measure of welfare gains developed in this chapter.

In the concluding section, I review the principles of intertemporal welfare measurement described in the chapter and identify some of the issues that arise when costs and benefits accrue to different generations.

INDIVIDUAL PREFERENCES AND INTERTEMPORAL CHOICE

The theory of preferences that we have been working with up to this point models only single-period choices and makes single-period utility a function of only the single-period consumption of marketed goods and services and environmental quality. With time introduced explicitly, the single-period utility in time period t is given by

$$u_t = u_t(X_t, q_t) \tag{1}$$

where X_t measures the expenditure on consumption in t (X_t is a numeraire good with a unit price) and q_t is the level of the environmental or resource service flow at t. This model has obvious limitations in that it cannot be used to analyze intertemporal choices and such things as saving and lending.

The alternative is a model of lifetime utility. This general representation of preferences would have an individual at any point in time experiencing a level of "lifetime" utility from the goods and services and environmental quality that person anticipates over his or her remaining lifetime:

$$u^* = u^*(X, Q) = u^*(X_1, ..., X_t, ..., X_T, q_1, ..., q_t, ..., q_T) \qquad (2)$$

where T is the number of years of remaining life. For simplicity, this formulation implies that the individual gains no satisfaction from making a bequest to others at the end of his or her life, but it is a straightforward matter to add a bequest motivation if that is an important part of the problem being analyzed (see, for example, Cropper and Sussman, 1988). Also in this analysis, we abstract from all considerations of uncertainty concerning the future.

The existence of single period preferences as described by equation (1) implies that the lifetime utility function (2) is separable, that is,

$$u = u[u_1(X_1, q_1), ..., u_t(X_t, q_t), ..., u_T(X_T, q_T)] \qquad (3)$$

It is common to assume that this expression is additively separable and that the single-period utility function is the same for all periods. Thus we can write

$$u^* = \sum_{t=1}^{T} D_t \cdot u(X_t, q_t)$$

where $u(\cdot)$ is invariant in t, $\partial u / \partial X_t$ is positive and decreasing in X_t, and where D_t is meant to capture the individual's time preference. For example, if an individual intrinsically preferred present consumption over future consumption, D_t would decrease with t and the marginal lifetime utility of X_t would be a decreasing function of t.

A useful way to characterize time preference is to let $D_t = 1/(1 + d)^t$ where d can be interpreted as a subjective rate of time preference analogous to an interest rate.[1] The relationship between d and time preference is as follows:

$d > 0$ represents positive time preference with the marginal utility of X_t decreasing with t;

---

[1]This form of geometrically decreasing D_t over time is necessary for intertemporal consistency of individual choice. See Strotz (1956) or Page (1977).

$d = 0$ represents neutral time preference; and

$d < 0$ represents negative time preference with the marginal utility of X_t increasing with t.

Olson and Bailey (1981) argue that empirical evidence points to $d > 0$ as the most common pattern of individual preferences. As they also usefully make clear, an individual's lifetime marginal utility of X_t ($t > 1$) could be less than the lifetime marginal utility of X_1 for either (or both) of two reasons: X_t is greater than X_1, or D_t is less than one ($d > 0$), or both.

Now assume that the individual knows with certainty that he or she will receive a stream of income payments, M_t, and a stream of environmental services, q_t, and that the individual can borrow or lend in a perfect capital market at an interest rate of $r\%$ per period. Assuming that the individual does not wish to leave a bequest and that all accounts must be settled in year T, the individual's lifetime budget constraint is

$$\sum_{t=1}^{T} X_t \cdot (1 + r)^{-t} = W^* = W_0 + \sum_{t=1}^{T} M_t \cdot (1 + r)^{-t} \qquad (4)$$

where W^* is lifetime wealth and W_0 is initial wealth.[2] Knowledge of the initial wealth, the streams of income and environmental services, and the interest rate is sufficient to determine the lifetime pattern of consumption of X that maximizes lifetime utility. In each period, if X_t is greater than (less than) M_t, the individual must borrow (lend) the difference.

The first-order conditions for maximizing equation (3) subject to the wealth constraint (4) take the form

$$\frac{\partial u^*/\partial X_t}{\partial u^*/\partial X_{t+1}} = 1 + r \qquad (5)$$

or, since $\partial u^*/\partial X_t = D_t \cdot \partial u/\partial X_t = (1 + d)^{-t} \partial u/\partial X_t$,

$$(1 + d) \frac{\partial u/\partial X_t}{\partial u/\partial X_{t+1}} = 1 + r \qquad \text{for all } t = 1, \ldots, t, \ldots, T\text{-}1 \qquad (6)$$

The individual's intertemporal marginal rate of substitution must be

_____

[2]Specifically, this formulation assumes that wealth is measured as of the beginning of the first period and that income and expenditures are made at the end of each period.

equal to one plus the interest rate. Thus the individual's intertemporal marginal rate of substitution can be inferred by observing the interest rate that governs these intertemporal trade-offs. The first-order conditions imply that if both r and d equal 0, the individual will equate the single-period marginal utility of consumption across all periods, thus choosing a constant stream of consumption over time. If $r = 0$ and $d > 0$, the single-period marginal utility of consumption in t must be less than that in $t + 1$, implying that the individual chooses to consume relatively more in period t. And an increase in r, other things being equal, requires adjustments to increase the marginal utility of consumption in earlier periods, thus implying a deferral of consumption to future time periods.

MEASURES OF WELFARE CHANGE

We turn now to the development of measures of intertemporal welfare change. First, consider a marginal increase in q_t. The individual would be willing to accept a reduced quantity of the numeraire in that period to hold utility constant. This compensating surplus measure of welfare change is the marginal rate of substitution between q_t and X_t, or

$$w_{q_t}^t = \frac{\partial u / \partial q_t}{\partial u / \partial X_t}$$

From equations (5) or (6), in intertemporal equilibrium the individual would be indifferent between paying $w_{q_t}^t$ in period t and paying

$$w_{q_t}^0 = w_{q_t}^t / (1 + r)^t$$

now. If the whole stream of future Q increases, the marginal willingness to pay now is the sum of the willingnesses to pay for each of the components of the increase, or

$$w_Q^0 = \sum_{t=1}^{T} w_{q_t}^t / (1 + r)^t$$

If the changes in Q are nonmarginal, aggregation over time is not so simple. As in chapter 3, we can define single-period compensating surplus (CS) and equivalent surplus (ES) welfare measures for changes in any q_t. These are changes in single-period income based on single-period utility. In an analogous fashion, we can use equations (2) or (3) to define compensating and equivalent changes in initial wealth that are based on lifetime utility. What is of interest is the relationship between these two alternative perspectives on the measurement of welfare change.

To keep it simple, consider a policy that would result in a change in q during only one period. Specifically, suppose that the policy results in $q_t'' > q_t'$.[3] What is the welfare gain to the individual of this change? The single period compensating surplus measure is the solution to

$$u\,(X_t - CS_t, q_t'') = u\,(X_t, q_t')$$ (7)

The present value of the single-period measure of gain is $(1 + r)^{-t}CS_t$.

In order to derive the compensating and equivalent wealth measures of lifetime welfare change, we first use the solution to the lifetime utility maximization problem to obtain the lifetime indirect utility function:

$$v^*(W^*, q_1, ..., q_t, ..., q_T)$$ (8)

From the perspective of lifetime utility, the individual is willing to make a payment at $t = 0$ that will equate the lifetime utilities with and without the change in q_t. Call this the compensating wealth, or CW. It is the solution to

$$v^*(W^* - CW, q_1, ..., q_t'', ..., q_T) = v^*(W^*, q_1, ..., q_t', ..., q_T)$$ (9)

Following the analysis by Blackorby, Donaldson, and Moloney (1984), it can be shown that the present-value welfare measures based on equations (7) and (9) are different except under certain special conditions.[4] Specifically, for the policy evaluated by equations (7) and (9):

$$|CW| \geq |(1 + r)^{-t}\,CS_t|$$ (10)

This result can be easily explained. If the individual pays CS in period t, then according to equation (7) single-period utility is restored to its

---

[3]It is straightforward to obtain similar results for changes in Q.

[4]Blackorby, Donaldson, and Moloney (1984) analyze the case where only prices and incomes change. In their model, the two welfare measures correspond only if prices are constant in the two situations, that is, if only incomes change.

initial level, and so is lifetime utility. But if the single-period marginal utility of consumption in t is affected by the change in q_t, then the first-order conditions for an optimum allocation of consumption over time will not be satisfied. The individual will wish to reallocate consumption across time through some pattern of additional borrowing and lending. Such reallocation of consumption will increase utility. This means that the individual could have made a larger single-period payment to keep lifetime utility at its original level. This establishes the inequality of equation (10).

The equality will hold only if the marginal utility of consumption is invariant to the combined changes in q_t and its offsetting CS_t. For example, suppose that the increase in q_t causes a decrease in the marginal utility of consumption in that period. The payment of CS_t will increase the marginal utility of consumption in period t. These two forces might exactly offset each other, leaving the marginal utility of consumption and the optimal intertemporal pattern of consumption unchanged. But if the single-period utility function is separable in q_t so that the marginal utility of consumption is unchanged, or if the increase in q_t leads to an increase in the marginal utility of consumption, then the net effect of the change in q_t and the offsetting compensation payment will be to increase the marginal utility of consumption in period t and to cause a reallocation of consumption across time.[5]

Similar inequalities can be derived for equivalent surplus and equivalent wealth (EW) measures of welfare gain and for both types of measures for welfare losses associated with a decrease in q_t. For example, suppose that q_t decreased. If the individual instead had to pay ES_t as the equivalent measure of welfare loss, and if this resulted in an increase in the marginal utility of consumption in t, then the individual could restore part of the loss of lifetime utility by reallocating spending toward period t, and ES_t would be an overestimate of the lifetime welfare loss.

These results can be summarized as follows:

For Gains (q_t increases)	For Losses (q_t decreases)
$CW \geq (1 + r)^{-t} CS_t$	$|CW| \leq |(1 + r)^{-t} CS_t|$
$EW \leq (1 + r)^{-t} ES_t$	$|EW| \geq |(1 + r)^{-t} ES_t|$

All of these expressions can be easily generalized to the case where there are changes in Q in many periods.

---

[5]See chapter 8 for more discussion of the effects of changes in q_t on the marginal utility of consumption.

These results have somewhat disturbing implications for applied welfare analysis. The typical practice in evaluating environmental policies that affect welfare over many years is to estimate single-period welfare measures on the basis of some indirect technique and to compute the present value of these single-period changes using some interest rate. But as shown here, if compensating surplus measures of single-period welfare change are used, this procedure will underestimate the true lifetime welfare benefit of the stream of improvements in Q and overstate the loss of decreases in Q. Conversely, if equivalent surplus single-period measures are used, the conventional practice leads to upward-biased estimates of total welfare gains and downward-biased estimates of lifetime losses.[6]

Of course, the important question is: How big is this bias? Keen (1990) has shed some light on this question. He showed that the difference between the single-period and the lifetime welfare measures was a term that he called the intertemporal compensating variation, or the maximum amount the individual would be willing to pay to be able to reallocate expenditure and utility across periods. Keen then derived expressions for this term and showed that in most instances the term would be of only second-order importance. In general, the error will be significant only if high values for the elasticity of intertemporal substitution combine with large values of CS_t relative to X_t and very uneven streams of CS_t over time.

Although most of the indirect methods for measuring welfare change produce only single-period measures, there are cases in which it is possible to obtain direct unbiased estimates of CW or EW. If changes in Q affect the market price of an asset, the change in its price might provide a basis for estimating CW or EW. For example, the hedonic price method is based on differences in the prices of assets such as houses. Since what is observed are differences in the market prices for the housing asset, price differentials between low- and high-quality houses reflect the compensating wealth change associated with the difference in the expected stream of environmental quality associated with the house. It is common practice in applied welfare analysis to convert the compensating wealth measure to an

---

[6]As Blackorby, Donaldson and Moloney (1984) point out, if the present value of the single-period CS_t is greater than zero, then the project unambiguously passes a Kaldor potential compensation test. Similarly, if the sum of the single-period ES_t is less than zero, the project unambiguously fails a Hicks potential compensation test. But projects with negative aggregate CS_t could still be efficient, while projects with positive aggregate present values of single-period ES_t might decrease aggregate welfare.

annual equivalent by using some interest rate or discount factor. Provided that the interest rate properly represents the individual's opportunities for intertemporal substitution, this annual equivalent can be used as an unbiased indicator of lifetime welfare change.

To summarize, if an individual can reallocate his stream of consumption over time through transactions at some interest rate r, that person will equate his intertemporal marginal rate of substitution with a discount factor equal to $1 + r$. This means that r can be interpreted as an intertemporal price or measure of intertemporal value, at least for marginal changes. For nonmarginal changes in Q, the correct measure of intertemporal welfare change is the compensating or equivalent change in initial wealth. But many of the available methods for measuring welfare change yield estimates of single-period values rather than lifetime values; the present value of these single-period welfare measures is, at best, only an approximation (an upper or lower bound) on the lifetime welfare measure. Given the limited availability of data, the approximation may be all that is obtainable.

WHICH INTEREST RATE IS THE RIGHT INTERTEMPORAL PRICE?

How can we estimate an individual's rate of time preference? Standard economic theory predicts that in a simple economy with no taxes and with perfect capital markets in which all individuals can borrow or lend at the market rate of interest, an individual will arrange her time pattern of consumption by borrowing and lending so that at each point in time her rate of time preference is just equal to the observed market rate of interest (Lind, 1982; Varian, 1990). In such a simple economy, this is the interest rate that should be used for the intertemporal aggregation of individuals' single-period welfare measures. But in fact there are taxes (certainly), inflation (at least sometimes), and various market imperfections. And we observe a multiplicity of market interest rates. The purpose of this section is to discuss briefly the choice of an interest rate for welfare evaluation when factors such as inflation and taxation are present, and to suggest some approaches to choosing an intertemporal price when faced with many rates of interest and tax rates.

Consider first the problem of the effects of taxes on income. As is well known, taxes can drive wedges between market prices and individuals' marginal rates of substitution. Suppose that an individual's interest income is taxed at the rate of $t\%$. Then money lent at $r\%$ will return only $r \cdot (1 - t)$% net of tax. It is this after-tax rate of return to which the optimizing lender will equate his intertemporal marginal rate of substitution. Similarly, if interest expense is treated as deductible in calculating tax liabilities, then the net cost of borrowing at a market rate of r is $r \cdot (1 - t)$%. Again, it is this after-tax net borrowing cost that will govern the individual's borrowing decisions. Discounting for welfare evaluation should be done at this after-tax interest rate.

A second problem is inflation. If there is price inflation, then the distinction between real and nominal interest rates becomes important. The individual would still arrange her time pattern of consumption so as to equate her rate of time preference with the expected real rate of interest. If inflation of i% is perfectly anticipated, then the observed market interest rate m and the real rate of interest r are related as follows:

$$\frac{(1 + m)}{(1 + i)} = 1 + r$$

or
$$m = r + i + r \cdot i \tag{11}$$

For reasonable rates of inflation and real interest rates, the third term in equation (11) will be small and can be ignored. Here the difference between the market interest rate and the inflation rate is an approximation of the real interest rate. If the estimates of individuals' single-period CS_t are based on the expectation of actual prices in that period, then discounting should be done at the market rate of interest, which will also reflect the anticipated rate of inflation, i. But the conventional practice in benefit measurement is to abstract from inflation by basing future period estimates of CS_t on the present price level. In that case, discounting should be done at the real rate of interest.

What is a reasonable value for this real rate of interest? One range of estimates comes from Lind (1982). After reviewing historical evidence on rates of return to a variety of investments, Lind concluded that rates of time preference could be as low as 0%, if historical real rates of return on U.S. Treasury bills were taken as the bench mark (p. 84). Over this same period of time, the real after-tax return on a broad portfolio of common stocks was about 4.6% (Lind, 1982, p. 83). This could be taken as an upper bound on the appropriate discount rate and would be the correct discount rate if the payments or receipts being discounted had the same risk characteristics as the market portfolio of common stocks.

Given the increasing openness of financial markets and the ease of moving funds between nations, world average interest rates may be more relevant than rates from any one nation's financial market. In a recent study, Barro and Martin (1990) presented estimates of an index of real expected and real actual short-term interest rates for nine major industrialized nations, including the United States. They calculated annual real rates of interest by subtracting either the expected or the observed rate of inflation from the market rate of interest on three-month government securities for each quarter, and averaged the four quarterly rates. The average pretax rates of interest they found for several intervals of time are presented in table 7-1. For more than two-thirds of the period studied, real rates of interest were below 2%, and they did not exceed 6%. These results are broadly consistent with Lind's conclusions. They also demonstrate the variability of real rates of interest over time. Specifically, real rates were higher for the period 1980–1989 than in earlier periods. Whether real rates will return to lower levels in the future cannot be foreseen.

Rather than look at market rates of interest, an alternative approach is to estimate the implicit interest rate that must be guiding individuals' observed choices over alternative temporal patterns of benefits and costs. For example, suppose an individual who has a choice between receiving $10 today and $15 in one year chooses the smaller immediate payment. This implies that the individual discounts future consumption at a rate of at least 50%. Empirical studies on choices of this sort often reveal unusually high implicit interest rates—more than 20% and up to and above 100%. Several people have estimated the discount rates implied by the choice to trade off price against lifetime operating costs when more energy-efficient appliances have higher purchase prices (Hausman, 1979; Gately, 1980; and Ruderman, Levine, and McMahon, 1986). These implied rates must be far higher than the effective interest rates available to the people making the choices. Also, experiments in which people

Table 7-1.   Real Expected and Actual Interest Rates for Nine Industrialized Nations

Time period	Real expected rate (%)	Real actual rate (%)
1959–1970	2.0	1.8
1971–1973	1.2	−1.0
1974–1979	0.0	−1.0
1980	2.4	1.8
1981–1986	4.2	5.3
1987–1988	2.3	2.8
1989	3.5	3.4

Source: Barro and Martin (1990)

are offered either hypothetical or real choices between payments of different sizes and dates consistently reveal implicit interest rates that are both high for short periods of time and declining as the interval lengthens. This evidence implies nongeometric discounting. For a discussion of some of the evidence, see Loewenstein and Thaler (1989) and Ainslie (1991).

On the other hand, Moore and Viscusi (1990) have estimated the discount rates implied by the way in which workers make trade-offs between higher wages now and loss of future years of life through accidental death on the job. They found implicit discount rates in the range of 2 to 12%, depending on the data source and the model specification. They concluded that *"Overall, we find no evidence of a significant discrepancy between rates of discount for health risks and financial rates of time preference"* (Moore and Viscusi, 1990, p. S–52; emphasis in the original).

Even if we have estimates of market rates of interest or implicit discount rates from revealed behavior, another problem arises in the calculation of welfare measures because of the fact that, for a variety of reasons, people face different interest rates. For example, individuals' real after-tax rates of interest will differ, since individuals face different marginal tax rates and may have different portfolios of investments. Differences in effective interest rates can also arise from market imperfections, transactions costs, and differences in the tax treatment of interest paid and interest received.[7] How should we deal with differences in effective interest rates across individuals? In principle, each individual's single period welfare changes should be discounted at the interest rate that person uses to make intertemporal allocation decisions. But in practice, we typically have single-period welfare measures that are aggregated across individuals who may face quite different rates of interest. Here I will sketch a procedure for calculating aggregate values in the second-best situation, where individuals face different effective interest rates at the margin.

Assume for the moment that no individual borrows and lends at the same time and that at any point in time each individual's transactions are entered into at a single interest rate that may vary across individuals. Consider an individual who at period t could borrow for one period at r_{bt} or could lend at another rate r_{st}, where both are real after-tax rates of interest. The individual, in effect, faces a kinked budget line. Whichever direction the individual chooses to move in, the interest rate governing that form of transaction will be the one

---

[7]Under current federal tax law most interest income is taxable, while interest paid on consumer loans is not tax-deductible.

with which that person equates his or her intertemporal marginal rate of substitution. That interest rate is the one which should be used to discount that individual's welfare measure over that interval of time.[8] So if the individual is a net lender between period t and period $t + 1$, and there is a change in q_{t+1}, the after-tax lending rate should be used to discount CS_{t+1} back to t.

Now consider aggregation across individuals who face different effective interest rates at any point in time because for example of differences in marginal income tax rates or differences in borrowing and lending behavior. Specifically, assume that the ith individual faces an effective real interest rate r_i that is constant over time. Assume then that we wish to calculate the sum of the compensating wealth payments that would leave each individual in his initial lifetime utility position. For individual i, the lifetime CW_i for a change in Q is approximated by the present value of the stream of single-period compensating surpluses:[9]

$$CW_i \simeq \sum_{t=1}^{T} \frac{CS_{it}}{(1 + r_i)^t}$$

The aggregate change for the n affected individuals is

$$\overline{CW} \simeq \sum_{i=1}^{n} \sum_{t=1}^{T} \frac{CS_{it}}{(1 + r_i)^t} \tag{12}$$

There is a simple way of calculating a weighted average discount factor that can be used with equation (12). First, each individual's share of each period's aggregate compensating surplus is calculated

$$s_{it} = \frac{CS_{it}}{\sum_{i=1}^{n} CS_{it}}$$

---

[8] If the individual neither borrows nor lends, there is no transaction from which to infer an interest rate.

[9] If r_i varies over time, then the individual's welfare measure is

$$CW_i \sum_{t=1}^{T} \frac{CS_{it}}{\prod_{t=1}^{t}(1 + r_{it})}$$

Individual discount factors for each period can be combined in the weighted averaging procedure described below.

Then this expression is used to substitute for CS_{it} in equation (12) to obtain

$$\overline{CW} = \sum_{i=1}^{n} \sum_{t=1}^{T} s_{it} \frac{\Sigma CS_{it}}{(1 + r_i)^t}$$

This can be simplified to

$$\overline{CW} = \sum_{t=1}^{T} \overline{D}_t \cdot \Sigma CS_{it}$$

where $\overline{D}_t$ is a weighted average discount factor:

$$\overline{D} = \sum_{i=1}^{n} s_{it} \cdot (1 + r_i)^{-t}$$

The preceding analysis has been based on the assumption that each individual enters into transactions at only one interest rate in any one period. But introspection and anecdotal evidence suggest that this assumption is often violated in practice. Many people simultaneously save by contributing to pension funds and borrow to buy new houses and consumer durables. And they may earn after-tax rates of return on savings instruments that are only a third or a quarter of their effective borrowing costs on outstanding credit card balances. This behavior raises two questions. First, how can this behavior be reconciled with the assumption of economic rationality? Second, what does this behavior imply for inferring intertemporal marginal rates of substitution and making intertemporal welfare measurements?

Lind (1990) has suggested that recent developments in behavioral economics can help to explain these practices. One key paper in this literature (Shefrin and Thaler, 1988) hypothesizes that people place different types of income, expenditures, and assets into different mental accounts and that they make different intertemporal allocation decisions for different sets of accounts. If it should prove possible to identify unique marginal rates of substitution or implicit discount rates for each mental account, and if it is possible to discover to which mental accounts people assign changes in environmental quality, then it may prove possible to develop a consistent discounting framework for intertemporal welfare measurement. But this possibility should be considered only as speculation and a possible area for future research.

In summary, the existence of multiple interest rates in the marketplace does not preclude the consistent intertemporal aggregation of individuals' welfare measures. In principle, each individual's set of welfare measures is aggregated across time, using that individual's intertemporal marginal rate of substitution as revealed by the interest rate at which that person transacts. In practice, there is a weighted average discount factor that reproduces this calculation. The case of individuals who simultaneously transact at different interest rates is more problematic. But the concept of mental accounts may help to explain this behavior and may also provide a basis for calculating the relevant intertemporal marginal rate of substitution.

THE CAPITAL COSTS OF ENVIRONMENTAL POLICIES

A major issue in the economic evaluation of public investment projects has been how to take account of the fact that financing the public investment through some combination of taxes and borrowing is likely to displace some private investment having a higher rate of return than the effective interest rate governing individuals' intertemporal substitutions. The divergence between the rate of return on private investment and the social discount rate can be attributed at least in part to the taxes on corporate and personal income. In this section, I briefly review the "shadow-cost-of-capital" approach to measuring the cost of public investment projects. I then discuss the implications of the shadow price of capital for the economic evaluation of public policies that require private investments that may displace other private investments.

Consider a public investment project with an initial capital cost of K_0. Assume that no further operating or maintenance costs are associated with the project. Let the project yield benefits of B per year in perpetuity. As before, let r represent the effective after-tax rate of interest governing all individuals' borrowing and lending decisions. Since r will reflect individuals' willingness to trade off present for future consumption, it should be used as the intertemporal price for welfare evaluation, given the individualist welfare perspective adopted in this book.[10] Let s represent the real marginal rate of return on investments in the private sector. Because of taxes on capital income at both the corporate and personal level, $s > r$.

---

[10]For an alternative perspective on the social rate of discount, see Marglin (1963b).

Using r, we can calculate the present value of the perpetual stream of benefits as a lower bound to the aggregate compensating wealth payment that is our desired lifetime welfare indicator. But if we find $B/r > K_0$, this does not necessarily signal that the project will improve welfare, because the relevant welfare comparison is not between the present value of the stream of benefits and the capital cost of the project. Rather, the relevant comparison is between the present value of the benefits of the project and the opportunity cost of the project as reflected in the stream of consumption forgone. In other words, the capital cost must be converted to its consumption equivalent for comparison with the benefit stream.

If K_0 were invested at the marginal rate of return in the private sector, it would produce a perpetual stream of $s \cdot K_0$ per year of future consumption. This is what is lost by diverting K_0 of capital resources from private investment to public investment, and this is the opportunity cost of the public investment. The present value of the stream of future consumption forgone is

$$\frac{s \cdot K_0}{r}$$

Thus in this simple example the shadow price or true social opportunity cost of one dollar of capital diverted from private investment is s/r, which is greater than one. In practice, the shadow price of capital will depend on a variety of factors reflecting the extent to which private investment is displaced, on net. If no private investment is displaced, the shadow price is one. More realistic formulations of the shadow price expression would take into account the facts that public investment will not necessarily displace private investment on a dollar-for-dollar basis, that public and private investments have less than infinite lifetimes, and that some portion of the future returns from displaced private investment could be reinvested for future consumption rather than being consumed immediately.[11] To summarize, the shadow price of capital is the present value (discounted at the individual rate of time preference, r) of the stream of future consumption forgone from one dollar of public investment.

The proper procedure for project evaluation is to discount benefits (and operating and maintenance costs, if any) at the social rate of discount. The present value of the benefit stream would then be

---

[11]For a review of some of these issues and some estimates of the magnitude of the shadow price, see Lind (1982). Earlier important references include Bradford (1975) and Marglin (1963a).

compared with the social cost of the capital investment. The latter would be calculated by multiplying the dollar cost of the investment by the appropriately calculated shadow price of capital.

Now suppose there is a pollution control regulation that, instead of requiring public investment, requires private firms to invest capital in pollution control. Suppose also that this regulation will yield environmental improvements with an aggregate value of B per year in perpetuity. Once again, the present value of the benefit stream from the regulation is B/r. Assume that it is estimated that firms can comply with the regulation by installing equipment with an aggregate capital cost of K_0, and that there are operating, maintenance, and repair costs of R per year and that the equipment will last forever. And suppose further that firms face infinitely elastic factor supply curves for all factor inputs and have constant returns-to-scale production functions. This assures that all of the costs of complying with the regulation will be passed on to consumers in the form of higher prices.

Because under these assumptions the aggregate supply of capital is infinitely elastic, the regulation does not result in any displacement of investment in the economy. The cost of the regulation comes in the form of higher prices to consumers. Prices must be raised sufficiently to amortize the investment of K_0 and to cover the operating costs of R per year. Ignoring the effects of higher prices on demand, the required price increase must generate additional revenues of $s \cdot K_0 + R$. This is the annual cost of the regulation to consumers.[12] To calculate the present value of this stream of costs, the effective consumers' interest rate, r, should be used. The present value of the social cost to consumers is approximated by $(s/r) \cdot K_0 + R/r$. In effect, the capital cost is first annualized using the pretax marginal rate of return on private investment (s) and then converted back to a present value using the consumption rate of interest (R).

Kolb and Scheraga (1990) show that essentially the same procedure should be followed even when the pollution control investment has a finite life and when other private investment is displaced.

---

[12]In order to take account of the effects of higher prices on demand and output, it is necessary to solve for that K_0 which is required to comply with the regulation after taking account of the effect of higher prices on output and the need for capital. For example, if some firms are expected to exit the industry, then fewer firms would have to make investments. The aggregate capital requirement would be smaller than if output and the number of firms were assumed to be unchanged. Also, the conceptually correct measure of cost is the CV of the consumers who must pay the higher prices.

The reason for this is that the displacement of private investment is not permanent. When prices are raised to recover the amortized cost of the pollution control capital, these funds are available for future investment. The net effect of the investment in pollution control capital is only to delay the stream of consumption benefits from postponed private investment.

SUMMARY

We have seen that the aggregation of an individual's single-period welfare measures over time can be done using a discount factor that represents the individual's marginal rate of intertemporal substitution or rate of time preference. For rational individuals this rate will be equal to the relevant after-tax real rate of interest. We have also reviewed evidence from financial markets which suggests that this rate (which can vary across individuals) lies in the range of 1 to 4%. This is much lower than the 10% discount rate mandated for use by federal agencies for many years (until 1992),[13] but it brackets the rate of 2% that the U.S. Congressional Budget Office is now using for its economic analyses (Hartman, 1990). I would feel comfortable using a rate of 2 to 3%, at least where the streams of benefits and costs accrue to people in the same generation.

However, as is well known, discounting future benefits or costs over long time horizons, especially over two or more generations, is controversial. Applying any discount rate to a future sum can make it look very small in present-value terms when the interval is 100 years or more. For example, at 2%, the present value of $1 in 100 years is about 14 cents; if the $1 comes in 200 years, its present value is only 1/2 cent at 2%.

The principal argument *for* discounting for economic welfare analysis also provides some insight into why one might choose not to make discounting calculations across generations. Recall that what is being discounted is individuals' willingness to pay or required compensation, either for a change (compensating measure) or to avoid a change (equivalent measure). The ethical justification for making these welfare calculations is their use in applying either the Kaldor or Hicks form of compensation test to determine whether it is possible to

---

[13]In October 1992, the Office of Management and Budget revised the discount rate downward to 7 percent (U.S. Office of Management and Budget, 1992). See Lyon (1990) for a more detailed discussion of discounting policies in the federal government.

compensate the losers so that no one is made worse off. When the gainers and losers are part of the same generation, actual compensation is feasible, and in any event both groups can participate in the decision. If a proposed policy would impose costs on a future generation, it is hard to imagine mechanisms for assuring that the compensation would be paid if it were thought desirable or necessary to obtain the consent of future generations.[14] And the future generation has no voice in present decisions. Similar problems arise in the case of a policy that would yield benefits to the future but impose costs on the present. The real problem with discounting across generations is not, then, that it results in very small numbers for the present values of future effects; rather, it is that it is hard to imagine a meaningful role for measures of compensation in the context of intergenerational resource reallocations.

REFERENCES

Ainslie, George. 1991. "Derivation of 'Rational' Economic Behavior from Hyperbolic Discount Curves," *American Economic Review* vol. 81, no. 2, pp. 334–340.

Barro, Robert J., and Xavier Sala i Martin. 1990. "World Real Interest Rates," *NBER Macroeconomics Annual* vol. 5, pp. 15–59.

Blackorby, Charles, David Donaldson, and David Moloney. 1984. "Consumer's Surplus and Welfare Change in a Simple Dynamic Model," *Review of Economic Studies* vol. 51, no. 164, pp. 171–176.

Bradford, David F. 1975. "Constraints on Government Investment Opportunities and the Choice of Discount Rate," *American Economic Review* vol. 65, no. 5, pp. 887–899.

Cropper, Maureen L., and Frances G. Sussman. 1988. "Families and the Economics of Risks to Life," *American Economic Review* vol. 78, no. 1, pp. 255–260.

Gately, Dermot. 1980. "Individual Discount Rates and the Purchase and Utilization of Energy-Using Durables: Comment," *Bell Journal of Economics* vol. 11, no. 1, pp. 373–374.

Hartman, Robert W. 1990. "One Thousand Points of Light Seeking a Number: A Case Study of CBO's Search for a Discount Rate Policy," *Journal of Environmental Economics and Management* vol. 18, no. 2, part 2, pp. S-3–S-7.

Hausman, Jerry. 1979. "Individual Discount Rates and the Purchase and Utilization of Energy-Using Durables," *Bell Journal of Economics* vol. 10, no. 1, pp. 33–54.

[14]See Lind (1990) for more on this point.

Keen, Michael. 1990. "Welfare Analysis and Intertemporal Substitution," *Journal of Public Economics* vol. 42, no. 3, pp. 47–66.

Kolb, Jeffrey A., and Joel D. Scheraga. 1990. "Discounting the Benefits and Costs of Environmental Regulations," *Journal of Policy Analysis and Management* vol. 9, no. 3, pp. 381–390.

Lind, Robert C. 1982. *Discounting for Time and Risk in Energy Policy.* Washington, D.C.: Resources for the Future.

Lind, Robert C. 1990. "Reassessing the Government's Discount Rate Policy in Light of New Theory and Data in a World Economy with a High Degree of Capital Mobility," *Journal of Environmental Economics and Management* vol. 18, no. 2, part 2, pp. S-8–S-28.

Loewenstein, George, and Richard H. Thaler. 1989. "Anomalies: Intertemporal Choice," *Journal of Economic Perspectives* vol. 3, no. 4, pp. 181–193.

Lyon, Randolph M. 1990. "Federal Discount Rate Policy, the Shadow Price of Capital, and Challenges for Reforms," *Journal of Environmental Economics and Management* vol. 18, no. 2, part 2, pp. S-29–S-50.

Marglin, Stephen A. 1963a. "The Opportunity Costs of Public Investment," *Quarterly Journal of Economics* vol. 77, no. 2, pp. 274–289.

Marglin, Stephen A. 1963b. "The Social Rate of Discount and the Optimal Rate of Investment," *Quarterly Journal of Economics* vol. 77, no. 1, pp. 95–111.

Moore, Michael J., and W. Kip Viscusi. 1990. "Discounting Environmental Health Risks: New Evidence and Policy Implications," *Journal of Environmental Economics and Management* vol. 18, no. 2, pp. S-51–S-62.

Olson, Mancur, and Martin J. Bailey. 1981. "Positive Time Preference," *Journal of Political Economy* vol. 89, no. 1, pp. 1–25.

Page, Talbot. 1977. *Conservation and Economic Efficiency: An Approach to Materials Policy.* Baltimore, Md.: The Johns Hopkins University Press for Resources for the Future.

Ruderman, Henry, Mark Levine, and James McMahon. 1986. "Energy-Efficiency Choice in the Purchase of Residential Appliances," in Willett Kempton and Max Neiman, eds., *Energy Efficiency: Perspectives on Individual Behavior.* Washington, D.C.: American Council for an Energy Efficient Economy.

Shefrin, Hersh M., and Richard H. Thaler. 1988. "The Behavioral Life-Cycle Hypothesis," *Economic Inquiry* vol. 26, no. 4, pp. 609–643.

Strotz, Robert. 1956. "Myopia and Inconsistency in Dynamic Utility Maximization," *Review of Economic Studies* vol. 23, pp. 165–180.

U.S. Office of Management and Budget. 1992. "Guidelines and Discount Rates for Benefit-Cost Analysis of Federal Programs," Circular no. A-94, revised. Washington, D.C.

Varian, Hal R. 1990. *Intermediate Microeconomics: A Modern Approach.* 2d ed. New York: Norton.

8

Values in an Uncertain World

U p to this point, the discussion in this book has been based at least implicitly on the assumption of perfect certainty. For example, individuals have been assumed to know what prices, income, and environmental quality will prevail over the relevant planning horizon. This clearly cannot be literally true. And for some questions, the essence of the matter is uncertainty about some event or condition, such as the probability of an accident or a natural disaster (say, a flood) occurring or the probability of contracting a serious illness following exposure to a chemical. The topic of this chapter is how to define and measure changes in economic welfare when uncertainty is an important characteristic of the economic world in which people are living and making choices.

The uncertainty faced by individuals can take many forms. For example, people could be uncertain about future prices of consumer goods, future income, or the returns on financial assets.[1] The uncertainty that is important in this chapter is that faced by individuals who are users or potential users of some environmental service flow or resource. Individuals might be uncertain as to whether a specific environmental service flow or resource will be available for their use in the future; they might be uncertain as to whether they will actually want to use some resource in the future; and they could be uncertain as to whether their exposures to environmental hazards and pollutants will actually result in illness or death. These uncertainties raise two kinds of questions for policymakers. The first is: What form of welfare criterion is appropriate under conditions of uncertainty, or more specifically, what modifications to the Hicks-Kaldor potential

---

[1]Since there is already an extensive literature on the welfare effects of price and income variability, I will not treat those problems here. See, for example, Helms (1985a, 1985b); Just, Hueth, and Schmitz (1982); and Rothschild and Stiglitz (1976).

compensation test are required? The second is: When public policies change the uncertainties facing people, how are the resulting welfare changes to be measured? These two questions are connected in that, as we shall see, the choice of a welfare criterion has important implications for how welfare changes are to be measured.

In the next section I develop the now standard model of individual preferences under uncertainty, which is based on the hypothesis that individuals maximize their expected utility. I identify and discuss the several alternative welfare indicators that can be derived from this model of preferences. In the second section I discuss alternative aggregate welfare criteria and their implications for the way individuals' welfare changes are measured. In the third section I present some models for the indirect measurement of individuals' willingness to pay for risk-reduction and risk-prevention policies. I also discuss the possibility that preferences are not consistent with the expected utility hypothesis, and examine the implications of nonexpected utility preferences for welfare measurement. In the fourth section I define the concepts of option value, quasi-option value, and margin of safety and discuss their relationship to the welfare measures developed in this chapter.

INDIVIDUAL UNCERTAINTY AND WELFARE

The term individual uncertainty refers to situations in which an individual is uncertain as to which of two or more alternative states of nature will be realized and is not indifferent as to which state actually occurs. I assume here that individuals can assign probabilities to these alternative states of nature and that the probabilities are correct in the sense of summing to one and incorporating all available information. Thus I am following the modern practice of making no distinction between risk and uncertainty. And I am abstracting from questions of risk perception, how individuals make assessments of probabilities, and how these assessments are revised in light of new information.[2]

---

[2]For a discussion of the often-cited distinction between risk (where probabilities are known) and uncertainty (where probabilities are unknowable) that is often attributed to Frank Knight, see LeRoy and Singell (1987). For introductions to some of the issues concerning risk perceptions and probability assessments and for references to the relevant literature in psychology and economics, see Arrow (1982), Viscusi (1985), and Machina (1990, especially pp. 169–175).

As a first step in this analysis I provide a more formal characterization of how risk is defined and the ways in which environmental policy can affect risk. I then turn to the characterization of individual preferences in situations involving risk, and ways of characterizing welfare changes.

Risk and Environmental Policy

Consider some environmental risk such as a chemical spill or pollution event, or some natural hazard such as a flood or an earthquake; other examples of such risks include tornadoes and hurricanes, accidents at nuclear reactors, explosions in chemical plants, releases of toxic materials from hazardous waste storage sites, and air pollution episodes associated with unusual meteorological conditions. Risks such as these can be described in terms of two characteristics: the range of possible adverse consequences and the probability distribution across consequences. In this analysis, adverse consequences are measured in units that reflect the consequences for people (for example, the number of buildings damaged and degree of destruction, or the number of days of illness) rather than in terms of measures such as height of flood stage, Richter scale reading, or maximum atmospheric concentration of a pollutant. To keep the exposition simple, I will consider only one possible adverse event and two states of nature: the event occurs with a given set of consequences, or the event does not occur.

Suppose there is some public policy action that has the effect of reducing the probability of the adverse event, reducing the severity of its consequences, or both. Examples of policy measures that reduce the magnitude of an adverse event include regulations requiring earthquake-resistant construction techniques and the building of public shelters for protection against tornadoes and hurricanes. These are often referred to as risk-reduction measures. Examples of policy measures that reduce the probability of an adverse event include regulations for nuclear reactor safety, standards for durable containment techniques at hazardous waste storage sites, and the construction of dams to control streamflow to reduce the probability of flooding. These are often referred to as risk-prevention measures.

It must be granted that the distinction between risk reduction and risk prevention may sometimes be arbitrary. For example, the benefits of wearing a seat belt could be modeled in terms of the belt's effect on the severity of injury associated with accidents of unchanged probability, or in terms of its effects on the probabilities of experiencing injuries of various severities per mile driven. The choice of a modeling strategy may depend upon such pragmatic considerations as availability of data.

Many risk-reduction and risk-prevention measures are public goods in that they have the characteristics of nonrivalry and nonexcludability. The public good character of these measures means that a private market system will fail to provide them in efficient quantities. Thus there is a case for government intervention to improve the efficiency of resource allocation. In order to determine whether risk-reduction and risk-prevention measures result in improvements in welfare, it is necessary to define and measure the benefits and costs of changes in risks.

Individual Preferences and Expected Utility

Turning to the characterization of individual preferences in situations involving risk, assume that an individual has a well-behaved preference ordering over bundles of goods X and that there is some adverse environmental event A over which that person has no control. The variable A measures the severity of the adverse event. Let $A = A^*$ represent the occurrence of the adverse event and $A = 0$ represent the absence of the adverse event. A takes the value A^* with probability π and 0 with the probability $1 - \pi$. For any given state of nature, this preference ordering can be represented from an ex post perspective by $u = u(X, A)$ with $u(X, 0) > u(X, A^*)$. I assume $u_A < 0$ and $u_X > 0$ where in this chapter subscripts indicate partial derivatives. To avoid unnecessary complexity, income M and prices P are assumed to be certain. I assume that $u(\cdot)$ is the same in both states, but that preferences over X are state-dependent. By this I mean that the consumption bundle actually chosen can depend on the state of nature, that is, on whether $A = 0$ or A^*. This means that the marginal utility of X can vary across states of nature.[3] The model can be easily generalized to many states and to make $u(\cdot)$ depend on the state of nature; see, for example, Graham (1981) and Freeman (1984b, 1985, 1989).

Given utility maximization, there is an ex post indirect utility function $u = v(M, P, A)$ that shows the maximum attainable utility

---

[3]An alternative formulation is to write two different utility functions, $u^*(X)$ and $u^\circ(X)$, to indicate the effect of A on utility. These two formulations are essentially equivalent in that in both cases A acts as a shifter of the marginal utility schedules for X and M.

given M, P, and A, where M is income and P is the vector of prices. It has the properties

$$v_M > 0 \qquad v_{MM} < 0$$
$$v_A < 0 \qquad v_{AA} < 0$$

From here on the P term is suppressed, since prices are assumed to be unchanged through the analysis.

I assume that individuals know the magnitude of the adverse event A^* and the probability of its occurrence. I also assume in this section that there are no opportunities for individual protective activities, that is, no ways of spending money to reduce either π or A^*. Ehrlich and Becker (1972) refer to such activities as self-protection (for reducing π) and self-insurance (for reducing A^*) and analyze individual behavior in the case of purely monetary losses. To the extent that self-protection and self-insurance activities are possible in the case of risks of nonmonetary losses, they may provide a basis for empirical estimation of the values of risk changes. I investigate this possibility in the section on indirect methods in this chapter.

In our analysis, let D represent the monetary value of the damages caused by the event given that the adverse event has occurred. D is the maximum sum of money the individual would give up to experience $A = 0$ rather than A^*. It is the solution to

$$v(M, A^*) = v(M - D, 0)$$

Thus D is a form of compensating surplus (CS) measure of welfare change where the reference level of utility is the utility realized if A^* occurs. From an ex post perspective, the value of reducing A^* to 0 is the expected value of the monetary equivalent of avoiding a sure A^*, or $\pi \cdot D$.

People often have to make choices before the state of nature is revealed, in other words, ex ante choices. Let us assume that individuals make these choices so as to maximize their expected utility where expected utility is defined as

$$E[u] \equiv \pi \cdot u(X, A^*) + (1 - \pi)u(X, 0)$$

The expected utility expression provides a basis for an alternative measure of the value of avoiding A^*, namely, the willingness to pay ex ante. This state-independent payment is sometimes referred to as option price, or OP. Option price is defined as the maximum payment the individual would make to change from the status quo risk to a situation in which A^* would not occur. It is also a form of compensating surplus, but one where the reference point is defined in terms of expected utility. OP is the solution to

$$\pi \cdot v(M, A^*) + (1 - \pi) v(M, 0) = v(M - OP, 0)$$

OP and $\pi \cdot D$ will not in general be equal.[4] This is because they are measuring two different things, or more precisely, measuring the monetary equivalents of two different forms of utility change, the ex post change in $v(\cdot)$ and the ex ante change in $E[v]$. Furthermore, these two measures are not the only possible ways of expressing a compensating measure of the welfare value of preventing the adverse event. This can best be seen with the aid of Graham's (1981) willingness-to-pay locus.

The Willingness-to-Pay Locus

Given the assumption that individuals maximize expected utility, their behavior can be described as the solution to the following problem:

$$\text{Max: } E[u] = \pi \cdot u(X, A^*) + (1 - \pi)u(X, 0) \qquad (1)$$
$$x$$

subject to the usual budget constraint that expenditure equal income. The solution to this problem is denoted as E^1.

Now consider a policy that would reduce the magnitude of the adverse event from A^* to 0. If the individual were required to pay D to avoid the adverse effects given that the event occurs and would pay nothing otherwise, the realized utility in each of the two states would be unchanged. Thus this set of payments would not change expected utility. Alternatively, if the individual were to pay OP before the uncertainty were resolved, by definition expected utility would be unchanged. These are only two of an infinite number of possible pairs of payments that would leave the individual indifferent, in expected utility terms, between the status quo and a situation in which the consequences of the adverse event were avoided with certainty. The

---

[4]The relationship between these two measures is the basis of an extensive literature on option value. For reviews of this literature, see Bishop (1982) and Smith (1983). I review the current status of this literature in the fourth section of this chapter.

set of these payments is denoted as:

t^* = payment given the state in which the adverse event occurs
t° = payment given the state in which it does not occur

These payments satisfy the following condition:

$$E^1 = \pi \cdot v(M - t^*, 0) + (1 - \pi)v(M - t^\circ, 0) \qquad (2)$$

This equation defines Graham's willingness-to-pay (WTP) locus, which is shown in figure 8-1.

Similarly, for a risk-prevention policy the WTP locus is given by the solution to

$$E^1 = E^2 \equiv \pi' \cdot v(M - t^*, A^*) + (1 - \pi')v(M - t^\circ, 0)$$

where π' is the probability of the adverse event with the risk-prevention policy in place. All of the subsequent analysis pertains to risk-reduction policies, but with suitable modification it can also be applied to the benefit-cost analysis of policies that change probabilities.

By setting the total differential of equation (2) equal to zero, we obtain an expression for the slope of the WTP locus:

$$\frac{dt^*}{dt^\circ} = -\frac{(1 - \pi)v_{M^\circ}}{\pi \cdot v_{M^*}} \qquad (3)$$

where v_{M^*} is the marginal utility of income evaluated at $M - t^*$ given the event has occurred, and v_{M° is evaluated at $M - t^\circ$ given no adverse event. If the individual is risk-averse as assumed above ($v_{MM} < 0$), then the WTP locus will be concave to the origin.[5] This is because increasing t^* (for example) raises v_{M^*} relative to v_{M°. Risk neutrality with its constant marginal utility of income would result in a linear WTP locus.

Three points on the WTP locus are of particular interest as possible welfare measures. The first is point D in figure 8-1. This point involves a payment of zero if the event does not occur and there are no realized damages, and a payment of t^{*D} if the event does occur. This sum represents the maximum payment the individual would make to experience the consequence 0 rather than A^* given that

---

[5]See Karni (1983) for the derivation of measures of risk aversion for state-dependent preferences.

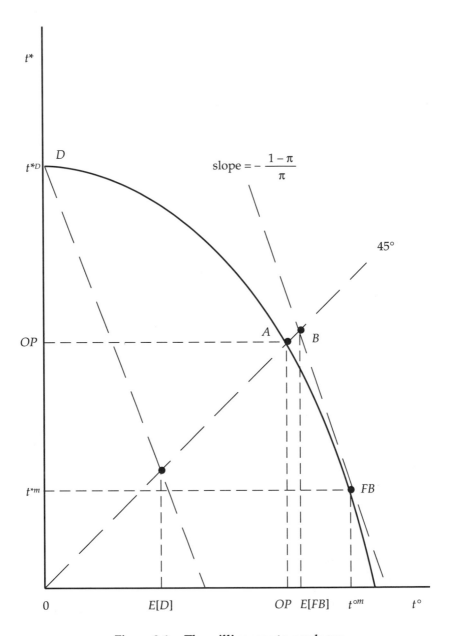

Figure 8-1.   The willingness-to-pay locus

the event occurs. This sum is also equal to the damages (D) from the event were it to occur. There is a line from point D with a slope of $-(1 - \pi)/\pi$ which represents the locus of all pairs of monetary values that have the same expected value as t^{*D}.[6] This line is one of a family of what I will call iso-expected payment lines. The point where the iso-expected payment line through point D intersects the 45° line can be used to find the expected value of damages $(E[D])$. This is the ex post expected CS measure of the welfare change of preventing A^*.

The second point of interest in figure 8-1 is A. It shows an alternative state-independent payment scheme in which $t^* = t^\circ = OP$. In this example, the state-independent payment is greater than the expected value of damages. But this will not always be the case. The difference between OP and $E[D]$ can be either positive or negative, and in some circumstances it can be quite large; see, for example, Graham (1981) and Freeman (1984b, 1985, 1989).

Finally, there is an iso-expected-payment line that is tangent to the willingness-to-pay locus at point FB in figure 8-1. This point represents that state-dependent payment scheme, t^{*m}, $t^{\circ m}$, which maximizes the expected value of the individual's payments $(E[FB])$. The tangency of the willingness-to-pay locus and the iso-expected payment line at point FB implies that the marginal utilities of income are equal in the two states.[7] This equality is the condition for the efficient distribution of risk and for the optimum purchase of contingent claims at actuarially fair prices.[8] In this sense, point FB represents an optimum contingent payment scheme. In Graham's (1981) terminology, this is the fair bet point.

As illustrated in figure 8-1, these three alternative payment schemes can give quite different summary measures of maximum

---

[6]For any t^*, these loci satisfy

$$E[t] = \pi \cdot t^* + (1 - \pi)t^\circ$$

Differentiating, this gives:

$$\frac{dt^*}{dt^\circ} = - \frac{(1 - \pi)}{\pi}$$

[7]Since the slopes of the WTP locus and iso-payment lines are equal, making use of equation (3) yields:

$$- \frac{(1 - \pi)}{\pi} = - \frac{(1 - \pi)v_{M^*}}{\pi \cdot v_{M^\circ}}$$

and $v_{M^*}/v_{M^\circ} = 1$

[8]A contingent claim is a contract specifying in advance a set of payments or receipts, or both, in which the amounts depend on the state of nature. An insurance policy is a contingent claim.

willingness to pay for the policy. This is because they differ with respect to the implied opportunities for insurance, that is, opportunities to alter the payments in different states of nature to reflect attitudes toward risks. In the example of figure 8-1, the fair bet point allows the individual to make a larger payment in the state of nature in which no adverse event occurs and in which, therefore, the individual is otherwise better off.

In order to see the relationship between state-dependent payments of this sort and insurance, let us suppose that actuarially fair insurance is available to the individual. Specifically, suppose that the individual could make a payment I before the uncertainty was resolved to purchase an insurance policy that would pay R if the adverse event occurred. This means that the individual can exchange income across states of nature according to $I/(R - I) = k$, the price of insurance. If the price of insurance is actuarially fair, $k = \pi/(1 - \pi)$. In figure 8-1, the insurance price line has the same slope as the iso-expected payment lines. As is well known, the condition for the optimal purchase of insurance is, in this case, the equality of the marginal utilities of income in the two states of nature.[9]

Suppose now that the individual is required to make the payment represented by point D if the adverse event occurs and pays nothing otherwise. Paying D is equivalent to experiencing damages equal to D in this state of nature. Suppose also that the individual can purchase insurance at fair prices. The iso-expected payment line through point D shows that the individual could purchase an insurance contract that would require the individual to pay $E[D]$ if the adverse event did not occur and would reimburse him so that his net payment would be $E[D]$ if the adverse event did occur. Since this point lies inside the WTP locus, it results in a higher expected utility than could be realized in the absence of the policy that would reduce A^* to zero. This means that if the individual can also purchase insurance so as to adjust his state-dependent payments, he would be willing to pay more than D for the policy that reduces A^* to zero. By

---

[9]The individual chooses I (and therefore R) so as to maximize

$$E[u] = \pi \cdot v(M + R - I, A^*) + (1 - \pi) v(M - I, 0)$$

subject to the constraint that $I/(R - I) = k$. Substituting the constraint into the objective function and differentiating gives the first order condition

$$\frac{\pi}{(1 - \pi)} \frac{v_{M^*}}{v_{M^\circ}} = k$$

If the insurance is fair, $k = \pi/(1 - \pi)$ and $v_{M^*} = v_{M^\circ}$.

failing to take into account the individual's opportunities for diversifying risk through insurance, the expected damage measure of welfare change underestimates the true willingness to pay of the individual.

Now suppose that there is some institutional barrier to imposing a payment scheme that varies across states of nature, but that the individual can still purchase contingent claims at fair prices. The maximum state-independent payment that could be extracted from the individual ex ante is actually represented by point B in figure 8-1. Since point B lies outside of the willingness-to-pay locus, the individual would be left worse off with this payment scheme in the absence of fair insurance. But with fair insurance, the individual can move down to the right and reach the fair bet point through the optimal purchase of insurance. Thus the availability of fair insurance makes the maximum expected payment of the fair bet point feasible even when varying payments across states of nature are not possible.

The existence of the willingness-to-pay locus and points such as B (in figure 8-1) that can be reached by purchasing contingent claims raises an important question. If welfare change is to be measured by the maximum payment that holds expected utility constant, which pair of state-dependent payments is the best welfare measure? We are not yet ready to answer this question, because how welfare is to be measured depends in part on the form of the social welfare criterion. The Potential Pareto Improvement (PPI) form of social welfare criterion requires a comparison of the welfare changes of those who gain and those who lose by the project. When losers face uncertainty about costs, the potential for state-dependent payments and compensation to redistribute risks between gainers and losers must be taken into account, along with the ability of each group to insure itself through the purchase of contingent claims. A full consideration of these issues must be left to the next section. But first we must show how uncertainty about costs can be represented in Graham's state-dependent payment model.

The Required Compensation Locus

The costs of government risk-management activities come ultimately in the form of reductions in the utilities of individuals. This is true whether the activity involves direct government spending or regulations imposing requirements on the private sector. Furthermore, these costs, as correctly measured, may be quite different from the direct expenditures of the government or regulated firms; see, for example, Hazilla and Kopp (1990). The measurement of these costs from an ex ante perspective entails first determining the reductions in expected utility for all of those who would bear the costs and then finding the minimum compensation required to restore each person to her original level of expected utility in the absence of the government activity.

Any individual who would bear part of the cost of a project has a reference level of expected utility in the absence of the project, say

$$E^2 = \pi \cdot v(M^*) + (1 - \pi) v(M^\circ)$$

The project would impose costs on the individual that could vary across states of nature. These costs could take several forms: a reduction in money income through taxation, the direct disutility of restrictions on nonmarket activities, or changes in product and factor prices.

If these costs vary across states, then compensation must be calculated so as to restore each cost-bearer's expected utility to the status quo level. Let c^* and c° represent all of the nonmonetary dimensions of costs and let M^{c^*} and M^{c° represent money income inclusive of project-induced income changes. So the problem can be analyzed graphically, suppose that the probabilities affecting costs and benefits depend on the same random process in nature, so that π and $1 - \pi$ represent the relevant probabilities. This assumption will be relaxed later.

Given these assumptions, the project would result in a level of expected utility in the absence of compensation of

$$E^2 \equiv \pi \cdot v(M^{c^*}, c^*) + (1 - \pi) v(M^{c^\circ}, c^\circ)$$

where M^{c^*} and M^{c° represent net income and c^* and c° represent the direct negative effects of the policy on utility in each state. There is an infinite number of pairs of state-dependent compensation payments, r^* and r°, that can be made to restore expected utility to E^2. These pairs are solutions to

$$E^2 = \pi \cdot v(M^{c^*} + r^*, c^*) + (1 - \pi) v(M^{c^\circ} + r^\circ, c^\circ) \tag{4}$$

The solution to this expression gives the required compensation (RC) locus for an individual who is harmed by the policy. It is shown in figure 8-2. This is the locus of all possible compensating payments that leave those who bear the costs no worse off in terms of expected utility than if the policy had not been undertaken. The slope of the RC

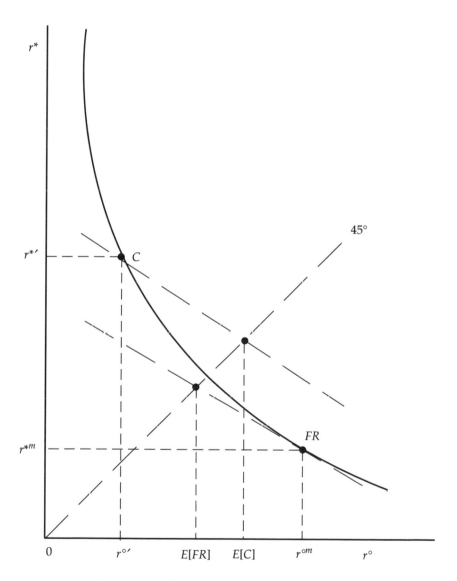

Figure 8-2.   The required compensation locus

locus is found by setting the total differential of equation (4) equal to zero:

$$\frac{dr^*}{dr^\circ} = - \frac{(1 - \pi)}{\pi} \frac{v_{M^\circ}}{v_{M^*}}$$

Diminishing marginal utility of income assures that the locus is convex to the origin. The probabilities of the two states also define a family of iso- expected payment lines. They can be used to find the set of compensations that optimally distributes the risk to the cost-bearer across states, that is, that combination of payments that equates the marginal utilities of income in the two states. This is shown as r^{*m}, $r^{\circ m}$, at point FR in figure 8-2. The intersection of the iso-expected payment line through this point with the 45° line gives the expected value of the fair compensation point, $E[FR]$.

It is also possible to define an expected value of the required compensation from an ex post perspective. Given the occurrence of the event, there is an $r^{*\prime}$ that satisfies

$$v(M^{c*} + r^{*\prime}, c^*) = v(M^*)$$

Similarly, $r^{\circ\prime}$ satisfies

$$v(M^{c\circ} + r^{\circ\prime}, c^\circ) = v(M^\circ)$$

These solutions define one point on the RC locus, labeled as C in figure 8-2. Expected cost is

$$E[C] \equiv \pi \cdot r^{*\prime} + (1 - \pi) r^{\circ\prime}$$

Finally, we can define a state-independent compensation analogous to OP (say, OR). Since it also holds expected utility constant, it is the solution to

$$E^1 = \pi \cdot v(M^{c*} + OR, c^*) + (1 - \pi)v(M^{c\circ} + OR, c^\circ)$$

But as will become clear in the next section, neither the expected value of cost nor OR will in general be useful as welfare measures for benefit-cost analysis.

AGGREGATION AND THE WELFARE CRITERION

A major issue in the literature on benefit-cost analysis under uncertainty, at least since Weisbrod's classic article (1964), has been what is the appropriate measure of the welfare change for an individual who benefits from a project.[10] The early option-value literature focused on comparing the expected value of consumer surplus with option price. Graham (1981) redirected the discussion by showing that there is an infinite number of alternative state-dependent payment vectors and that the choice of a vector depends at least in part on the availability of contingent claims and individuals' opportunities to redistribute risk. Graham added a third candidate to the list of potential welfare measures, the expected value of the fair bet point; and he showed under what conditions the third candidate would be preferable to expected surplus and option price.

The literature since Graham has continued to focus on the question of the availability of contingent claims and whether risks are individual or collective; see, for example, Mendelsohn and Strang (1984), Graham (1984), Cory and Saliba (1987), Colby and Cory (1989), and Smith (1990). For the most part this literature has ignored the cost side of the problem or treated it in a simplified fashion. In particular, most of this literature has neglected explicit treatment of uncertainty in costs, the possibility for changing the nature of risk through state-dependent compensation schemes, and the implications of the latter possibility for the form of the Potential Pareto Improvement criterion.[11] Also, for the most part this literature has dealt with only the simplest form of uncertainty—that where only two alternative states of nature are possible.[12] The simple two-state model and its graphic treatment can be maintained only if both cost and benefit uncertainty result from the same random process and there are only two possible outcomes. If there are n possible benefit outcomes and m possible cost outcomes, the total number of alternative states of nature is $n \cdot m$.

In this section I first discuss alternative forms of the PPI criterion under uncertainty. I then consider a simple model with one gainer and one loser so that the problems of aggregation and collective as opposed to individual risk are not important. I show that in this simple model none of the three candidates for welfare measurement is a reliable screen for potential Pareto improvements. This is because all three candidates ignore the opportunity for redistribution of risk between the gainer and the loser through state-dependent payments

---

[10]This section is adapted, with permission, from Freeman (1991b).

[11]An exception is the work of Meier and Randall (1991).

[12]Exceptions include Graham (1981) and Graham (1992). The latter paper presents a much more formal, abstract treatment of some of the problems considered here.

and compensation. I then extend the analysis to a many-person economy. In those cases where the collective nature of risk precludes contingent claims markets, the importance of opportunities for redistributing risk between gainers and losers through state-dependent payment and compensation remains. Neither state-independent payments and compensation nor expected values of fair bet points are reliable indicators of potentially Pareto improving projects.

Potential Pareto Improvements Under Uncertainty

Recall that the foundation of the Potential Pareto Improvement criterion is that a project passes only if there is some way to make redistributive payments so that no one is made worse off by the project and some people are made better off. It is generally agreed that in the context of risk and uncertainty "better off" and "worse off" are defined in terms of expected utility.[13] But there are several ways of formulating potential compensation tests in terms of expected utility. As shown above, there is an infinite number of potential payments and compensations for each individual that will leave expected utility unchanged. However, not all of these payment and compensation vectors are feasible in the aggregate, in the sense that whatever state of nature occurs, the potential payment equals or exceeds the required compensation. In other words, not all of the payment and compensation vectors pass an ex post balanced budget test. Here I develop the PPI criterion in a general form that is consistent with the balanced budget test. Then in the next subsection I illustrate the implications of this criterion for project selection and for welfare measurement in the simple two state model. The generalization to many states is straightforward. I then discuss the aggregation of individual welfare measures across many gainers and losers.

A proposed policy can be deemed an improvement in PPI terms only if there is at least one set of state-dependent payments that will finance (in the sense of balancing the budget in all states of nature) a set of state-dependent compensations without making anyone worse off and leaving at least one person better off in expected utility terms. More formally,

let $i = 1, ..., n$ index alternative outcomes of the random process affecting benefits;

---

[13]Bishop (1986), for example, considers two forms of ex post compensation tests. He rejects them both on the grounds that they are overly restrictive, that is, would fail too many projects, and that they ignore individuals' attitudes toward risk that are reflected in their ex ante expected utility functions.

let $j = 1, ..., m$ index alternative outcomes of the random process
affecting costs;

if T and R represent points on the n and m dimensional WTP and RC
loci respectively, a project passes the PPI criterion if (and only if)
there is at least one T and R such that $t_i \geq r_j$ for all i, j with the
strict inequality holding for at least one pair.

Graphically, a project represents a PPI only if the WTP locus in the
$n \cdot m$ dimensional space intersects the RC locus.[14] Notice that since
this expression of the PPI criterion compares t_i and r_j independently
of the probabilities associated with alternative outcomes, expected
values are irrelevant in screening for PPI projects.

This version of the PPI criterion can pass projects that fail more
restrictive forms of the criterion. This is because the criterion takes
into account the possible benefits of redistributing risks between
gainers and losers through state-dependent payments and compensation.[15] With this form of the criterion, project evaluation requires
consideration of both the real effects of the project and project financing, because it is the latter that determines the benefits, if any, of
redistributing risks.

Consider a project that would affect only two people. One person would gain, at least in one state of nature. The other would bear
the costs. In this simplification, the problem of how to aggregate
benefits and costs within the beneficiary and cost-bearer groups is
avoided. Assume that it is not possible to buy contingent claims on
income across the two states of nature that are relevant. Should the
project be built? The answer depends on which version of the PPI
compensation test one adopts.

One form of compensation test was offered by Bishop (1986),
who called it an Ex Ante Compensation Test. For the two-person
economy considered here, this test asks whether the option price of
the gainer exceeds the required compensation of the loser in every
state of nature. Ready (1988) also discussed this criterion; he called it
a "weak" form of potential Pareto improvement. The implications of
this form of compensation test can be seen in figure 8-3. Point A on
the willingness-to-pay locus shows the option price of the gainer. The

---

[14]An equivalent procedure is to aggregate the positive and negative payments into a net WTP locus. If this locus has a positive segment in the $n \cdot m$
dimensional space, the project passes the PPI test.

[15]Cook and Graham (1977) were apparently the first to recognize the
potential risk redistribution benefits of public investments in risk prevention
and reduction.

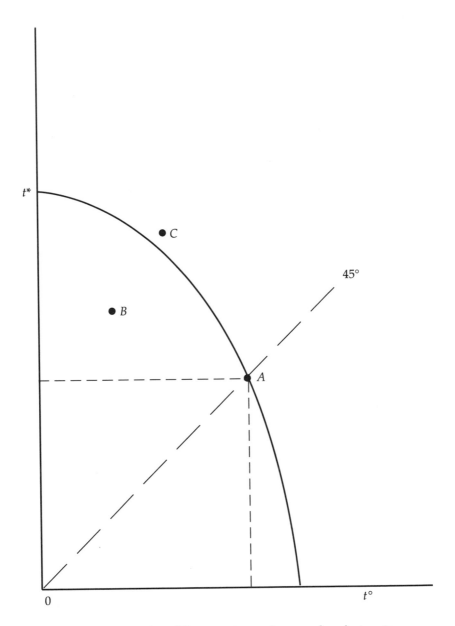

Figure 8-3. The willingness-to-pay locus and project costs

ex ante compensation or weak PPI criterion will pass only those projects whose costs lie below and to the left of point *A*. Points *B* and *C* represent projects with uncertain costs. Project *B* would not pass the test because if the adverse event occurred, the person bearing the cost would require compensation that exceeded the amount collected in the form of option price.

Ready also proposed an alternative, which he called the "strong" potential Pareto criterion. A project passes the strong form of this criterion if there exists some set of state-dependent payments that exceed the state-dependent project costs in every state. The strong form of PPI criterion would pass a project with uncertain costs represented by point *B* in figure 8–3, but it would reject a project whose costs were represented by point *C*. The strong form of the PPI criterion justifies the use of the expected value of the fair bet point as a benefit measure. If $E[FB]$ exceeds $E[FR]$, then the project passes.

Note that although both weak and strong forms of the criterion allow for cost uncertainty, they both treat costs as a point in the payment and compensation space. And they both ignore opportunities to redistribute risk among cost bearers or between gainers and losers through state-dependent payments *and compensation.*

Recall that according to the general form of the criterion stated above, a proposed policy is an improvement only if there is at least one set of balanced-budget state-dependent payments that will finance a set of state-dependent compensations without making anyone worse off and leaving at least one person better off. In terms of figure 8–3, the general form of PPI criterion would pass a project with costs represented by point *C* if the required compensation locus through point *C* intersected the WTP locus. As noted above, this form of the criterion takes account of both the real effects of the project and the redistribution of risks through project financing.

Applying the Criterion in a Two-Person World

Again consider the simplest case of a project that would affect only two people and where it is not possible to buy contingent claims on income across the two states of nature. To provide a specific example, suppose that the project is to divert water from a river through an existing generator to produce hydroelectric power for a beneficiary. The cost would be borne by a farmer who would otherwise use the water for irrigation. In a rainy year, the benefit would be large because of the large flow through the generator. The cost to the farmer would be small because the rainfall would substitute for the irrigation water. Similarly, in a dry year the benefit would be small and the cost to the farmer large.

Should the project be built? Since the benefits and the costs are subject to the same random process, the WTP and RC loci can be combined in a single diagram; figure 8-4 shows the case. Points B and C show the state-dependent benefits and costs. The WTP and RC loci pass through these two points. Another way of stating the PPI criterion is that it asks whether there is a set of state-dependent transfers from the beneficiary to the cost-bearer such that neither party is made worse off and at least one of the parties gains. Any state-dependent transfer is represented by one point in the WTP–RC space. Any such point on or below the WTP locus leaves the beneficiary no worse off; if the point is on or above the RC locus, the cost-bearer is no worse off. Thus any set of transfers that lies in the lens-shaped area between the two loci will successfully finance this project. The project passes the PPI test.

There are several important points to note about this conclusion. First, neither the expected value of cost nor the expected value of benefit is relevant. In the example of figure 8-4, the expected value of cost substantially exceeds the expected value of benefit. But the project still passes the PPI test because there are feasible payment schemes that redistribute the risks associated with the states of nature in a way that can make both parties better off.[16]

The irrelevance of expected benefit and expected cost holds even if benefits and costs are certain, as long as individuals are risk-averse. For example, suppose that points C and B in figure 8-4 are both on the 45° line, with C above B. It is still possible that the aggregate WTP and RC loci through these points could intersect over some range. This demonstrates the importance of considering the manner in which the payment and compensation scheme redistributes risk in project evaluation. For example, suppose that both individuals faced substantial uncertainty about their incomes. The beneficiary would prefer a payment scheme with the larger payment in the state of nature in which income was high. The cost-bearer would prefer to receive more compensation in the low-income state. Project financing provides both

---

[16]Thus Graham (1981) is incorrect when he says that if the project cost is represented by a point such as C in figure 8-3, "... which lies above the aggregate locus, then there does not exist a Pareto improving way of providing the [project]" (Graham 1981, p. 718). It is possible for the aggregate RC locus through his point G to lie below the aggregate WTP locus over some range, indicating payments and compensation that would pass the Pareto test.

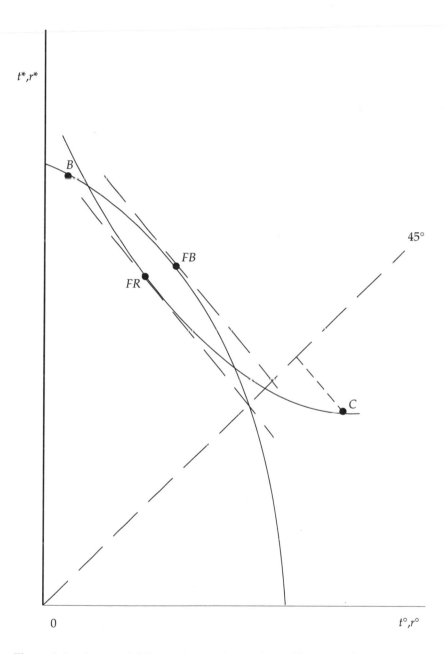

Figure 8-4.   A potential Pareto-improving project with expected costs exceeding expected benefits

individuals with a means of hedging the income uncertainty. The risk of the project also has to be evaluated in the context of both individuals' total risk portfolios.

The second point about the conclusion above is that option prices are not reliable indicators of PPI projects. Again, the example in figure 8-4 shows that it is possible for the state-independent compensation to exceed the state-independent willingness to pay and yet have the project pass the PPI test because of the ability of state-dependent payments and compensation to redistribute risk in a more favorable manner.

The third point is that neither the expected value of the fair bet point nor the expected value of fair compensation is relevant to project evaluation. Although the potential Pareto improving project in figure 8-4 has an expected value of the fair bet point exceeding the expected value of compensation, this condition is not sufficient to assure that the project is potentially Pareto improving. A case in which the expected value of the fair bet point exceeds the expected value of fair compensation is shown in figure 8-5. Yet because the WTP locus lies below the RC locus everywhere, there is no state-dependent transfer scheme that would make this project potentially Pareto improving. Although on average (that is, in an expected value sense) WTP exceeds RC, there are some states of nature in which the funds for the required compensation could not be collected from the beneficiary.

These results can be summarized in the form of three statements about necessary and sufficient conditions for identifying PPI projects.

1. The expected value of benefits being greater than the expected value of costs is neither necessary nor sufficient for a PPI project.
2. The option price of beneficiaries being greater than the option price of cost-bearers is sufficient but not necessary for a PPI project.
3. The expected value of the fair bet point exceeding the expected value of the fair compensation point is necessary but not sufficient to identify a PPI project.

Up to this point I have not considered the feasibility of collecting or paying out the necessary state-dependent sums. If a social welfare judgment is made not to require payment and compensation, the feasibility of a state-dependent payment and compensation scheme is not an issue. However, if a social welfare judgment is made to require the payment and compensation, then the potential Pareto improve-

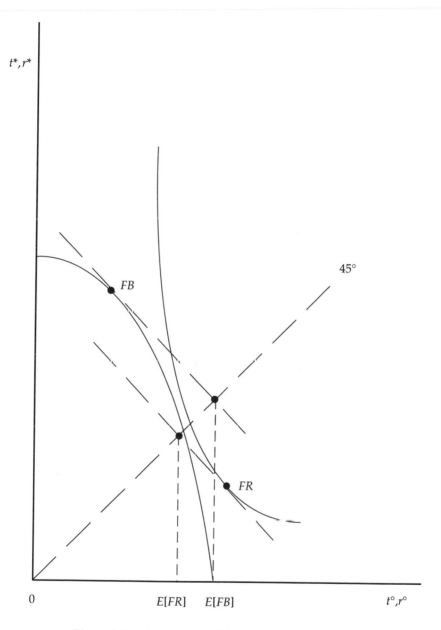

Figure 8-5. A non-potential Pareto-improving project

ment test must be based on feasible state-dependent payments and compensations. Graham (1981) and Cory and Saliba (1987) both provide examples showing that some combinations of state-independent payments and state-dependent payments and rebates can be used to obtain any required state-dependent payment pattern. As Graham puts it, "the trick, of course, is to collect the right payment in every state, but this may well be no more difficult than collecting the same payment in every state" (1984, p. 1102).

However, if the social welfare judgment requires payment and compensation and it proves to be impossible to devise a state-dependent scheme, then the potential Pareto improvement test must be based on option price measures, as pointed out by Graham (1981, p. 722). But this is not because option price is a superior welfare indicator in general; rather, it is because an actual compensation welfare criterion is more restrictive than a potential compensation criterion.

Aggregation in the Many-Person Economy

In the many-person economy the principal issue to be addressed is how the WTP and RC loci of individuals should be aggregated for the purposes of social welfare analysis. The answer depends on the nature of the risks faced by gainers and losers, that is, whether risks are collective or individual in nature. By collective risk I mean the case in which, if the event affecting benefits and costs occurs, all of the potentially affected individuals experience the event. Individual risk refers to a case in which the risks facing different individuals are independent.

Consider the case of individual risk first. With individual risks, the analysis of aggregate benefits and costs is much simplified. If one individual who would be a beneficiary has a probability of π^b of experiencing the adverse event, then $\pi^b \cdot N$ individuals will experience the event with virtual certainty, where N is the number of potential beneficiaries. Similarly, $\pi^c \cdot K$ of the K bearers of the costs will experience the cost state associated with the event with virtual certainty. Although individuals are uncertain as to their own outcomes, aggregate payments and compensation can be calculated with virtual certainty.

Assume for the moment that all people are identical. If those cost-bearers who experience the event are compensated by $r^{*\prime}$ and the rest of the cost-bearers are compensated by $r^{\circ\prime}$, then the aggregate compensation is

$$\pi \cdot K \cdot r^{*\prime} + (1 - \pi)K \cdot r^{\circ\prime} = K[\pi \cdot r^{*\prime} + (1 - \pi)r^{\circ\prime}]$$

The aggregate compensation is minimized by finding the expected fair compensation for an individual and multiplying by K. The same procedure is used to find the maximum aggregate willingness to pay. If

$$N[\pi^c \cdot t^{*m} + (1 - \pi^b)t^{\circ m}] > K[\pi^c \cdot r^{*m} + (1 - \pi^c)r^{\circ m}]$$

then the project passes the PPI test. If individuals are different, all individuals' fair bets and compensations must be found and summed. In general, if each gainer makes the state-dependent payments represented by her fair bet point, the aggregate payment is the sum of the expected values of the fair bet points. Similarly for the losers. And there is a virtual certainty that the gainers can compensate the losers and still be better off.

As long as a social welfare judgment is made not to require actual payment and compensation, this is all that needs to be said on the matter. However, if a social welfare judgment is made to require payment and compensation, then one must address the question of the feasibility of the required state-dependent payment and compensation schemes. If it should turn out that state-dependent payment and compensation is not feasible, then individual option prices are relevant, unless contingent claims could be purchased at actuarially fair prices. In the case of individual risks, it is more likely that such contingent claims will in fact be available.

In the case of collective risk, the full aggregate WTP and RC loci are required for welfare analysis. The procedure for aggregating individuals' WTP loci is explained in Graham (1981, pp. 718–719). For each possible marginal rate of substitution (MRS) between contingent payments in different states as measured by dt^*/dt°, each individual's payment vector is determined and all of these payment vectors are summed to obtain

$$T^{*m} = \sum_i^N t_i^{*m} \text{ and } T^{\circ m} = \sum_i^N t_i^{\circ m}$$

where i indexes the N beneficiaries and the superscript m indicates that all individuals' payments are at their fair bet points for a given common MRS. This procedure gives one point on the aggregate willingness-to-pay locus for each MRS. Each point also represents an efficient distribution of risk within the group of beneficiaries since all individuals have the same MRS for contingent payments. A similar procedure of aggregating individuals' required compensation at different marginal rates of substitution will yield the aggregate RC locus. Each point on the RC locus corresponds to an efficient distribution of cost risk among cost-bearers.

A project passes the potential Pareto improvement test if there exist payment and compensation points, $T^* = R^*$ and $T^\circ = R^\circ$, that lie both on or above the aggregate RC locus and on or below the aggregate WTP locus. The graphic analysis is similar to that in the two-person world shown in figures 8-4 and 8-5. Aggregate option price is sufficient but not necessary as a screen for PPI projects; and the aggregate fair bet points are necessary but not sufficient as indicators of PPI projects.

There are two exceptions to these conclusions. First, if contingent claims could be purchased at actuarially fair prices, the expected values of the fair bet and fair compensation points do provide a sufficient test for potential Pareto improvement. This is because individuals' actual payments and compensations can be made at the expected values of the fair bet and fair compensation points. Individuals can then adjust their own positions through contingent claims purchases to reach their individual willingness-to-pay or required compensation locus at their fair bet point. However, the nature of collective risks is likely to preclude a market for contingent claims at actuarially fair prices.

Second, if a social welfare judgment is made to require compensation, and if state-dependent payments and compensations are not feasible, then as is the case in the two-person world the welfare criterion must be based on option price measures. If individuals are identical, the appropriate test is based on the aggregate OP^B and OP^R because they are equal to the sums of the individual OPs and ORs. But if individuals are different, then the sums of the individuals' option price measures will be less than the aggregate measures. This is because the former do not provide for the efficient distribution of risk across individuals within each group. The appropriate welfare test becomes

$$\sum_i^N OP_i > \sum_j^M OR_j$$

In this section I have examined some of the consequences of extending the model of benefit-cost analysis for uncertain projects to reflect the symmetry of benefits and costs as changes in the expected utilities of affected individuals. One of the implications of this analysis is that there is much work still to be done on the development of measures of benefits and costs in order to implement the appropriate welfare criteria. Specifically, since presently available methods of benefit and cost estimation under uncertainty generally give only expected cost and benefit and option price measures, it is necessary to develop new methods for estimating the WTP and RC loci.

INDIRECT METHODS FOR MEASURING VALUES

When individuals have opportunities to adjust to risky positions through transactions in related private goods markets, it may be possible to use one of the indirect methods described in chapter 4 to infer individuals' values for risk changes. In this section I describe how models of averting behavior and hedonic prices can be used to obtain measures of value that are based on observable behavior. The welfare measures I develop are based on the hypothesis of expected utility maximization. I then show how these models can be extended to take account of other assumptions about individuals' preferences. It must be pointed out that since most observable transactions involve ex ante state-independent payments, the models described here are useful only in estimating option price measures of value.

Expressions for Option Price and Marginal Willingness to Pay

Now suppose that public policies are under consideration that will reduce either the magnitude of an adverse event or its probability. The option price for a policy that reduces the magnitude of an adverse event is the set of state-independent payments that results in the same level of expected utility as would have occurred in the original uncertain situation. Denote this payment as OP^A to indicate that it refers to a risk-reduction policy. Making use of the indirect utility function, OP^A for a policy that reduces A from A^* to zero is the solution to

$$\pi \cdot v(M, A^*) + (1 - \pi)v(M, 0) = \pi \cdot v(M - OP^{A^0}, 0)$$
$$+ (1 - \pi)v(M - OP^{A^0}, 0)$$
$$= v(M - OP^{A^0}, 0) \tag{5}$$

Similarly, the value of reducing A from A^* to $A' > 0$ is the solution to

$$\pi \cdot v(M, A^*) + (1 - \pi)v(M, 0) = \pi \cdot v(M - OP^{A'}, A')$$
$$+ (1 - \pi)v(M - OP^{A'}, 0)$$

For some purposes, it will be more useful to deal with marginal values. The option price for a marginal reduction from A^* is the marginal change in income that holds expected utility constant. This can be found by taking the total differential of the expression for expected utility in its indirect utility function form, setting it equal to zero, letting $d\pi = 0$, letting $dA = 0$ for $A = 0$, and rearranging terms. We get the individual's marginal willingness to pay, ex ante (w^{A*}), or

$$w^{A*} = \frac{dM}{dA^*} = -\frac{\pi \cdot v_{A*}}{[\pi \cdot v_{M*} + (1 - \pi) v_{M\circ}]} \quad (6)$$

where $*$ and $\circ$ indicate the values at which the partial derivatives are evaluated. Recalling that $v_A < 0$, this expression is positive for reductions in A^* ($dA^* < 0$), indicating a positive willingness to pay to reduce the magnitude of the uncertain event.

Equation (6) shows a variant of a standard result from welfare theory. The marginal willingness to pay for a change in A^* is equal to the marginal disutility of A^* converted to a money measure by using the marginal utility of income. But since this is an ex ante willingness to pay, the marginal utility terms are the expected values of the relevant marginal utilities, that is, the weighted averages of the marginal utility terms in the two states of nature where the weights are the probabilities of the two states.

The option price for reducing the probability of the event to zero is found by solving the following expression for $OP^{\pi\circ}$:

$$\pi \cdot v(M, A^*) + (1 - \pi)v(M, 0) = 0 \cdot v(M - OP^{\pi\circ}, A^*)$$
$$+ 1 \cdot v(M - OP^{\pi\circ}, 0) \quad (7)$$
$$= v(M - OP^{\pi\circ}, 0)$$

Comparison of this expression with equation (5) shows that the option price for reducing the probability of an event to zero is equal to the option price for reducing its severity or magnitude to zero. Thus in this case there is no difference between the value of risk reduction and the value of risk prevention, since in both cases the policy being valued eliminates the risk entirely.

The option price for a reduction in the probability of the event from π to π' is found by solving the following expression for $OP^{\pi'}$:

$$\pi \cdot v(M, A^*) + (1 - \pi)v(M, 0) = \pi' \cdot v(M - OP^{\pi'}, A^*) \quad (8)$$
$$+ (1 - \pi')v(M - OP^{\pi'}, 0)$$

The marginal value for a change in π can be derived by taking the total differential of equation (8), setting it equal to zero, and holding

dA^* at zero. The result is:[17]

$$\frac{dM}{d\pi} = \frac{v(M, 0) - v(M, A^*)}{[\pi \cdot v_{M^*} + (1 - \pi)v_{M^\circ}]} \tag{9}$$

This expression is positive, indicating a positive willingness to pay for reductions in π ($d\pi < 0$). It can be rewritten to show the dependence of the change in M on the magnitude of the change in π:

$$w^\pi = dM = \frac{v(M, 0) - v(M, A^*)}{[\pi \cdot v_{M^*} + (1 - \pi)v_{M^\circ}]} \cdot d\pi \tag{10}$$

where w^π is the willingness to pay ex ante for a change in the probability of A^*.

A comparison of equations (10) and (6) is instructive. Whereas (6) is based on the marginal disutility, as given by the term v_{A^*}, (10) is based on the nonmarginal difference in utility between the two states of nature, converted to monetary units by a weighted average of the marginal utilities of income in the two states of nature.

Averting Behavior

Observations of an individual's averting behavior might provide a basis for inferring the values of changes in risk. Suppose that, ex ante, an individual can select a level of private spending R, that will reduce the magnitude of A^* given that the event occurs according to the relationship $A^* = A(R, G)$, where G is the level of government protective spending. An example would be purchasing and wearing a seat belt to reduce the severity of injury given the occurrence of an accident. Assume that this function has the following properties:

$$A(0, 0) = A^*, \qquad A_R^* < 0, \qquad \text{and } A_G^* < 0$$

The individual chooses R, given G so as to maximize expected utility:

$$E[u] = \pi \cdot v[M - R, A(R, G)] + (1 - \pi)v[M - R, 0] \tag{11}$$

The first-order conditions include

$$\frac{1}{A_R^*} = \frac{\pi \cdot v_A}{\pi \cdot v_{M^*} + (1 - \pi)v_{M^\circ}} \tag{12}$$

---

[17] This is a standard result in the analysis of the value of risk prevention. See, for example, Jones-Lee (1974), Cook and Graham (1977), Machina (1983), and Smith and Desvousges (1988).

where v_{M^*} is the marginal utility of income evaluated at the level of A associated with the given level of G. The term $1/A_R^*$ is the reciprocal of the marginal productivity of expenditure on risk reduction, or equivalently, the marginal private cost of reducing A^*. The right-hand side is the marginal value of reducing A^* that was derived above.

The relevant value for policymakers is the value to the individual of an increase in public spending on risk reduction, dG. To find the option price measure, we take the total differential of equation (11), set it equal to zero, and substitute the first-order condition for the choice of private protective spending into this expression. After some simplification, we have

$$\frac{dM}{dG} = - \frac{A_G^*}{A_R^*} = \frac{\partial R}{\partial G} \tag{13}$$

This means that the individual's marginal willingness to pay for a small increase in government spending is the ratio of the marginal productivities of private spending and public spending in reducing A^* or the marginal rate of technical substitution between R and G in reducing A^*. This measure can be calculated if the technical relationship $A(R, G)$ is known. The relationship is observable in principle. The welfare change is also given by the marginal rate of substitution between private and public spending, holding expected utility constant. This is not the same thing as the observed change in private spending. For example, if G increases, the individual will reduce R but will also attain a higher level of expected utility. As discussed in chapter 4, if R enters the utility function directly, then the welfare measure will include unobservable marginal utility terms, and the measure derived here will be an underestimate (overestimate) if R provides positive (negative) utility.

Similar results can be obtained for the case where individual ex ante spending has the effect of reducing the probability of the adverse event. Now let the production function relating private and public expenditures to the probability of the adverse event be

$$\pi = \pi(R, G) \text{ where}$$

$$\pi(0, 0) = \pi^*$$

$$\pi_R < 0, \pi_G < 0$$

The individual chooses R so as to maximize expected utility given by

$$E[u] = \pi(R, G) v(M - R, A^*) + [1 - \pi(R, G)] v(M - R, 0) \tag{14}$$

The first-order conditions include

$$\frac{1}{\pi_R} = - \frac{v(M - R, 0) - v(M - R, A^*)}{\pi(R, G)v_{M^*} + [1 - \pi(R, G)]v_{M^\circ}} \tag{15}$$

Again, the left-hand side is the reciprocal of the marginal productivity of private expenditure on reducing the probability, or, equivalently, the marginal private cost of reducing π. The right-hand side is the marginal value of reducing π as given by equation (10).

The marginal value of an increase in public spending to reduce the probability of the event is found by totally differentiating equation (14), setting the result equal to zero, and substituting the first-order condition where appropriate. The result is:[18]

$$\frac{dM}{dG} = - \frac{\pi_G}{\pi_R} = \frac{\partial R}{\partial G} \tag{16}$$

This result is similar to the case of public spending to reduce A^*. The individual's marginal willingness to pay for public spending is equal to the ratio of the marginal productivities of private and public spending to reduce π, or to the marginal rate of technical substitution between R and G, holding A^* constant. Again, this result is analogous to those derived in the existing literature on protective spending in the absence of uncertainty.

Unfortunately, these results do not carry over to the case where the averting activity jointly produces reductions in π and A^*. Repeating the steps described above but making both A^* and π functions of R and G leads to the following expression:

$$\frac{dM}{dG} = - \frac{\pi \cdot v_{A^*} \cdot A_G + \pi_G \cdot (v^* - v^\circ)}{\pi \cdot v_{A^*} \cdot A_R + \pi_R \cdot (v^* - v^\circ)}$$

The unobservable utility terms do not cancel out of this expression, so marginal willingness to pay cannot be inferred from information on the averting technology.[19] As shown in chapter 4, the inability to use the averting behavior model in this case is due to the jointness of the implicit production technology. It is not a consequence of introducing risk into the analysis.

---

[18]Berger and coauthors (1987) obtain an similar expression from their analysis of air pollution, averting behavior, and the risk of illness.

[19]Shogren and Crocker (1991) establish this result.

Hedonic Prices

If either the probability or the magnitude of a risk (or both) is a characteristic of heterogeneous goods such as housing, hedonic price estimation can be used to obtain the relevant ex ante marginal values for risk changes. Suppose that the magnitude of the adverse event varies across the space used for residential housing. For example, the dose of a toxic chemical from an accidental release would depend on the distance from the source of the release. If people are aware of this spatial variation, then they should be willing to pay more for houses in those areas with lower-magnitude risks. Competition for these more attractive houses would result in a systematic inverse relationship between the price of housing, P_h and A_i^*, where i indexes the spatial location of the house. For simplicity, suppose that the magnitude of the event is the only relevant characteristic of housing. Then the price of a house at location i can be found from the hedonic price function $P_h (A_i^*)$. Given income, the probability of the event, and the magnitude of the event, the individual chooses a location so as to maximize expected utility:

$$E [u] = \pi \cdot v [M - P_h (A_i^*), A_i^*] + (1 - \pi)v [M - P_h (A_i^*), 0] \quad (17)$$

The first-order condition is

$$-\frac{\partial P_h}{\partial A_i^*} = \frac{\pi \cdot v_{A_i^*}}{\pi \cdot v_{M^*} + (1 - \pi)v_{M^\circ}} \quad (18)$$

Since the right-hand side of this condition is the ex ante marginal value of a reduction in A_{i^*} (see equation [6] again), this condition says that expected utility maximization calls for setting the marginal value of risk reduction equal to its marginal implicit price, the slope of the hedonic price function. Thus if individuals and the housing market are in equilibrium, the estimated marginal implicit price of risk reduction for each individual reveals each individual's marginal ex ante valuation for risk reduction. However, since a house is a long-lived asset, and P_h is an asset price, equation (18) yields a compensating wealth measure of the lifetime welfare change associated with a permanent change in π.[20]

If the relevant housing characteristic that varies across space is the probability of the adverse event, and housing prices reflect differences

---

[20]See Brookshire and coauthors (1985) for an empirical application to the valuation of reductions in earthquake risks.

in π_i, the results are similar. The expression for expected utility is

$$E[u] = \pi_i \cdot v(M - P_h(\pi_i), A^*) + (1 - \pi_i)v(M - P_h(\pi_i), 0) \quad (19)$$

and the first-order condition is

$$\frac{\partial P_h}{\partial \pi_i} = -\frac{v[M - P_h(\pi_i), 0] - v[M - P_h(\pi_i), A^*]}{\pi_i \cdot v_{M^*} + (1 - \pi_i)v_{M^\circ}} \quad (20)$$

Again, the right-hand side is the ex ante marginal value of the probability change in equation (10). Thus the observed implicit price of probability reduction also reveals the individual's marginal ex ante value of risk reduction.[21] If both the probability and the magnitude of the event vary independently across space, housing prices will be a function of both characteristics. Both equations (18) and (20) must be satisfied in equilibrium. Hedonic price functions that do not include both characteristics as explanatory variables will be misspecified.

Welfare Change with Non-Expected Utility Preferences

The models described above for indirect benefit measurement are based on expected utility as a representation of individuals' preferences under uncertainty. There is substantial evidence that individuals' choices frequently violate expected utility theory. For examples of this evidence and discussions of its implications for the theory of preferences under uncertainty, see Kahneman and Tversky (1979), Arrow (1982), Grether and Plott (1979), Machina (1987), and Thaler (1987). An important question, therefore, is whether models of value of the sort described here can be modified for use with non-expected utility preferences.

Revising the models in this way is straightforward and involves no additional complications, at least in certain circumstances.[22] This follows from the key features of the indirect methods for estimating individuals' values from data on behavior. Generally speaking, the models involve first deriving the expression for welfare change, finding the first-order conditions for optimization, and substituting them into the expression for welfare change. Given the assumptions of the models described here, the substitution allows for the cancelling out

---

[21]Smith (1985) has also shown this result. See MacDonald, Murdoch, and White (1987) for an empirical application to flood risks.

[22]This section is adapted, with permission, from Freeman (1991a).

of any observable utility terms. It turns out that at least for many non-expected utility representations of preferences, the same result occurs, so that the derived observable welfare measures are independent of the particular form of preferences. This is a straightforward consequence of the envelope theorem. In what follows I first show this for the general case and then for two specific forms of non-expected utility preferences.

Let I be some general index of preferences where the preferences depend on income, prices (implicitly), the probabilities of different states of nature and the magnitudes of the adverse event in different states. Thus:

$$I = f(M, A, \pi) \tag{21}$$

I assume that equation (21) is convex and twice differentiable. This expression could be nonlinear in the probabilities or incorporate regret and rejoice terms or other deviations from the standard expected utility function, or both. Expected utility preferences also fit this general formulation.

Consider the averting behavior model where $A^* = A(R, G)$. The first-order condition for the optimum R is

$$\frac{\partial I}{\partial R} = -f_{M^*} + f_{A^*} \cdot A_R^* = 0$$

or

$$f_M = f_{A^*} \cdot A_R^*$$

To find the marginal welfare measure for a policy that reduces A^*, totally differentiate equation (21), rearrange terms, and substitute in the first-order condition to obtain

$$dI = f_{M^*} \cdot dM + (f_{A^*} \cdot A_R^* - f_M)dR + f_{A^*} \cdot A_G^* \cdot dG = 0$$

$$\frac{dM}{dG} = -\frac{f_{A^*} \cdot A_G^*}{f_{M^*}}$$

$$= -\frac{A_G^*}{A_R^*}$$

Thus the marginal willingness to pay for publicly supplied risk reduction is equal to the marginal rate of technical substitution between public and private risk reduction.

Suppose that in the hedonic model it is probabilities that vary across space. Then the general index of preferences would be

$$I = f[M - P_h(\pi_i), A, \pi_i] \tag{22}$$

The first-order condition for the selection of the risk characteristic of housing is

$$\frac{\partial I}{\partial \pi_i} = -f_M \cdot \frac{\partial P_h}{\partial \pi_i} + f_{\pi_i} = 0$$

or

$$f_{\pi_i} = f_M \cdot \frac{\partial P_h}{\partial \pi_i}$$

Totally differentiating equation (22) to obtain the welfare measure for the change in π and substituting the first-order condition gives

$$dI = f_M \cdot dM + \left(f_{\pi_i} - f_M \cdot \frac{\partial P_h}{\partial \pi_i}\right) d\pi_i + f_{\pi_i} \cdot dG = 0$$

$$\frac{dM}{dG} = -\frac{f_{\pi_i}}{f_M}$$

$$= \frac{\partial P_h}{\partial \pi_i}$$

The marginal willingness to pay for publicly supplied risk prevention is equal to the observable marginal implicit price of the risk characteristic of housing.

To illustrate this general result, consider the prospect theory model of Kahneman and Tversky (1979). In the two-state model based on the indirect utility function, the index of preferences takes the following form:

$$I = g(\pi)v(M, A^*) + g(1 - \pi) \cdot v(M, 0) \tag{23}$$

where

$$g(0) = 0, g(1) = 1$$

and

$$g(\pi) + g(1 - \pi) < 1 \text{ for } 0 < \pi < 1$$

Assume that the magnitude of the adverse event depends both on the level of expenditure on a private averting activity R and public expenditure G. The value of reducing A^* is

$$\frac{dM^*}{dA^*} = \frac{g(\pi)v_{A^*}}{g(\pi)v_{M^*} + g(1 - \pi)v_{M^\circ}} \tag{24}$$

which is not directly observable because of the utility and probability weighting terms. But it can be inferred. Given the level of G, the individual's optimal level of the private averting activity is given by

$$\frac{\partial I}{\partial R} = -g(\pi)v_{M^*} + g(\pi) \cdot v_{A^*} \cdot A_R^* - g(1 - \pi)v_{M^\circ} = 0$$

Thus

$$\frac{1}{A_R^*} = \frac{g(\pi)}{g(\pi)v_{M^*} + g(1 - \pi)v_{M^\circ}} \cdot v_{A^*} \tag{25}$$

The marginal value to the individual of a change in G is found by totally differentiating equation (23), setting it equal to zero, and solving for:

$$\frac{dM}{dG} = \frac{dR}{dG} - \frac{g(\pi)}{g(\pi)v_{M^*} + g(1 - \pi)v_{M^\circ}} \cdot v_{A^*} \cdot A_R^* \cdot \frac{dR}{dG}$$

$$- \frac{g(\pi)}{g(\pi) \cdot v_{M^*} + g(1 - \pi) \cdot v_{M^\circ}} \cdot v_{A^*} \cdot A_G$$

After substituting in equation (25), this becomes

$$\frac{dM}{dG} = -\frac{A_G^*}{A_R^*}$$

Similarly, if it is π that can be reduced by private and public expenditure, the value of a reduction in π is

$$\frac{dM}{d\pi} = \frac{g(1 - \pi) \cdot v(M - R, 0) - g(\pi) \cdot v(M - R, A^*)}{g(\pi) \cdot v_{M^*} + g(1 - \pi) \cdot v_{M^\circ}}$$

The first-order condition for private averting expenditure is

$$-\frac{1}{\pi_R} = \frac{v(M - R, 0) - v(M - R, A^*)}{g(\pi) \cdot v_{M^*} + g(1 - \pi) \cdot v_{M^\circ}}$$

And after substitution, the value of the public risk prevention expenditure is

$$\frac{dM}{dG} = -\frac{\pi_G}{\pi_R}$$

Similar results can be derived for other forms of preferences, for example the regret theory of Loomes and Sugden (1982); see Freeman (1991a).

Extending these results to the case of weak complementarity is straightforward. Suppose that the consumption of some market good x_i increases the probability of occurrence of some adverse event for the purchaser. Suppose further that there is some public policy action represented by G that can reduce the risk associated with consuming x_i for all consumers. So:

$$\pi = \pi(x_i, G)$$

with $$\partial\pi/\partial x_i > 0$$

and $$\partial\pi/\partial G < 0$$

This policy will increase the general preference index for all consumers of x_i and will cause the demand curve for x_i to shift out. Formally, the general preference index can be written as

$$I = f(M, A, \pi, p_i)$$

where p_i is the price of the complementary good. The general form of the expenditure function is

$$e = g(p_i, A, \pi, I^*)$$

Weak complementarity requires that there be a choke price for x_i and that at the choke price

$$\frac{\partial e}{\partial \pi} = 0$$

As is shown in chapter 4, when the conditions for weak complementarity are satisfied, the monetary equivalent of the increase in well-being is the area between the compensated demand curves for the good before and after the public policy change. Similarly, if the market good affects the severity of the adverse event, weak complementarity requires that $\partial e/\partial A^*$ or $\partial I/\partial A^*$ be zero when the good is not purchased. Since these results are independent of any particular specification of the preference function (other than the conditions of weak complementarity), they will hold for expected utility and non-expected utility preferences as well as in the case of certainty.

I have shown that the three broad classes of models that have been developed to measure the benefits of environmental change from revealed preferences under certainty can be easily generalized to apply to valuing changes in risk. This generalization of the models does not require that individual preferences take the expected utility form. The principal requirement is that individuals be maximizing some objective function. Then by the envelope theorem, welfare measures that contain unobservable preference terms can be reduced to functions of observable relationships by substitution of the first-order conditions for preference maximization. Thus if the conditions for utilizing these models are satisfied, there is no particular need to be concerned with how people make their choices under uncertainty.

ADDITIONAL TOPICS IN THE WELFARE ECONOMICS OF UNCERTAINTY

In this section, I briefly discuss four other topics that are related to the welfare economics of uncertainty and that bear on environmental policymaking. The first topic is the relationship between option price and the expected damage associated with some risk. The latter has a long history of use in applied benefit-cost analysis. The question is whether it is a useful proxy for the preferred ex ante welfare measure. The second topic is option value and whether option value is something that can be measured and added to other types of values when welfare measures are computed. The third topic is quasi-option value, its source, and its validity as a welfare measure. The fourth topic is the margin of safety provided by a particular environmental regulation and whether the welfare economics of uncertainty can provide any guidance as to the "adequacy" of a specific margin of safety.

Comparing Option Price and Expected Damage Measures

There is a long history of using ex post expected damage measures in benefit-cost analysis.[23] Expected damage measures have the virtue of

---

[23]Perhaps the earliest example of this approach to measuring the value of risk changes is the U.S. Army Corps of Engineers' method for estimating the benefits of flood control projects. The Corps' method involves calculating the

being relatively easy to calculate from experience with risky events. For example, for most river basins in the United States there is a large body of data on hydrology, land use, and the distribution of structures that can be used to calculate expected flood damages under alternative proposed flood control projects. But even if these losses (including lost utility) were accurately measured, we have shown that they will not in general correctly measure welfare change from the ex ante perspective.

Since at least in some circumstances expected damage measures may be relatively easy to obtain, an important question is whether they can be taken as useful approximations of the other forms of ex ante values when policy-makers would prefer to have the latter. Are the differences between the two types of measure likely to be small or large?[24] Is the expected demand measure likely to be larger or smaller than the option price measure? What factors determine the sign of the difference and its magnitude?

The derivation of an expected damage measure for risk reduction is straightforward. The first task is to establish a measure of the value of avoiding an event given that A^* will occur with certainty. The willingness to pay to avoid the consequences of the event given its occurrence is a compensating surplus (CS) measure of damage, D. It is found as the solution to

$$v(M, A^*) = v(M - CS, 0) \qquad (26)$$

The expected damage measure of the value of avoiding the event that has probability of π of occurring is the mathematical expectation of the willingness to pay given by equation (26), that is, $\pi \cdot D$. The marginal value of reducing A^* given that the event has occurred can be found by totally differentiating the left-hand side of (26), setting it equal to zero, and solving for

$$\frac{dM}{dA^*} = -\frac{v_{A^*}}{v_{M^*}} \qquad (27)$$

---

reduction in the costs of replacement, repair, and cleaning up after a flood for each possible flood stage, multiplying the monetary damages by the probability associated with that level of flooding, and summing across all possible flood stages; see, for example, Eckstein (1961). Of course, we would now add to the repair and cleaning costs some monetary measure of the loss of utility associated with the flooding events.

[24]See Anderson (1979) and Scotchmer (1989) for earlier analyses of the differences between ex ante and ex post welfare measures for changes in prices.

Let this be denoted by d^A to indicate that it is a marginal value and that it refers to changes in A. Equation (27) is a standard result showing that ex post the marginal value is the marginal utility of A^* converted to a money measure by using the marginal utility of income. Taking the expectation yields the ex post value of a marginal reduction in A^*:

$$E[d^A] = \frac{dM}{dA^*} = -\pi \frac{v_{A^*}}{v_{M^*}} \tag{28}$$

The percentage difference between the option price and expected damage measures is defined according to the following expression, using equations (6) and (28):

$$\% \text{ difference} \equiv \frac{w^{A^*} - E[d^A]}{w^{A^*}} = (1 - \pi) - (1 - \pi)\frac{v_{M^\circ}}{v_{M^*}} \tag{29}$$

This expression shows that if the marginal utility of income is independent of the event, so that $v_{M^\circ} = v_{M^*}$, the difference in the measures is zero. This would be the case, for example, if the individual were risk-neutral or could purchase actuarially fair insurance.

If the adverse event reduces the marginal utility of income so that v_{M^*} is less than v_{M°, then the expected demand measure will exceed the option price measure and the difference will be negative.[25] On the other hand, if the adverse event increases the marginal utility of income, the option price measure will be larger than the expected damage measure. In both cases the magnitude of the difference will be large when the difference in the marginal utilities of income is large and when the probability of the adverse event is low. To summarize, the sign of the difference between the option price and expected damage measures depends on the way in which the marginal utility of income varies with A. And the magnitude of the difference depends on how much it varies and on the probability of the event.

In Freeman (1989) I computed these differences for several explicit forms of von Neuman-Morgenstern cardinal utility functions

---

[25]See Viscusi and Evans (1990) for evidence that occupational injuries reduce the marginal utility of income.

with different degrees of risk aversion.[26] For some functional forms, the marginal utility of income is higher given the adverse event. One way of interpreting this last characteristic is that income and the absence of A^* are substitutes in consumption such that if A^* does not occur, the marginal utility of the consumption that M allows is diminished. This might be the case if the event increased the marginal utilities of ameliorating activities such as cleaning up and repairing damages after a flood. Option price values for reducing A^* will be greater than expected damages, and more so for larger losses occurring with lower probabilities.

The findings for utility functions with this characteristic can be summarized by the following statements:

1. Since the differences are positive as predicted by equation (29), empirical valuation measures based on changes in expected damages will understate the option prices.

2. The difference is an increasing function of CS, since a larger CS means a larger difference between v_{M^*} and v_{M°.

3. The differences are for the most part trivially small for small losses (equivalent to up to 1 percent of income). Only for high degrees of risk aversion does the difference approach 10 percent for low-probability losses.

4. For relatively large losses (equivalent to, say, 10 percent of income), the differences range between 5 percent and 20 percent, depending on the probability and for moderate degrees of risk aversion. For relatively high degrees of risk aversion, the differences range above 60 percent.

5. For the catastrophic loss (equivalent to 50 percent of income), the differences are large for all but the least risk-averse form of utility function and most likely event. For the more risk-averse forms of utility function, this means that option prices for risk reduction can exceed expected damages by factors of 2.5 to more than 100.

––––––––––––

[26]Examples include

(i) $v = (M - A^*)^b$ with $0 < b < 1$
(ii) $v = -e^{-b(M - A^*)}$ with $b > 0$
(iii) $v = -e^{-bM(1 - A^*)}$
(iv) $v = \ln[(1 - A^*)M]$

For (i) and (ii), the price flexibility of income for A^* is zero, meaning that the marginal value of a change in A is independent of M. Risks of this sort are essentially equivalent to financial risks. And an arbitrary choice of units for measuring A^* can be made such that A^* is equal to the ex post D.

This last conclusion might be a possible explanation for a phenomenon that some observers have noted, that is, that the public seems unwilling to accept some very low-probability, high-consequence risks that experts have judged to be acceptable. If the experts' judgments are based on ex post valuation measures and if people are relatively risk-averse and their preferences take this form, the experts could be vastly underestimating the true potential welfare costs that these risks would impose on people.

For other forms of utility functions, v_{M° is greater than v_{M^*}.[27] One way of interpreting this characteristic is that income and the absence of A^* are complements in consumption such that if the event does not occur, the marginal utility of the consumption that M allows is high. But if the event does occur, the capacity for consumption to generate utility is diminished. For example, the event "broken leg" decreases the marginal utility of expenditures for hiking and skiing trips. For these functions, expected damages exceed option prices.

Calculations with these functions show that differences are negative as predicted and are less than 10 percent except for low probabilities of very large losses (50 percent of income). For most functional forms, the magnitude of the difference is larger for the most risk-averse utility functions. But for one functional form the difference is largest when the degree of risk aversion is zero. And for one functional form, even the sign of the difference depends on the degree of risk aversion.

As these results demonstrate, option price and expected damage measures of the value of risk reduction are likely to be different. Expected damage measures are very unreliable proxies for option price measures. The sign of the difference depends on specific features of the functions chosen to represent preferences. The degree of risk aversion is not a reliable predictor of the size or even the sign of the differences. Values for risk reduction based on expected damages are not likely to be useful, and could be seriously misleading as guides for risk management decisions.

I now turn to the value of risk prevention. The expected damage measure of the value of reducing the probability of a loss is the reduction in the expected value of the compensating surplus associated with the loss. As explained above, the expected loss, $E[D]$, is $\pi \cdot D$.

---

[27]Examples include:

$$v = (1 - A^*)^c M^b \qquad \text{with } 0 < b, c, < 1, \text{ and}$$
$$v = (1 - A^*) \ln M$$

By differentiation, we have

$$dE\,[D] = d\,(\pi \cdot CS) = CS \cdot d\pi \qquad (30)$$

In order to examine the relationship between the option price and expected damage measures of the value of a probability change, we need to compare the right-hand sides of equations (10) and (30). But they are not easily comparable. The option price measure starts with the difference between two utility levels and converts it to a money measure by using the weighted average of the marginal utilities of income in the two states. The expected damage measure is already in monetary units; specifically, it is the difference in money expenditure necessary to achieve the same level of utility under two different sets of conditions.

Again, in Freeman (1989) I assumed specific functional forms and parameters and calculated differences. The results are difficult to summarize except to say that differences can be positive or negative, large or small, and can be large even in the absence of risk aversion.

In estimating the values of reducing the probability of adverse events, expected damage measures are unreliable indicators of the desired option prices. Expected damage measures can be either underestimates or overestimates, and the errors involved in using them as proxies for option prices can be large. But they cannot be predicted without detailed knowledge of the specific characteristics of individuals' preferences.

Is Option Value a Value?

The concept of option value was introduced by Weisbrod in a much-cited paper more than twenty-five years ago (Weisbrod, 1964). This concept has had an interesting history since its birth, both in terms of its development as a theoretical construct and because of its relationship to and role in policy discussions concerning environmental resources. Weisbrod argued that an individual who was unsure of whether he would visit a site such as a national park would be willing to pay a sum over and above his expected consumer surplus to guarantee that the site would be available should he wish to visit it. Weisbrod called this extra sum the option value of the site. Option value was seen to arise when an individual was uncertain as to whether he would demand a good in some future period and was faced with uncertainty about the availability of that good. If option price (OP) is defined as the maximum sum the individual would be willing to pay to preserve the option to visit the site before his own demand uncertainty is resolved, then the excess of option price over expected consumer surplus can be called option value (*OV*). It was thought that option value should be measured, if it were possible, and added to expected consumer surplus in order to obtain the full measure of the value of providing an environmental service.

Weisbrod apparently viewed the existence of positive option value as being intuitively obvious. Indeed, there is no formal mathematical or logical proof of the existence of option value in Weisbrod's paper. But as the subsequent literature has shown, there are a number of subtleties and complications to the concept, as well as traps for the unwary investigator. For example, Cicchetti and I thought we had proved that option value was positive for risk-averse individuals (Cicchetti and Freeman, 1971), but our proof was wrong (see Schmalensee 1972; Anderson, 1981; and Bishop, 1982). Schmalensee showed that even for a risk-averse individual, option value could be greater than, equal to, or less than zero depending upon the particular circumstances.[28] As Bohm (1972) pointed out in a comment on Schmalensee, what matters is the relationship between the marginal utilities of income in the different states of nature.

The matter seemed to rest there until Hartman and Plummer (1987) and I (Freeman, 1984b) provided specific characterizations of the nature of demand uncertainty and examined the implications of different types of demand uncertainty for the relationship between the marginal utilities of income for different states of nature. Hartman and Plummer, for example, showed that if an individual was uncertain about future income and the demand for the good in question was a positive function of income, option value is unambiguously negative for risk-averse individuals. I showed that in this case risk lovers would have positive option values, and that one plausible form of state-dependent preferences assured positive option values for risk-averse individuals.

At about the same time Bishop (1982) suggested that useful insights might be obtained by considering the simpler case in which an individual is certain of demand but faces uncertain supply of the good. He showed that option value is greater than zero for a project that eliminates the uncertainty of supply.[29] Thus option value ap-

---

[28]This is essentially a matter of the relationship between the expected damage (or surplus) measure and the option price measure discussed in a preceding section of this chapter.

[29]However, Freeman was able to show that Bishop's conclusion did not

peared to be resurrected as a form of benefit associated with a guaranteed future supply of goods, such as national parks.

This, after all, seems to have been the motivation for the original investigation of option value. Friedman (1962) in *Capitalism and Freedom* had argued that since there were no externalities associated with uses of national parks, all of the relevant economic values could be captured by the owners of parks through admission fees. Thus the allocation of land to national parks should be subjected to a market test. Weisbrod's paper (1964) was clearly at least in part a response to this argument. And Krutilla (1967) included option value in his list of reasons why markets might fail to achieve allocative efficiency for unique environmental resources. Option value was presented as an economic value over and above the expected use values; and it could not be captured through admission fees. The preservation of the option to visit the park was a form of public good that might justify the preservation of a natural area even when expected use values were less than opportunity costs.

Schmalensee's (1972) theoretical proof that the sign of option value is ambiguous seems to have reduced interest in option value as a policy-relevant concept, at least temporarily. But in 1983 the Environmental Protection Agency (EPA) listed option value as one form of intrinsic benefit that could be included in the benefit-cost analyses of proposed regulations required under the terms of Executive Order 12291 (U.S. Environmental Protection Agency, 1983). Measurement of option value has also been part of the work to be done in several EPA-sponsored research projects. In light of the state of the theory, this continuing effort to measure option value and to use it in policymaking amounts to an interesting comment on the power of the idea.

The idea of option value may have been most useful through stimulating the more rigorous analysis of the theory of welfare measurement under uncertainty. Because of the theoretical contributions of authors such as Schmalensee (1972), Graham (1981), Bishop (1982), and Smith (1987a, 1987b), we can now see that what has been called an option value is really just the algebraic difference between the expected values of two different points on a willingness-to-pay locus. Specifically, it is the algebraic difference between the expected value of the consumer surplus and the state-independent willingness to pay (option price). Since these two points represent alternative ways

---

hold for all possible forms of reduction in uncertainty of supply. What I called supply-side option value could also be either positive or negative. See Freeman (1985) and Wilman (1987).

of measuring the same welfare change, the difference between their expected values cannot be a separate component of value. Furthermore, option value cannot be measured separately; it can only be calculated if we have enough information on preferences to calculate both option price and expected surplus. And finally, as I have shown in this chapter, neither of these points on the willingness-to-pay locus has any particular claim as a superior welfare measure. I think it is time to expunge option value from the list of possible benefits associated with environmental protection.

Planners' Uncertainty and Quasi-option Value

When policymakers are uncertain about the magnitude of the benefits or the costs of alternative courses of action, or both, decision rules and procedures should be modified to reflect this uncertainty. One possible modification entails altering the time sequence of choices so as to take advantage of information that might become available in the future. The term quasi-option value was coined by Arrow and Fisher (1974) to describe the welfare gain associated with delaying a decision when there is uncertainty about the payoffs of alternative choices, and when at least one of the choices involves an irreversible commitment of resources. They showed that quasi-option value is not dependent on risk aversion. It can be present even when decision makers make choices on the basis of expected values of monetary measures of benefits and costs.

Most of the literature on the role of this concept in environmental decision making has concluded that consideration of quasi-option value would lead to relatively less irreversible development and relatively more preservation of natural environments. However, the conclusion that there is a quasi-option value benefit to preserving a natural area or to delaying its development springs from a specific feature of the models used by Arrow and Fisher (1974), Krutilla and Fisher (1975), Conrad (1980), Miller and Lad (1984), Fisher and Hanemann (1987), and others. In these models quasi-option value stems from the value of the information gained by delaying an irreversible decision to develop a natural area. But it is not difficult to imagine situations where the relevant information to guide future decisions can be gained only by undertaking now at least a little development. For example, suppose there is uncertainty about the magnitude of a mineral deposit underlying a wilderness area. Perhaps the only way that the uncertainty about the magnitude of the benefits of development relative to preservation can be resolved is through exploratory drilling, that is, through a little bit of development. In such cases there can be positive quasi-option value to development, or, what is the same thing, a negative quasi-option value to preservation (Miller and Lad, 1984; Freeman, 1984a).

Whether quasi-option value exists or whether it is positive or negative for preservation depends on the nature of the uncertainty, the opportunities for gaining information, and the structure of the decision problem. Quasi-option value is not a component of the values individuals attach to resource changes. Even if individuals' utility functions were known, quasi-option value could not be estimated separately and added into a benefit-cost calculation. Quasi-option value is a benefit of adopting better decision-making procedures. Its magnitude can only be revealed by comparing two strategies where one of the strategies involves optimal sequential decision-making to take advantage of information obtained by delaying irreversible resource commitments. The decision maker who knows how to use an optimal sequential decision-making strategy has no reason to calculate quasi-option value. The calculation would be redundant, since the best decision is already known.

The Margin of Safety

Section 109 of the Clean Air Act of 1970 calls for setting primary national ambient air quality standards so as to protect human health, "allowing an *adequate margin of safety*" (emphasis added). The question to be addressed in this section is whether the phrase "adequate margin of safety" can be given a meaningful interpretation in the context of the conventional theory of welfare measurement under uncertainty. A basic assumption that underlies the process for setting air quality standards as established by Congress is that there are threshold levels of pollution below which there would be no observable adverse effects on human health. The existence of a threshold means that the risk to health associated with air pollution is zero even at some positive pollution levels. Under the act, protection of human health was to be achieved by setting the standard for air quality at or below the threshold for each pollutant. The margin of safety was intended to take into account uncertainties in determining the exact position of the threshold.

In considering the question of meaningful interpretation of "adequate margin of safety," the framework developed by Lichtenberg and Zilberman (1988) is useful. Suppose that some standard of performance has been established as a policy goal; in the case of the Clean Air Act, the standard is zero excess adverse health effects caused by air pollution. Further, suppose that uncertainty concerning the parameters of the function relating ambient air quality to incidence of disease or mortality makes it impossible to determine with certainty the minimum levels of air pollution that are necessary to achieve this target. This means that for any given air quality standard, there is some non-zero probability that the target of zero excess adverse health effects will not actually be achieved. The probability of failure to meet the target can be reduced by setting a more stringent air quality standard.

The margin of safety can be interpreted as one minus the probability of failing to achieve the target. The question of what is an adequate margin of safety can be restated as: What is the maximum acceptable probability of failure to meet the target? From an economic perspective, the answer to this question is found by comparing the additional benefits to people of reducing the probability of failure with the costs. The costs are what it takes to meet the higher air quality standard that is consistent with the lower probability of failure. The benefits are the willingness to pay to reduce this probability. The optimum is where the marginal benefits just equal the additional costs. In order to implement this economic approach to determining the adequacy of the margin of safety, we need to have a way of measuring people's willingness to pay for reduced probability of failure.

A key feature of the approach described here is that for any pollution level the individual's risk of mortality or illness is not known with certainty. Rather, there is some probability distribution of risks, with greater uncertainty about risks being reflected in a higher variance of that probability distribution. In contrast, the models of individual choice described in this chapter are based on the assumption that the risk is known with certainty. Thus in the case of certainty there is no need for policy-makers to consider a margin of safety. What would be required to give economic content to the term "adequate margin of safety" in the context of placing values on reductions in risk is to build more sophisticated models of individual choice. One approach is to assume that people have preferences over probability distributions of risk. Given distributions with the same mean, people would prefer distributions with smaller variances. Given distributions with the same variance, people would prefer distributions with lower means for risk. An improvement in air quality that reduced the probability of failing to achieve the target of no excess adverse health effects would shift each individual's probability distribution of risk, thereby conveying benefits to each person. None of the empirical models for estimating values of risk reduction take into account this added level of complexity.

SUMMARY

In this chapter I have explored the various ways in which welfare changes can be defined and measured for changes in risks and how these measures relate to the Potential Pareto Improvement criterion when it is employed in benefit-cost analyses of policies dealing with risk. The methods presented here are based on an important maintained assumption, which is that individuals are capable of assigning probabilities to alternative outcomes that are consistent with probability theory and contain all relevant information. However, there is a substantial body of evidence that people find it hard to think in probabilistic terms and often have serious misperceptions about the magnitude of important environmental risks.[30] And, as noted above, people often make choices that are not consistent with expected utility theory.

The observation that our models of behavior may not be very good descriptions of how people actually behave toward risk raises some interesting and difficult questions concerning the evaluation of public policies dealing with risks. Two of these questions are: Whose probabilities should be used in policy analysis, those of the experts or those of the people facing the risks? And, whose preferences should count, those of the rational expert, or those of the people facing the risks? Portney (1992) posed these questions in the starkest manner with his hypothetical example of Happyville's contaminated water supply, in which the experts are utterly convinced that the contaminating chemical is benign but the citizens believe it is responsible for the above-average incidence of cancer in their town. Each citizen is willing to pay $1,000 for what the experts say will be unnecessary and costly treatment.

Arrow (1983) and Machina (1990) have both considered these questions in their more general form. They agree that the probabilities are part of the description of the policy problem and the effect of the policy, and that it is important that this description be accurate. Thus they recommend reliance on the experts' probability assessments. Machina also goes on to recommend that individuals' preferences be recognized even if they are not consistent with expected utility theory.

---

[30]For references to some of this evidence and discussions of its implications for economic analysis, see Machina (1990) and Arrow (1982). Other important references include Slovic, Fischhoff, and Lichtenstein (1979, 1980, and 1982).

I think that this is good advice, but many problems come in a form in which it may not be possible to follow that advice. Consider Portney's Happyville example. On the information given, it is not possible to determine whether the people have misperceived the risk of cancer (errors in probabilities) or whether they place a very high value on preventing a very small but catastrophic risk (preferences). All we know is that each citizen is willing to pay $1,000 for treatment of the water supply. If that will cover the cost, it would seem to be welfare-enhancing to build the treatment plant.

When indirect measurement methods are employed to estimate the values people place on changes in risks, these questions of probabilities and preferences may be less important. Suppose for the sake of argument that individuals assign probabilities to adverse outcomes that are consistently higher than the probabilities assessed by risk experts. Which probabilities should be used, those of the experts or those of the people? Here I think that the answer is pretty clear. As long as the objective of the research is to learn about people's preferences and willingness to pay, we have to know what they think they are getting in return for their expenditures on risk reduction and risk prevention. This means relying on their own subjective probabilities. In any case, when indirect methods such as averting behavior and hedonic price analysis are used to infer values of risk changes, it is not generally possible for the researcher to determine whether a high marginal value results from the subject's overestimate of the risk change or from a high valuation of a correctly perceived change. All that is observed is the risk-dollar trade-off.

REFERENCES

Anderson, James E. 1979. "On the Measurement of Welfare Cost Under Uncertainty," *Southern Economic Journal* vol. 45, no. 4, pp. 1160–1171.

Anderson, Robert J. 1981. "A Note on Option Value and the Expected Value of Consumer's Surplus," *Journal of Environmental Economics and Management* vol. 8, no. 2, pp. 187–191.

Arrow, Kenneth J. 1982. "Risk Perception in Psychology and Economics," *Economic Inquiry* vol. 20, no. 1, pp. 1–9.

Arrow, Kenneth J. 1983. "Behavior Under Uncertainty and its Implications for Policy," in B. T. Stigum and F. Wenstop, eds., *Foundations of Utility and Risk Theory with Applications*. Dordrecht, Germany: Reidel.

Arrow, Kenneth J., and Anthony C. Fisher. 1974. "Environmental Preservation, Uncertainty, and Irreversibility," *Quarterly Journal of Economics* vol. 88, no. 1, pp. 312–319.

Berger, Mark C., Glenn C. Blomquist, Don Kenkel, and George S. Tolley. 1987. "Valuing Changes in Health Risks: A Comparison of Alternative Measures," *Southern Economic Journal* vol. 53, no. 4, pp. 967–984.

Bishop, Richard C. 1982. "Option Value: An Exposition and Extension," *Land Economics* vol. 58, no. 1, pp. 1–15.

Bishop, Richard C. 1986. "Resource Evaluation Under Uncertainty: Theoretical Principles for Empirical Research," in V. Kerry Smith, ed., *Advances in Applied Microeconomics*, vol. 4. Greenwich, Conn.: JAI Press.

Bohm, Peter. 1972. "Option Demand and Consumer's Surplus: Comment," *American Economic Review* vol. 65, no. 3, pp. 233–236.

Brookshire, David S., Mark A. Thayer, John Tschirhart, and William D. Schulze. 1985. "A Test of the Expected Utility Model: Evidence from Earthquake Risks," *Journal of Political Economy* vol. 93, no. 2, pp. 369–389.

Cicchetti, Charles J., and A. Myrick Freeman III. 1971. "Option Demand and Consumer's Surplus: Further Comment," *Quarterly Journal of Economics* vol. 85, no. 3, pp. 528–539.

Colby, Bonnie E., and Dennis C. Cory. 1989. "Valuing Amenity Resources Under Uncertainty: Does the Existence of Fair Contingent Claims Markets Matter?" *Journal of Environmental Economics and Management* vol. 16, no. 1, pp. 149–155.

Conrad, Jon M. 1980. "Quasi-Option Value and the Expected Value of Information," *Quarterly Journal of Economics* vol. 94, no. 2, pp. 813–820.

Cook, Philip J., and Daniel A. Graham. 1977. "The Demand for Insurance and Protection: The Case of Irreplaceable Commodities," *Quarterly Journal of Economics* vol. 91, no. 1, pp. 143–156.

Cory, Dennis C., and Bonnie Colby Saliba. 1987. "Requiem for Option Value," *Land Economics* vol. 63, no. 1, pp. 1–10.

Eckstein, Otto. 1961. *Water Resources Development: The Economics of Project Evaluation*. Cambridge, Mass.: Harvard University Press.

Ehrlich, Isaac, and Gary S. Becker. 1972. "Market Insurance, Self Insurance and Self Protection," *Journal of Political Economy* vol. 80, no. 4, pp. 623–648.

Fisher, Anthony C., and W. Michael Hanemann. 1987. "Quasi-Option Value: Some Misconceptions Dispelled," *Journal of Environmental Economics and Management* vol. 14, no. 2, pp. 183–190.

Freeman, A. Myrick, III. 1984a. "The Quasi-Option Value of Irreversible Development," *Journal of Environmental Economics and Management* vol. 11, no. 3, pp. 292–295.

Freeman, A. Myrick, III. 1984b. "The Sign and Size of Option Value," *Land Economics* vol. 60, no. 1, pp. 1–13.

Freeman, A. Myrick, III. 1985. "Supply Uncertainty, Option Price, and Option Value," *Land Economics* vol. 61, no. 2, pp. 176–181.

Freeman, A. Myrick, III. 1989. "Ex Ante and Ex Post Values for Changes in Risks," *Risk Analysis* vol. 9, no. 3, pp. 309–317.

Freeman, A. Myrick, III. 1991a. "Indirect Methods for Valuing Changes in Environmental Risks with Non-expected Utility Preferences," *Journal of Risk and Uncertainty* vol. 4, no. 2, pp. 153–165.

Freeman, A. Myrick, III. 1991b. "Welfare Measurement and the Benefit-Cost Analysis of Projects Affecting Risks," *Southern Economic Journal* vol. 58, no. 1, pp. 65–76.

Friedman, Milton. 1962. *Capitalism and Freedom.* Chicago, Ill.: University of Chicago Press.

Graham, Daniel A. 1981. "Cost-Benefit Analysis Under Uncertainty," *American Economic Review* vol. 71, no. 4, pp. 715–725.

Graham, Daniel A. 1984. "Cost-Benefit Analysis Under Uncertainty: Reply," *American Economic Review* vol. 74, no. 5, pp. 1100–1102.

Graham, Daniel A. 1992. "Public Expenditure Under Uncertainty: The Net Benefit Criteria," *American Economic Review* vol. 82, no. 4, pp. 882–846.

Grether, David M., and Charles R. Plott. 1979. "The Economic Theory of Choice and the Preference Reversal Phenomenal," *American Economic Review* vol. 69, no. 4, pp. 623–638.

Hartman, Richard, and Mark Plummer. 1987. "Option Value Under Income and Price Uncertainty," *Journal of Environmental Economics and Management* vol. 14, no. 3, pp. 212–225.

Hazilla, Michael, and Raymond J. Kopp. 1990. "The Social Cost of Environmental Quality Regulations: A General Equilibrium Analysis," *Journal of Political Economy* vol. 98, no. 4, pp. 853–873.

Helms, L. Jay. 1985a. "Errors in the Numerical Assessment of the Benefits of Price Stabilization," *American Journal of Agricultural Economics* vol. 69, no. 1, pp. 93–100.

Helms, L. Jay. 1985b. "Expected Consumer's Surplus and the Welfare Effects of Price Stabilization," *International Economic Review* vol. 26, no. 3, pp. 603–617.

Jones-Lee, Michael W. 1974. "The Value of Changes in the Probability of Death or Injury," *Journal of Political Economy* vol. 99, no. 4, pp. 835–849.

Just, Richard E., Darrell L. Hueth, and Andrew Schmitz. 1982. *Applied Welfare Economics and Public Policy.* Englewood Cliffs, N.J.: Prentice-Hall, Inc.

Kahneman, Daniel, and Amos Tversky. 1979. "Prospect Theory: An Analysis of Decisions Under Risk," *Econometrica* vol. 47, no. 1, pp. 263–291.

Karni, Edi. 1983. "Risk Aversion for State-dependent Utility Functions: Measurement and Applications," *International Economic Review* vol. 24, no. 3, pp. 637–647.

Krutilla, John V. 1967. "Conservation Reconsidered," *American Economic Review* vol. 57, no. 4, pp. 777–786.

Krutilla, John V., and Anthony C. Fisher. 1975. *The Economics of Natural Environments: Studies in the Valuation of Commodity and Amenity Resources.* Baltimore, Md.: The Johns Hopkins University Press.

LeRoy, Stephen F., and Larry D. Singell, Jr. 1987. "Knight on Risk and Uncertainty," *Journal of Political Economy* vol. 95, no. 2, pp. 394–406.

Lichtenberg, Erik, and David Zilberman. 1988. "Efficient Regulation of Environmental Health Risks," *Quarterly Journal of Economics* vol. 103, no. 2, pp. 167–178.

Loomes, Graham, and Robert Sugden. 1982. "Regret Theory: An Alternative Theory of Rationale Choice Under Uncertainty," *Economic Journal* vol. 92, no. 368, pp. 805–824.

MacDonald, Don N., James C. Murdoch, and Harry L. White. 1987. "Uncertain Hazards, Insurance, and Consumer Choice: Evidence from Housing Markets," *Land Economics* vol. 63, no. 4, pp. 361–371.

Machina, Mark J. 1983. "Generalized Expected Utility Analysis and the Nature of Observed Violations of the Independence Axiom," in B. T. Stigum and F. Wenstop, eds., *New Foundations of Utility and Risk Theory with Applications.* Dordrecht, Germany: Reidel.

Machina, Mark J. 1987. "Choice Under Uncertainty: Problems Solved and Unsolved," *Economic Perspectives* vol. 1, no. 1, pp. 121–154.

Machina, Mark J. 1990. "Choice Under Uncertainty: Problems Solved and Unsolved," in P. Brett Hammond and Rob Coppock, eds., *Valuing Health Risks, Costs, and Benefits for Environmental Decision Making.* Washington, D.C.: National Academy Press.

Meier, Charles E., and Alan Randall. 1991. "Use Value Under Uncertainty," *Land Economics* vol 67, no. 4, pp. 379–389.

Mendelsohn, Robert, and William J. Strang. 1984. "Cost-Benefit Analysis Under Uncertainty: Comment," *American Economic Review* vol. 74, no. 5, pp. 1096–1099.

Miller, Jon R., and Frank Lad. 1984. "Flexibility, Learning and Irreversibility in Environmental Decisions: A Bayesian Approach," *Journal of Environmental Economics and Management* vol. 11, no. 2, pp. 161–172.

Portney, Paul R. 1992. "Trouble in Happyville," *Journal of Policy Analysis and Management* vol. 11, no. 1, pp. 131–132.

Ready, Richard C. 1988. "Cost-Benefit Analysis Under Temporal Uncertainty," doctoral dissertation, Univerity of Wisconsin, Madison.

Rothschild, M., and J. E. Stiglitz. 1976. "Equilibrium in Competitive Insurance Markets," *Quarterly Journal of Economics* vol. 90, no. 4, pp. 629–650.

Schmalensee, Richard. 1972. "Option Demand and Consumer's Surplus: Valuing Price Changes Under Uncertainty, *American Economic Review* vol. 62, no. 5, pp. 813–824.

Scotchmer, Suzanne. 1989. "Equivalent Variation and Uncertain Prices," *Economics Letters* vol. 29, no. 2, pp. 127–128.

Shogren, Jason F., and Thomas D. Crocker. 1991. "Risk, Self-Protection, and Ex Ante Economic Value," *Journal of Environmental Economics and Management* vol. 20, no. 1, pp. 1–15.

Slovic, Paul, Baruch Fischhoff, and Sarah Lichtenstein. 1979. "Rating the Risks," *Environment* vol. 21, pp. 14–20, 36–39.

Slovic, Paul, Baruch Fischhoff, and Sarah Lichtenstein. 1980. "Facts versus Fears: Understanding Perceived Risk," in W. A. Albers, ed., *Societal Risk Assessment: How Safe is Safe Enough?* New York: Plenum.

Slovic, Paul, Baruch Fischhoff, and Sarah Lichtenstein. 1982. "Response Mode, Framing, and Information-Processing Effects in Risk Assessment," in Robin Hogarth, ed., *Question Framing and Response Consistency*. San Francisco, Calif.: Jossey-Bass.

Smith, V. Kerry. 1983. "Option Value: A Conceptual Overview," *Southern Economic Journal* vol. 49, no. 3, pp. 654–668.

Smith, V. Kerry. 1985. "Supply Uncertainty, Option Price, and Indirect Benefit Estimation," *Land Economics* vol. 61, no. 3, pp. 303–307.

Smith, V. Kerry. 1987a. "Non-Use Values in Benefit-Cost Analysis," *Southern Economic Journal* vol. 54, no. 1, pp. 19–26.

Smith, V. Kerry. 1987b. "Uncertainty, Benefit-Cost Analysis, and the Treatment of Option Value," *Journal of Environmental Economics and Management* vol. 14, no. 3, pp. 283–292.

Smith, V. Kerry. 1990. "Valuing Amenity Resources Under Uncertainty: A Skeptical View of Recent Resolutions," *Journal of Environmental Economics and Management* vol. 19, no. 2, pp. 193–202.

Smith, V. Kerry, and William H. Desvousges. 1988. "The Valuation of Environmental Risks and Hazardous Wastes Policy," *Land Economics* vol. 64, no. 3, pp. 211–219.

Thaler, Richard G. 1987. "The Psychology of Choice and the Assumptions of Economics," in Alvin E. Roth, ed., *Laboratory Experimentation In Economics: Six Points of View*. Cambridge: Cambridge University Press.

U.S. Environmental Protection Agency. 1983. *Guidelines for Performing Regulatory Impact Analyses*. Washington, D.C.: Office of Policy Analysis.

Viscusi, W. Kip. 1985. "Are Individuals Bayesian Decision Makers?" *American Economic Review* vol. 75, no. 2, pp. 381–385.

Viscusi, W. Kip, and William N. Evans. 1990. "Utility Functions that Depend on Health Status: Estimates and Economic Implications," *American Economic Review* vol. 80, no. 3, pp. 353–374.

Weisbrod, Burton A. 1964. "Collective Consumption Services of Individual Consumption Goods," *Quarterly Journal of Economics* vol. 77, no. 3, pp. 71–77.

Wilman, Elizabeth A. 1987. "A Note on Supply-Side Option Value," *Land Economics* vol. 63, no. 3, pp. 284–289.

9

Environmental Quality as a Factor Input

I n addition to providing services directly to individuals as consumers, environmental and resource systems can affect the costs and output levels in the production sector of an economy. Costs and output levels can be affected, for example, by changes in the flow of minerals and petroleum from the ground, the negative effects of air pollution on the flow of food and fiber from agriculture, and the impact of pollution on the costs of manufactured goods through requirements for more frequent cleaning, repair, and replacement of materials. The effects of these changes will be transmitted to individuals through the price system in the form of changes in the costs and prices of final goods and services and changes in factor prices and incomes. All of these examples involve a common economic mechanism. Improvements in the resource base or environmental quality lower costs and prices and increase the quantities of marketed goods, leading to increases in consumers' and perhaps producers' surpluses. Similarly, increases in pollution can cause economic harm to producers and consumers by decreasing their surpluses.

Most of the early studies of the effects of air and water pollution on producers were based on the damage function approach.[1] This approach involves (1) estimating a dose/damage function that relates some measure of pollution to a physical measure of damage; (2) applying this function to estimates of the inventory of materials exposed or at risk; and (3) multiplying the result by some unit value.

For example, in the case of damages to materials and structures, the physical damage might be corrosion, soiling, or loss of thickness of paint. This measure would have to be translated into an estimate of

---

[1]For reviews of many of these studies, see Freeman (1979, pp. 244–246) and Freeman (1982, chapters 5 and 9).

the increase in the frequency of some repair or replacement activity. The unit cost of this activity would provide the basis for estimating a monetary damage. The problem with this approach is that paint thickness and corrosion rates are not the economically relevant impacts. Rather, what is required is some understanding of the responses of producers in terms of increased frequency of repair, degree of degradation of performance, and so forth, all of which translate into increases in the cost of production. Similarly, in the case of agriculture the damage function approach focuses on reduction in harvestable yield and multiplies this by a market price. But this approach ignores effects on consumers resulting from possible changes in market price and changes in the cost of production of a crop.

Properly specified economic models of the effects of pollution on producers make use of cost functions or production functions to link the physical effects of changes in environmental quality to changes in market prices and quantities, and ultimately to changes in consumers' and producers' surpluses. Either directly in the case of the production function approach, or implicitly in the case of the cost function approach, these models incorporate the whole range of possible producers' responses to changes in pollution levels (through, for example, material substitution, increased protection activities, and changes in maintenance and repair schedules).

The basic theory for estimation of productivity benefits is outlined in chapter 4. In the first section of this chapter, I lay out the theory in more detail for the case of single-product firms in a competitive industry. In the second section I develop the analysis for the more likely case of multiproduct firms. I then examine (in the third section) how the welfare effects on factor owners and consumers are passed through vertically linked markets for inputs and intermediate products when there are several intermediate stages of production. In the fourth section I consider the effects that market distortions such as monopoly power and regulated prices have on welfare measures. In the fifth section I show how the methods presented in this chapter can be used to value changes in the productivity of natural resource systems such as commercially exploited forests and fisheries.

BASIC THEORY

The purpose of this section is to derive welfare measures for changes in some parameter, q, that enters directly into the production functions of single-product firms. This parameter can be interpreted as a measure of environmental quality or the quality of a resource input into production.[2] Alternatively, it could be a measure of innovation or technological change. Whatever the interpretation, an increase in this variable is assumed to increase the output attainable with any given set of inputs. Just, Hueth, and Schmitz (1982) have provided a rigorous analysis of how to measure changes in welfare due to price distortions in factor and product markets; their analysis expands and generalizes on the work of Just and Hueth (1979).[3] These models provide a basis for analyzing the effects of productivity induced changes in product and factor prices. As we will see, in the case of single-product firms the results that they derive for price distortions carry over in a straightforward manner to the case of parametric shifts of cost and supply functions caused by environmental changes.

In this section I derive two alternative measures of welfare change for both marginal and nonmarginal changes in q for the case of single-product firms. One measure is the value of the marginal product of q derived from the production function. The other measure is based on the aggregate cost function for the industry and can be interpreted as the area between the old and new supply curves. The analysis in this section is essentially short-run, focusing on changes in quasi-rents to firms and on consumers surpluses. In the long run, quasi-rents are competed away, except for those accruing to specialized factors owned by the firms. These rents can be viewed as increases in the prices of such factors.

Consider a competitive industry with n firms producing a single good, x. The ith firm is assumed to have a production function $x_i = x_i(V_i, k_i, q)$, $(i = 1, ..., n)$, where V_i is a vector of variable factor inputs to firm i, v_{ij} $(j = 1, ..., m)$, k_i is a fixed factor, and q measures environmental or resource quality. For example, q could be a measure of air or water quality or natural soil fertility. Alternatively, q could be some measure of technical efficiency such as the rate at which an electrical generating station converts thermal energy to electricity. In this case, the model developed here would measure the benefits of investments in increasing technical efficiency. To simplify the notation, the k_i terms will be suppressed in all that follows. Both V_i and q have positive marginal products; but q is given exogenously to the industry.

---

[2]This section is adapted, with permission, from Freeman and Harrington (1990).

[3]See also the earlier contributions of Anderson (1976) and Schmalensee (1976).

Assume that the industry faces perfectly elastic supplies for all factor inputs at prices f_j, or in vector notation, F.[4] Let aggregate industry output be denoted by $y = \Sigma x_i$. The industry faces an inverse demand function for its product, $p = p(y)$, where p is the market price, and income and all other prices are assumed constant. Define an aggregate production function

$$y = y(v_{11}, \ldots, v_{nm}, q)$$

to be the sum of the n firm production functions.

The Production Function Approach

Now assume that demand functions are compensated so that consumer welfare changes can be measured by the appropriate areas. Then the social welfare W associated with producing y is the area under the demand curve for y, less the cost of the inputs:

$$W(v_{11}, \ldots, v_{nm}, q) = \int_0^y p(u)du - \sum_i F \cdot V_i$$

The first-order conditions for choosing the v_{ij} to maximize social welfare are

$$\frac{\partial W}{\partial v_{ij}} = \frac{\partial y}{\partial v_{ij}} \cdot p(y) - f_j = 0, \qquad \text{for all } i,j$$

Since each firm is a price taker, this welfare optimum is also the competitive equilibrium. These first-order conditions define input demand functions $v_{ij}^*(f_j, q)$, and, in turn, an output function $y^*(q) = y[v_{ij}^*(f_j, q)]$ and a social welfare function $W(q) = W[v_{ij}^*(f_j, q)]$. The asterisks indicate optimally chosen quantities. In what follows, the f_j will not appear as function arguments except where factor prices are assumed to be variable.

Using the envelope theorem, we have

$$\partial W/\partial q = p(y^*) \cdot \partial y[v^*(q), q] \cdot /\partial q \tag{1}$$

---

[4] If factor supplies are less than perfectly elastic, the changes in rents to factor owners as factor prices change must also be included in the social welfare measure. Also, as Just, Hueth, and Schmitz (1982, chapter 9 and appendix D) show in their analysis of multimarket changes, general equilibrium supply and demand curves must be used.

The net welfare gain is, in effect, the value of the marginal product of q in the production function. Note that $\partial y / \partial q$ is not the observed increase in y^*. Rather, it is the increase in y that would occur holding all other inputs constant. Knowledge of the production function is required to implement this measure.

The Cost Function Approach

The variable cost functions for firms can be added to obtain an aggregate variable cost function for the industry:

$$C = C\,[y^*\,(q),\,q]$$

Assuming profit maximization with price equal to marginal cost, this also defines a market supply curve. The total contribution to social welfare made by the production and consumption of this good is the sum of consumers' surpluses and quasi-rents to firms:

$$W = \int_0^y p\,(u)\,du - C\,[y_t^*q]$$

This, of course, is maximized when output is set where price equals marginal cost,

$$p\,(y) = \partial C\,(y^*,\,q)/\partial y \tag{2}$$

and, as before, the optimal output is a function of the parameter q.

Again applying the envelope theorem, we have the following expression for welfare change associated with marginal changes in q:

$$\partial W/\partial q = -\,\partial C\,(y^*,\,q)/\partial q \tag{3}$$

This marginal value can be calculated if the cost function is known. The duality of the cost function and the production function ensure that the two measures of marginal welfare change given by equations (1) and (3) are equivalent.

The change in total welfare is positive if increasing q reduces costs. However, this fact does not necessarily imply that both producers' surplus and consumers' surplus must increase. Consumers' surplus (CS) is given by the area under the compensated demand curve less the actual expenditure, or

$$CS(q) = \int_0^{y^*} p(u)\, du - p(y^*) \cdot y^*(q)$$

The marginal effect on consumers' surplus of a change in q is the derivative of this expression, or

$$\partial CS/\partial q = - \frac{\partial p}{\partial y^*} \cdot \frac{\partial y^*}{\partial q} \cdot y^* \tag{4}$$

Of the three terms in equation (4), $\partial p/\partial y^* \leq 0$ by the law of demand and $y^* > 0$. Thus consumer surplus will decrease if and only if equilibrium output y^* decreases with an increase in q. This possibility cannot be ruled out.[5] It is possible for the marginal cost curve to increase over the relevant range, leading to an increase in price, even though total cost is reduced. Therefore, it is possible for consumers to be made worse off by an increase in q.

The marginal change in producers' quasi-rent, R, is the marginal welfare change given by equation (3) less the change in consumers' surplus given by equation (4), that is,

$$\frac{\partial R}{\partial q} = - \frac{\partial C}{\partial q} + \frac{\partial p}{\partial y^*} \cdot \frac{\partial y^*}{\partial q} \cdot y^* \tag{5}$$

The first term is positive, while the second term is usually negative. Thus quasi-rent could decrease only if the second term were larger in absolute value. The second term can also be expressed in terms of the price elasticity of demand:

$$\epsilon = \frac{\partial y^*}{\partial p} \cdot \frac{p}{y^*} < 0$$

When this expression is substituted into equation (5), we see that the

---

[5]To determine the sign of $\partial y^*/\partial q$, differentiate the first-order condition (2):

$$(\partial p/\partial y^*) \cdot (\partial y^*/\partial q) = (\partial^2 C/\partial y^{*2}) \cdot (\partial y^*/\partial q) + (\partial^2 C/\partial y^* \cdot \partial q)$$

and rearrange to obtain

$$\partial y^*/\partial q = (\partial^2 C/\partial y^* \partial q)/(\partial p/\partial y^* - \partial^2 C/\partial y^{*2})$$

Since the denominator of this expression is always negative, $y^*(q)$ is decreasing whenever $\partial^2 C/\partial y^* \partial q$ is positive.

magnitude of the second term varies inversely with the elasticity of demand. For sufficiently inelastic demands, the fall in price brought about by the increase in quantity could actually harm producers by reducing their quasi-rents.

Nonmarginal Changes in q

For nonmarginal changes in q, say a change from q^0 to q^1, the aggregate benefit can be found by integrating either equation (1) or equation (3).

$$W_q = \int_{q^0}^{q^1} p(y^*) \cdot \left\{ \partial y [V^*(q), q]/\partial q \right\} dq$$

$$= -\int_{q^0}^{q^1} \left\{ \partial C [y^*(q), q]/\partial q \right\} dq \qquad (6)$$

As equation (6) indicates, both of these measures require integration along a path. To implement them, not only are we required to know either the production function or the cost function, but we also must know how equilibrium output changes with q or how the levels of inputs change with q, or both.

However, W_q may be calculated directly from the demand curve and the cost function if the initial and final output levels y^{*0} and y^{*1} are known. This calculation is the change in the area bounded by the demand curve and cost curve:

$$W_q = \int_0^{y^1} p(y) \, dy - C(y^1, q^1) - \int_0^{y^0} p(y) \, dy + C(y^0, q^0) \qquad (7)$$

This is shown in figure 9-1, where $p_y(y)$ is the inverse demand curve and $MC_y(q)$ indicates the industry marginal cost curves as a function of y^* and q. Graphically, equation (7) is equivalent to

$$W_q = (a + b + c + d + e) - (c + e) - (a + b + c) + (b + c)$$
$$= b + d$$

The welfare change can be represented diagramatically by the area between the old and new cost curves bounded by the demand curve. The division of the gain between producers and consumers is also shown graphically in figure 9-2. Consumers gain the areas $w + v$. Producers lose area w but gain u. The net social gain is $u + v$. If the cost and the demand functions are known, these areas can be calculated by taking the appropriate integrals.

Price ($)

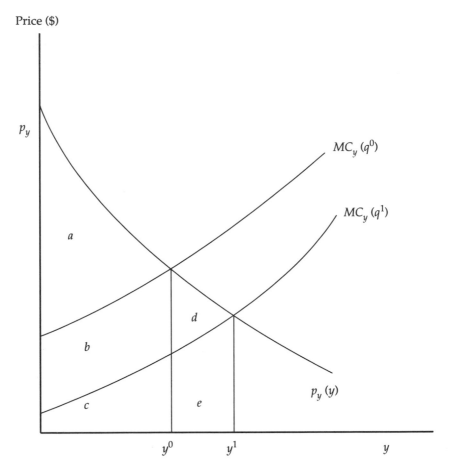

Figure 9-1.   The welfare change for single-product firms

Conclusions

In this section, two alternative measures have been identified for the welfare change associated with a change in a nonmarket input such as q. One is based on shifts in the cost or supply curves (the cost function approach); the other is based on the direct impact of the change in q on the production function (the production function approach). Of course, because of duality, they are both based on essentially the same information and must give the same results. As a practical matter, the choice of a measure depends on the availability of data.

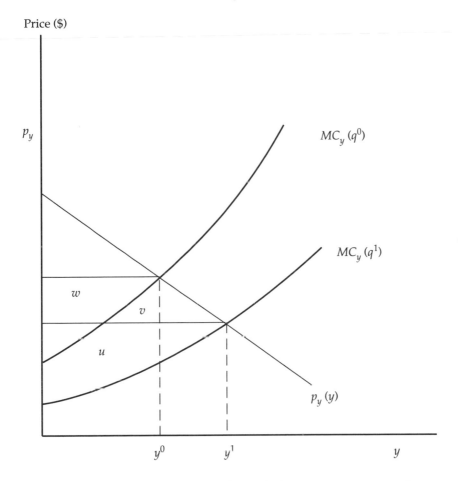

Figure 9-2.   The division of the welfare gain between consumers and producers

In some cases it may be possible to estimate the production function from data on physical inputs and outputs and environmental quality. For example, the availability of experimentally derived crop-loss functions for ozone (see Heck and coauthors, 1983) makes it feasible to use the production function approach to measure the benefits to agriculture of reducing ozone air pollution. Alternatively, one could estimate the cost function for an industry (Mathtech, 1982) or the rent or profit function (Mjelde and coauthors, 1984).

The production function and cost function methods are based on observations of the optimizing behavior of producers. An alternative to observing this behavior is to simulate it through formal optimization models. One example of this approach is the model used by Adams, Hamilton, and McCarl (1984) and Adams and McCarl (1985) to estimate the effects on the agricultural sector of controlling ozone air pollution. They combined a mathematical programming model of representative farms in the five Corn Belt states with an experimentally derived damage function to simulate producers' responses to changes in ozone levels. In another study of the economics of controlling ozone air pollution and agriculture, Kopp and coauthors (1985) constructed supply functions from region-specific farm budget and farm practice data. This model lacks the formal optimization characteristics of the Adams and McCarl model, but it allows for more detailed treatment of differences in farm economics across regions.

MULTIPRODUCT FIRMS

In the case of multiproduct firms with joint production technologies, things are not so simple.[6] For marginal changes, the production function and cost function approaches can still be applied, with appropriate modification. But for nonmarginal changes, it is not correct simply to add up areas between observed shifts of marginal cost or supply curves. When production is characterized by jointness, the model used for measurement must take account of the interconnectedness of the marginal cost or supply functions for the various outputs. The gain in welfare for an increase in q is measured by the sum of the changes in consumers' surpluses in the markets for all of the affected products plus the aggregate change in quasi-rents to the affected firms. However, in measuring the changes in quasi-rents, it is necessary to take account of how changes in the output of one good affect the marginal cost curves of the other goods being jointly produced.

For notational simplicity, we assume here only two products. The other notation is unchanged. The production function for the ith firm is given by

$$x_i (x_{1i}, x_{2i}, v_{il}, \ldots, v_{im}; q) \geq 0$$

---

[6]It is the presence of joint production and the associated economies of scope that explain the existence of multiproduct firms (Baumol, Panzer, and Willig, 1982).

That is, $x_i \geq 0$ is feasible production and $x_i = 0$ is efficient production. We also assume $\partial x_i / \partial q > 0$, which assures that increasing q will increase the welfare associated with the production of x_{1i} and x_{2i}.

Both the aggregate production function and cost function approaches still provide a straightforward basis for measuring the marginal welfare change. As before, assume that each firm is a price taker in all of its product and factor markets and that the industry faces infinitely elastic factor supply curves. In a fashion similar to the analysis leading to equations (1) and (3), we can define the aggregate production function as the sum of the n firm production functions:

$$y (y_1, y_2, V, q) = 0$$

where $y_1 = \Sigma x_{1i}$, $y_2 = \Sigma x_{2i}$, and $V = (v_{11}, ..., v_{nm})$ is the vector of all inputs for all firms. As before, the social welfare function is

$$W (V, q) = \int_0^{y_1} p_1 (u_1) \, du_1 + \int_0^{y_2} p_2 (u_2) \, du_2 - \Sigma F \cdot V$$

$$\text{subject to } y (y_1, y_2, V, q) = 0$$

The first order conditions for a maximum of social welfare are

$$W_1 = p_1 (y_1) - \lambda \partial y / \partial y_1 = 0$$
$$W_2 = p_2 (y_2) - \lambda \partial y / \partial y_2 = 0$$
$$W_{v_j} = -f_j - \lambda \partial y / \partial v_j = 0$$

where λ is the Lagrangian multiplier on the production constraint. These conditions define optimal $y_1^*(q)$, $y_2^*(q)$, and $V^*(q)$ in the usual way. By the envelope theorem,

$$\partial W / \partial q = -\lambda \cdot \partial y / \partial q \qquad (8)$$

If we now differentiate the production function with respect to q, we have

$$\frac{\partial y}{\partial y_1} \cdot \frac{\partial y_1^*}{\partial q} + \frac{\partial y}{\partial y_2} \cdot \frac{\partial y_2^*}{\partial q} + \frac{\partial y}{\partial q} + \Sigma_j \frac{\partial y}{\partial v_j} \cdot \frac{\partial v_j^*}{\partial q} = 0$$

Combining this with equation (8) yields

$$\partial W / \partial q = -\lambda \cdot \partial y / \partial q$$

$$= \lambda \cdot \frac{\partial y}{\partial y_1} \cdot \frac{\partial y_1^*}{\partial q} + \lambda \cdot \frac{\partial y}{\partial y_2} \cdot \frac{\partial y_2^*}{\partial q} + \frac{\partial y}{\partial q} + \Sigma_j \lambda \cdot \frac{\partial y}{\partial v_j} \cdot \frac{\partial v_j^*}{\partial q} = 0$$

Substituting terms from the first-order conditions yields

$$\partial W/\partial q = p_1(y_1^*) \cdot \frac{\partial y_1^*}{\partial q} + p_2(y_2^*) \cdot \frac{\partial y_2^*}{\partial q} - \sum_j f_j \frac{\partial v_j^*}{\partial q}$$

Alternatively, in terms of the aggregate cost function, $C(y_1^*, y_2^*, q)$, the benefits are

$$\partial W/\partial q = -\partial C(y_1^*, y_2^*, q)/\partial q$$

In the case of nonmarginal changes in q, it is tempting to say that all that needs to be done is to add up the areas between the new and old supply curves for each of the products of the multiproduct firms. But that would be wrong. Because of the interdependencies of the marginal cost curves under joint production, adding up these areas will not give a correct measure of the change in quasi-rents. The benefit of a nonmarginal change is the increase in the social values of the outputs, net of any changes in the joint cost of production. Again, for a change from q^0 to q^1 we have

$$W_q = \int_{y_1^0}^{y_1^1} p_1(y_1^*)\, dy_1 + \int_{y_2^0}^{y_2^1} p_2(y_2^*)\, dy_2$$
$$+ [C(y_1^0, y_2^0, q^0) - C(y_1^1, y_2^1, q^1)] \tag{9}$$

The change in joint costs, ΔC, can be decomposed into three steps:

$$\Delta C = C(y_1^0, y_2^0, q^0) - C(y_1^0, y_2^0, q^1)$$
$$+ C(y_1^0, y_2^0, q^1) - C(y_1^1\ y_2^0, q^1)$$
$$+ C(y_1^1, y_2^0, q^1) - C(y_1^1, y_2^1, q^1)$$

Substituting this into (9) gives

$$W_q = \int_{y_1^0}^{y_1^1} p_1(y_1^*)dy_1 + \int_{y_2^0}^{y_2^1} p_2(y_2^*)dy_2 - \int_{q^0}^{q^1} [\partial C(y_1^0, y_2^0, q)/\partial q]dq$$

$$- \int_{y_1^0}^{y_1^1} [\partial C(y_1, y_2^0, q^1)/\partial q]dy_1 - \int_{y_2^0}^{y_2^1} [\partial C(y_1^1, y_2, q^1)/\partial q]dy_2 \tag{10}$$

It should be noted that in each of the markets, one of the two marginal cost curves used in defining the welfare measure presented here is not actually observed. Implementation of equation (10) requires full knowledge of the joint cost function. Basing measures on comparisons of observed marginal cost curves before and after the change will result in error, the sign and magnitude of which will depend on the specific characteristics of the joint technology.

A graphic interpretation of equation (10) may prove to be helpful. First consider the case where y_1 and y_2 are substitutes in production in the sense that $\partial^2 C/\partial y_1 \partial y_2 = \partial^2 C/\partial y_2 \partial y_1 > 0$. This case is shown in figure 9-3. The solid cost curves represent the pre- and post-change observed marginal cost curves. The dashed cost curve in panel A corresponds to the fourth integral in (10), where q has changed holding y_2 at y_2^0.

The geometric areas in panels A and B corresponding to the five integrals in equation (10) are

$$W_q = (a + b + c) + (f + g) + (d + e)$$
$$- c - g$$
$$= a + b + d + e + f$$

The welfare change is measured by the area between the two cost curves for y_1 holding y_2 at y_2^0 plus the observed welfare triangle f in the market for y_2 (panel B). An empirical measure based on areas between observed cost curves would yield an underestimate of the true welfare change.

Figure 9-4 shows the case where y_1 and y_2 are complements in production, that is, $\partial^2 C/\partial y_1 \partial y_2 = \partial^2 C/\partial y_2 \partial y_1 < 0$. The geometric equivalent to equation (10) is

$$\Delta W = (a + b + c) + (f + g) + d$$
$$- (b + c + e) - g$$
$$= a + d - e + f$$

Again, measuring the first term requires knowledge of the unobserved cost curve for y_1 after the change in q but holding y_2 at y_2^0.

It is of interest to know the magnitude of the error that would result from failure to recognize the existence of joint costs. To get an approximation of the error, form a second-order Taylor polynomial expanded around (y_1^0, y_2^0, q^0). The approximate change in cost is

$$\Delta C = C(y_1^1 \, y_2^1, q^1) - C(y_1^0, y_2^0, q^0) = \Delta C^1 + \Delta C^2 + \Delta C^J$$

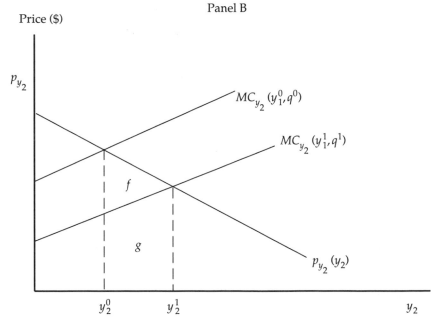

Figure 9-3.   The welfare measure for multiproduct firms when outputs are substitutes in production

Panel A

Panel B

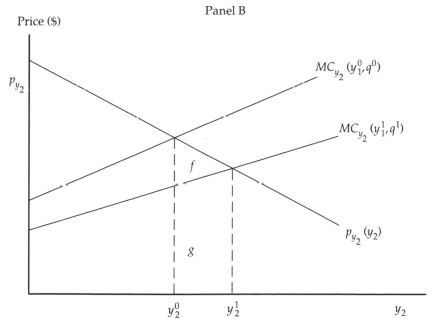

Figure 9-4.   The welfare measure for multiproduct firms when outputs are complements in production

where

$$\Delta C^1 = (y_1^1 - y_1^0) \cdot \frac{\partial C}{\partial y_1} + \frac{1}{2}(y_1^1 - y_1^0)^2 \cdot \frac{\partial^2 C}{\partial y_1^2} + (y_1^1 - y_1^0)(q^1 - q^0)\frac{\partial^2 C}{\partial y_1 \partial q}$$

$$\Delta C^2 = (y_2^1 - y_2^0) \cdot \frac{\partial C}{\partial y_2} + \frac{1}{2}(y_2^1 - y_2^0)^2 \cdot \frac{\partial^2 C}{\partial y_2} + (y_2^1 - y_2^0)(q^1 - q^0)\frac{\partial^2 C}{\partial y_2 \partial q}$$

$$\Delta C^J = (y_1^1 - y_1^0)(y_2^1 - y_2^0)\frac{\partial^2 C}{\partial y_1 \partial y_2}$$

The third term ΔC^J is the contribution to the change in cost by the jointness in y_1 and y_2.

There is another way of looking at the difficulties involved here. Recall that one component of the welfare change is the change in quasi-rents to firms. With joint production, in general it is not possible to measure the quasi-rent of a multiproduct firm from data from just one market. However, Just, Hueth, and Schmitz (1982) have shown that in the case of price changes there are circumstances in which it is possible to measure the change in quasi-rents in either a single-factor market or a single-product market. Their analysis can be extended to the case of parametric shifts in the production technology. What is required is that there be either a necessary input or a necessary output. A necessary output is one for which there is some positive minimum price at which the firm will choose to stop producing not only that output but all other products as well; in other words, production will shut down completely. All of the quasi-rent to a firm can be attributed to the necessary output and can be measured by integrating above the supply curve for that output from the shut-down price to the current market price. Alternatively, the quasi-rent can be measured by the area under the demand curve for any necessary factor input. An input is deemed to be necessary if there is some price for that input at which its derived demand falls to zero and if all of the firm's outputs fall to zero when that input is set at zero.

The measurement of changes in quasi-rents can be shown graphically for the case of two necessary outputs, y_1 and y_2, and one necessary input, v_1. Suppose that, as shown in figure 9-5, an increase in q shifts the supply functions for both outputs and the demand function for the necessary input to the right. The consumer surplus component of the welfare change is the areas $w_1 + v_1 + w_2 + v_2$ in panels B and C. The increase in the quasi-rents to all firms can be measured alternatively by $u_1 - w_1$ in panel B, or $u_2 - w_2$, or b in panel A.

To see this, first suppose that y_1 is the necessary output. At q^0 if p_1 is at p_1^0, other things being equal, firms shut down and quasi-rents

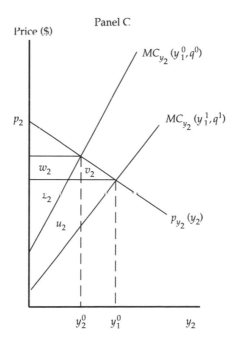

Figure 9-5.   The welfare gain for multiproduct firms with necessary outputs or a necessary input

are zero. If p_1 increases to p_1^1, firms produce both goods in positive quantities; but all of the quasi-rents are attributable to the increase in p_1 and are measured by the area $w_1 + z_1$. Comparing quasi-rents at q^1 and q^0 gives $\Delta R = R^1 - R^0 = u_1 + z_1 - w_1 - z_1 = u_1 - w_1$. For the necessary input v_1, if its price is f_1', firms shut down and quasi-rents are zero. If f_1 falls, firms produce both goods. All of the quasi-rents are attributable to the increased use of v_1 and are measured by a.

The three alternative measures of the total welfare gain are:

y_1 is necessary:

$$W_{q1} = w_1 + v_1 + w_2 + v_2 + u_1 - w_1$$
$$= u_1 + v_1 + w_2 + v_2$$

y_2 is necessary:

$$W_{q2} = w_1 + v_1 + w_2 + v_2 + u_2 - w_2$$
$$= w_1 + v_1 + u_1 + v_2$$

v_1 is necessary:

$$W_{qv} = w_1 + v_1 + w_2 + v_2 + b$$

These alternative measures can be compared to the measures obtained by adding up areas between supply curves. The difference between these measures can be seen as follows:

$$\text{True } W_q = u_1 + v_1 + w_2 + v_2$$

$$\text{Adding areas} = u_1 + v_1 + u_2 + v_2$$

$$\text{Difference} = w_2 - u_2$$

In the two-product case, the difference is a difference between two areas in one of the product markets, depending upon which product market is chosen for the measurement of the change in quasi-rents to firms. In the n product case, the difference between the two measures would be the sum of $n - 1$ such differences in areas.

Where empirical measures of producers' benefits have been reported in the literature, production has typically been modeled either implicitly or explicitly as being nonjoint, even in those cases where multiproduct firms are typical. For example, Adams, Hamilton, and McCarl (1984), Adams and McCarl (1985), Kopp and coauthors (1985), and Adams, Crocker, and Katz (1984) have all provided estimates of the benefits to agriculture of controlling ozone air pollution, using variations of the cost function approach. The first two studies used supply functions derived from models of farm behavior rather than from econometric estimates. Independence of the marginal cost or supply curves by crop was a characteristic of these models. Adams, Crocker, and Katz (1984) used econometric estimates of supply functions for four crops; but the estimating equations imposed nonjointness on the production technology even though many farms produced at least two of the four crops modeled (corn and soybeans). Other researchers (Mjelde and coauthors, 1984, Garcia and coauthors, 1986, and Mathtech, 1982) have aggregated across products to obtain industry profit or cost functions. But consistent aggregation of this sort is possible only if marginal cost functions are independent, a condition that is not satisfied when there is joint production. The analysis presented here suggests that in the future it will be important to confront the question of joint production directly. If in fact the industries being studied are characterized by joint production, the empirical models of production must reflect this; and the measures of welfare change must take this into account in the way just described.

VERTICALLY LINKED MARKETS

The next case to consider is that of vertically linked markets in which the output of one set of firms is purchased as an input by another set of firms. Assume that q affects costs and prices in one industry that is part of a set of vertically linked industries. The model used to derive welfare measures in this case must take account of the fact that every price change affects both buyers and sellers, but in different directions; that is, for a price increase, the buyer loses while the seller gains, and vice versa for price decreases. For simplicity, in what follows assume that all firms are single-product firms. Some firms are producers that purchase primary factor inputs and sell an intermediate product y_w. Other firms buy the intermediate product and factor inputs and sell y_r at retail to consumers. Assume that all factor supply curves but one are infinitely elastic; the exception is the supply curve for labor used in the wholesale industry, v_w. Assume also that the compensated supply curve for v_w is upward-sloping and that y_w and y_r are produced and sold in perfectly competitive markets. Figure 9-6 shows the case.

Assume now that q affects the producers of y_w. Suppose that at the initial level of q the supply function for y_w is MC_w^0, and the demand curve is D_w^0, as shown in panel B of figure 9-6. The first effect of an increase in q is to shift MC_w down to MC_w^1. This results in a similar downward shift of MC_r and a decrease in the retail price, as shown in

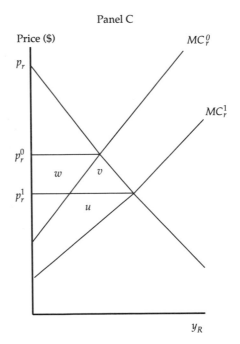

Figure 9-6. **Measuring welfare gains in vertically linked markets**

panel C of figure 9-6. In the wholesale market, the demand curve for y_w is derived on the assumption that the retail price is constant. When the retail price falls, the derived demand for y_w shifts to the left. The wholesale producers respond to the changes in demand along a path that reflects the adjustments in price in the retail market. The path of this adjustment can be described by what Just, Hueth, and Schmitz (1982) call a general equilibrium demand curve. This demand curve shows the maximum willingness to pay by the y_r industry given its costs and the retail price at which it must sell its product, and assuming that all other input prices are constant. This is D_w^* in figure 9-6.

At the same time, the demand curve for v_w in panel A of figure 9-6 is shifting, for two reasons. First, the decrease in p_w pushes the demand curve for v_w to the left. Second, the change in q also affects the productivity of and demand for v_w. The direction of this effect depends on the specific characteristics of the production function and on the elasticity of demand for y_w. As the price of v_w changes, this shifts MC_w in the same direction. There are also two general equilibrium supply functions, S^{*0} and S^{*1}, in panel B, one for each level of q. They describe the supply adjustments to changing factor prices. They show the minimum supply price for y_w for each level of q, taking account of changes in factor prices as the derived demand for v_w shifts. The net result is an equilibrium in this market at the intersection of D_w^* and S^{*1}, as shown in figure 9-6 panel B.

Three groups of economic agents are affected by the change in q so there are three components to the measure of welfare gain. The first group, the consumers of y_r receive an increase in consumer surplus, shown as $w + v$ in figure 9-6 panel C. The second group, the owners of v_w, experience a change in their rents because of the change in its price. They may either gain or lose depending upon how the demand for v_w changes; in figure 9-6, panel A, they lose the area $c + d$. The third group consists of the firms in the wholesale and retail industries that experience changes in their quasi-rents. The change in quasi-rents in the retail industry can be measured directly in the retail market by the area $u - w$. As discussed above, if v_w is a necessary input, then the change in quasi-rents to the wholesale industry can be measured in the factor market by the area $c - a$. Considering all of the changes, and netting out transfers between buyers and sellers resulting from price changes, we have

$$W_q = \Delta PS_v + \Delta R_w + \Delta R_r + \Delta CS_r$$
$$= -(c + d) + (c - a) + (u - w) + (w + v)$$
$$= -(d + a) + (u + v)$$

Following Just, Hueth, and Schmitz (1982), it can be shown that the welfare gain can also be measured in the directly affected wholesale market, provided that the measurements are based on the general equilibrium supply and demand curves. These curves reflect the adjustments of prices in linked markets that transfer gains and losses among buyers and sellers. Specifically,

$$W_q = \Delta CS_w^* + \Delta R_w^* \tag{11}$$

where CS_w^* ($= \Delta R_r + \Delta CS_r$) and R_w^* ($= \Delta R_w + \Delta PS_v$) are the areas behind the general equilibrium demand and supply curves respectively. So from equation (11),

$$W_q = f + g + j - f$$
$$= g + j$$

Thus if general equilibrium demand and supply functions for the directly impacted market can be estimated, it may be more convenient to measure welfare gains using this expression than to attempt to measure separately the changes in quasi-rents and surpluses in all of the remaining vertically linked markets.

Now consider the case of multiproduct firms that are part of a chain of vertically linked markets. As I have just shown, the welfare gain can be measured in the directly affected market. But the welfare measure must be modified to take account of the multiproduct characteristic of the producing industry. Specifically, if there is any jointness in production, it is not correct simply to add the areas $g + j$ across all products. Rather, the correct measure is the sum of the consumer surplus changes as measured by areas like $f + g$, and the change in producing firms' quasi-rents as measured by $j - f$, in the market for one essential output.

MARKET DISTORTIONS

The welfare measures derived in the preceding sections come from models in which all markets are perfectly competitive. In this section I briefly examine two kinds of market distortions. The first is distortion caused by monopoly power in the output market; the second is the distortion caused by price support and commodity production control programs such as those affecting U.S. agricultural markets.

Monopoly

The correct welfare measure in the case of monopoly can be derived graphically in a straightforward manner. Measurement is another question. Figure 9-7 shows the demand, marginal revenue, and marginal cost curves for a monopolist. If an increase in q shifts the marginal cost curve outward, the welfare measure is the sum of the increases in consumer surplus and monopoly quasi-rents:

$$W_q = \Delta CS + \Delta R$$

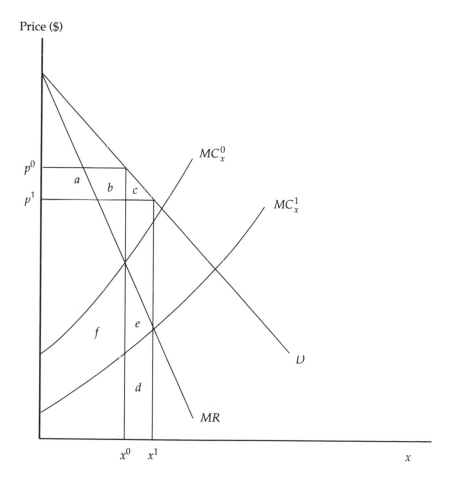

Figure 9-7.   The welfare change for a monoply firm

where

$$\Delta CS = a + b + c$$
$$\Delta R \ \ = \Delta TR - \Delta C$$
$$\ \ \ \ \ \ \ = d + e - (d - f)$$
$$\ \ \ \ \ \ \ = e + f$$

Thus,

$$W_q = a + b + c + e + f$$

As in the case of perfect competition, consumers always gain because of the price reduction. But in this case the monopolist also always gains from an increase in q. The monopolist's profit is

$$I = p(x) \cdot x - C(x, q)$$

$$\frac{\partial I}{\partial q} = \left(p + x \frac{dp}{dx} - \partial C / \partial x \right) \frac{dx}{dq} - \partial C / \partial q$$

The first-order condition for profit maximization requires that the term in parentheses be equal to zero. So

$$\frac{\partial I}{\partial q} = - \partial C / \partial q > 0$$

The problem posed for measurement is that the marginal cost curve of the monopolist cannot be observed from market data on optimally chosen prices and quantities. One possible approach to deriving marginal cost functions is to construct models of the firm based on engineering or technological data (see, for example, Russell and Vaughan, 1976).

Price Supports: The Case of Agriculture

In the United States, the federal government intervenes in the markets for several major crops so as to maintain a price above the market clearing price and to support the incomes of farmers. This intervention and distortion of the markets for crops complicates the analysis of the benefits of controlling air pollution-induced damages to agriculture, the benefits of technological change in agriculture, and the costs of regulations to protect other environmental values (for example, the costs of controls on pesticide applications and the costs of changes in farming practices to control erosion and nonpoint source pollution). Here I will use a fairly simple example to illustrate the issues that must be dealt with in order to arrive at a defensible measure of welfare change in the case of market distortions in agriculture.

As described in McGartland (1987), there are three major mechanisms through which the government intervenes in agricultural markets. The first is the "target price/deficiency payment," through which farmers receive a subsidy equal to the difference between a target price and the actual market clearing price for each unit of the crop they produce. The target price, which is the effective average revenue for farmers, determines the quantity produced according to the supply function for the crop. The second mechanism is the loan rate program. Under this program farmers can turn over part of their production to the government as collateral against a loan made at the loan rate per unit of the crop. Since there is no stigma attached to forfeiting on the loan, the loan rate becomes, in effect, the price at which the government is willing to purchase surplus production on the market. Under the third mechanism, for some crops government-sanctioned marketing orders place limits on the amount of the crop that farmers can place on the open market. This induced restriction on market supply serves to raise the price above the competitive market equilibrium. To illustrate the issues involved, I will describe the measurement of welfare change for the simple case of a target price/deficiency payment program.[7]

Assume that the government has established p_t as the target price as shown in figure 9-8. The supply curve for the crop is initially S^0; farmers therefore produce x^0. When this quantity is placed on the market, the equilibrium price is p_m^0.[8] The government makes up the difference between the target price and the market price resulting in an aggregate deficiency payment to farmers equal to the area $p_m^0 FDp_t$.

---

[7]For a description of the measurement of welfare changes for the case of a combination of a deficiency payment and loan rate program, see McGartland (1987) and Kopp and Krupnick (1987).

[8]Models for welfare measurement are sometimes constructed by assuming a particular functional form and elasticity for the demand function, and selecting parameter values so as to "force" the demand curve through some assumed price and quantity equilibrium such as point F in figure 9-8. For examples, see Adams, Hamilton, and McCarl (1984), National Academy of Sciences (1983), and Kopp and coauthors (1985). Lichtenberg and Zilberman (1986) and McGartland (1987) point out some of the problems that arise if the calculations and calibration process do not accurately reflect the form of the government intervention in the market.

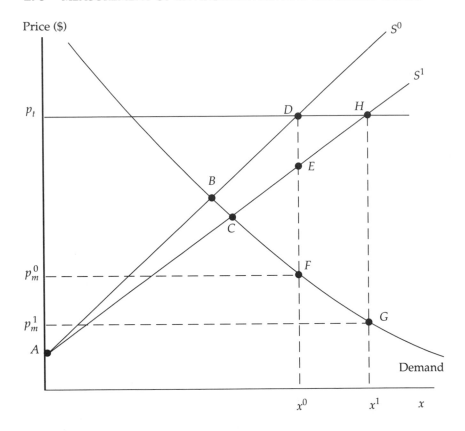

Figure 9-8. Welfare measures in price-supported markets: the case of agriculture

Since the efficient level of output would be at the intersection of the demand curve and the supply curve (point B), the excess production results in a deadweight loss equal to the area BDF. Note that the deficiency payment is a transfer from taxpayers to producers, and that as such it does not enter into the welfare calculation.

Now suppose that an air pollution control program increases productivity for this crop, resulting in the new supply curve S^1. Producers respond by increasing output to x^1, given the unchanged target price p_t. What is the social value of this extra output? The most conservative answer is that the social value is zero, and that, therefore, the "benefit" (negative) of the pollution control is the extra cost of producing the additional output, measured by the area x^0EHx^1. This "conservative" answer would make sense in some circumstances—for example, if the additional output were simply destroyed to keep it off the market. But under the target price/deficiency payment system, the increment to output is placed on the market and drives the market clearing price down to p_m^1. The value to consumers of the extra output is the area x^0FGx^1. But the value to consumers is less than the cost of this increment to production. Although the additional output costs more than it is worth to consumers (the difference being the area $FEHG$), the policy also reduces the cost of producing the original output level, x^0. This is clearly a benefit and is measured by the area ADE. The net welfare gain is the cost reduction (ADE) minus the excess cost of the increase in output ($FEHG$). This can be either positive or negative.

The proper welfare measure can also be seen in another way. If the market were in equilibrium, the net gain in producers' and consumers' surpluses would be the area ABC. But given the policy, there is also an increase in the deadweight loss from BDF to CHG. From this perspective, the net welfare change is the efficiency gain (ABC) less the increase in deadweight loss (CHG minus BDF). As can be seen from figure 9-8, these two ways of describing the welfare change are geometrically equivalent.

The welfare measure provided here is based on two key assumptions. The first is that farmers actually respond to the reduction in pollution as predicted by the new supply curve. The second is that there are no changes in government policy in response to the increase in agricultural productivity. Concerning the first assumption, Kopp and Krupnick (1987) have argued that farmers might not fully perceive the increase in productivity that lies behind the downward shift in the supply curve, especially given the natural variability of yields due to vagaries of weather, rainfall, and so forth. Since the supply curve is a long-run relationship, it incorporates the predicted increases in the inputs of labor, capital, and land on the part of farmers. But if farmers do not perceive the increase in productivity, they will not make these increases in factor inputs.

With respect to the second assumption, Madariaga (1988) has argued that government policymakers are likely to respond to any increase in agricultural output with adjustments in policy to curtail at least somewhat the surplus in production. Adjustments in policy could include reducing the target price and increasing the acreage reduction requirements. Kopp and Krupnick (1987) have presented calculations of social welfare gains for a 10 percent reduction in ambient ozone for four crops under alternative assumptions about policy responses. In all cases, the benefits of the cost reduction outweighed the increases in the deadweight loss. They show that benefit estimates can vary by up to 30 percent, depending upon the assumptions about policy responses.

In conclusion, the form that an agricultural price support policy takes can have a substantial impact on the benefits actually realized from any change in agricultural productivity. Measures of welfare change should be based on realistic assumptions concerning the relevant features of agricultural policy. Welfare measures should be interpreted as being conditional upon the existence of the agricultural policy being assumed. Given present agricultural policy, the benefits of pollution control are partly dissipated by the costs of the additional excess production of farm crops. Conversely, the costs associated with environmental regulations that raise the cost of agricultural production tend to be partly offset by the reduction in the deadweight losses due to excess production.

VALUING CHANGES IN THE PRODUCTIVITY OF NATURAL RESOURCE SYSTEMS

The productivity of commercially exploited natural biological systems can depend on such things as the flow of nutrients into the system, the population of a predator species, or the level of a pollutant. Any of these factors might be subject to human manipulation. So it would be useful to have an economic framework for evaluating the welfare consequences of policies to change these things. Since the outputs of such resource systems are traded in markets, the framework presented in this chapter is applicable. The welfare consequences come in the form of changes in producers' and consumers' surpluses.

In this section, I provide a brief description of how an environmental quality variable can be introduced into the standard economic models of two types of natural resource systems, the commercial forest and the commercial fishery. In the case of the forest, an investment in environmental quality may alter the optimum time of harvest, so the intertemporal features of the forest optimization model must be examined. Moreover, an environmental quality variable might affect the economic value of a nonmarket output such as recreation. Thus the interaction between the nonmarket and market outputs must be analyzed in a multiple-use framework.

In the case of the fishery, there are two alternative bio-economic models in the literature, the Schaefer–Gordon growth model and the Beverton–Holt stock-recruitment model. Environmental parameters can be easily introduced into both models. But the economic implications of a change in environmental quality are more transparent in the case of the Schaefer–Gordon model, as we will see. In addition, the economic value of an environmental quality change will depend upon the institutions for ownership or management of the fishery. I will show the results of a comparison of the economic value of an environmental change under the alternatives of optimum management and open-access exploitation.

Commercial Forests and the Role of Time

The models described in the preceding sections were timeless in the sense that environmental changes and changes in prices and quantities were contemporaneous. There was no need to model explicitly any intertemporal effects. Where the linkages between environmental changes and market changes are not contemporaneous, time must be built into the model explicitly. An example is the case of the effects of changes in air quality on commercial forest productivity.

One of the things that makes the case of the effects of environmental change on commercial forests interesting from an economic perspective is that a change affecting a stand of young trees today will not have an effect on marketed outputs for perhaps forty years, when the trees are harvested and sold. If there were an active market in forest land, any increase in the growth rate of young trees now would have an immediate impact on the market price of land where there were standing, growing, trees. But even so, any consumer surplus benefit from lower prices of forest products would not be realized until the increased harvests actually took place. In what follows, I ignore the product price effects on consumers and focus only on rents accruing to forests and owners.

Suppose that the trees on a plot of land grow in net value over time according to

$$G_t = G\,(t,\,q)$$

where G_t is the stumpage value, that is, the market price of the harvested volume at age t, net of harvest costs and transportation to the mill. This is a very general formulation that allows for growth in harvestable volume as well as for changes in the price per unit of volume because of changes in quality. I also assume that replanting and management costs are zero.

Assuming no economically relevant alternative uses for a unit of land, the landowner's economic problem is to choose a sequence of harvesting dates to maximize the present value of the stream of net receipts at each harvest, that is,

$$\text{Max:} V = \sum_{h=1}^{\infty} e^{-h \cdot r \cdot t} G(t, q)$$

$$= \frac{G(t, q)e^{-r \cdot t}}{1 - e^{-r \cdot t}} \tag{12}$$

where t is the age of the stand at the time of harvest, r is the interest rate, and h indexes the generation of the stand. For simplicity, I assume that there are no costs for planting, thinning, and other management activities during the rotation.[9] The stand should be harvested at the age that satisfies

$$\frac{\partial V}{\partial t} = \frac{\dfrac{\partial G}{\partial t} - r \cdot G(\cdot) - \dfrac{r \cdot G(\cdot)e^{-r \cdot t}}{1 - e^{-r \cdot t}}}{1 - e^{-r \cdot t}} = 0$$

or

$$\frac{\partial G}{\partial t} = r \cdot G(\cdot) + \frac{r \cdot G(\cdot)e^{-r \cdot t}}{1 - e^{-r \cdot t}} \tag{13}$$

The stand should be allowed to grow as long as the marginal gain in value through growth ($\partial G/\partial t$) exceeds the interest forgone by not realizing G through harvest ($r \cdot G$) plus the opportunity cost of postponing the stream of returns from future rotations. Harvest should occur when the marginal gain from waiting just equals the marginal opportunity cost.

Two questions of interest are the effects of changes in q on V and on the optimum rotation length. From equation (12)

$$\frac{\partial V}{\partial q} = \frac{\partial G}{\partial q} \cdot \frac{e^{-r \cdot t}}{1 - e^{-r \cdot t}}$$

The increase in q increases G at the time of the next harvest; and the second term in this expression gives the present value of the stream of these increases over the infinite future. So if the growth function were known, including the effect of q, then calculating the effect of changes in the steady-state level of q on the stream of rents would be

[9]For a more complete treatment of the forestry optimization problem, see Samuelson (1976), Hyde (1980), and Bowes and Krutilla (1989).

straightforward. But where q varies over time, because, for example, of trends in pollution associated with economic growth, such calculations would require more detailed knowledge of the growth function than is currently available.

Answering the second question appears to be more difficult. If t^* is the solution to equation (13), it appears to be impossible to determine unambiguously the sign of $\partial t^*/\partial q$. Callaway has reported that for one form of growth function, a decrease in q shortens the optimal rotation at low interest rates but increases it at higher interest rates.[10]

Environmental Quality and Multiple-Use Management of Forests

Suppose that in addition to the periodic harvest of marketable products, the forest unit provides a flow of nonmarket services such as recreation that depends on the age of the forest and on q. Let the value of the nonmarket service be given by $R(t,q)$. Over one harvest cycle, the present value of the nonmarket service flow is

$$V_n = \int_0^{t^*} R(t,q)dt$$

The objective function becomes:[11]

$$\text{Max: } V = \sum_{h=1}^{\infty} e^{-h \cdot r \cdot t} \left[G(t,q) + \int_0^t R(t,q)dt \right]$$

As Hartman (1976) first showed, including the value from the standing forest in the objective function will either increase the optimum age at harvest (t^*) or make harvesting uneconomic entirely. This is because the standing forest adds a second term to the left-hand side of equation (13), the marginal benefit of delaying the harvest. The optimum harvest (if it exists) occurs when

$$\frac{\partial G}{\partial t} + R(\cdot) = r \cdot G(\cdot) + \frac{r \cdot G(\cdot) e^{-r \cdot t}}{1 - e^{-r \cdot t}}$$

An increase in q can be valuable both because of its increase in the

---

[10]John M. Callaway, Jr., personal communication, March 17, 1987.

[11]See chapter 12 for a discussion of models for measuring the value of recreational resources, and chapter 5 for a discussion of the possibility of nonuse values arising from natural resources.

present value of the flow of harvestable product and because of its effect on the value of the flow of services from the standing forest.[12]

One benefit of modeling exercises of this sort is to provide guidance for the natural science research that is required to support future economic analyses. As these models suggest, it is important to know how changes in q affect the whole time pattern of the growth of trees and the flow of nonmarket services over the life of the forest.

Environmental Quality and Commercial Fisheries

The economic analysis of fisheries rests on a foundation of a biological model of the growth and mortality of a species. Two alternative biological models have dominated the literature on the economics of fisheries. The first is the so-called Schaefer–Gordon model, which makes the growth rate of the aggregate biomass of the species at any point in time a function of the current level of biomass. The alternative Beverton–Holt model explicitly describes both the number of fish in each age cohort and their weight or size. The Beverton–Holt model may be more realistic from a biological perspective; but it is also more complex from an economic perspective.

THE SCHAEFER–GORDON MODEL

In this model the relationship between the growth of the aggregate stock, g, and the aggregate size of the stock, z, takes the following form:

$$g = b \cdot z \left(1 - \frac{z}{k}\right)$$
$$= b \cdot z - \frac{b}{k} \cdot z^2$$

where b is the intrinsic growth rate and k is the carrying capacity of the environment.[13] This quadratic growth function results in a logistic time path for the stock in the absence of harvest, with z approaching the carrying capacity asymptotically.

Environmental parameters have been incorporated into the Schaefer–Gordon growth model in several empirical studies of fish-

---

[12]For further discussion of the Hartman result and multiple-use forest management in general, see Bowes and Krutilla (1989).

[13]When the stock has grown to the carrying capacity ($z = k$), the growth rate is zero.

eries. Either the intrinsic growth rate or the carrying capacity (or both) can be made a function of an environmental quality parameter, q. For example, in his study of the North American lobster fishery, Bell (1972) included seawater temperature as one variable helping to explain the annual harvest. Lynne, Conroy, and Prochaska (1981) examined the effects of changes in the acreage of marine wetlands on annual harvests of blue crabs in the Florida Gulf Coast fishery. And Bell (1989) performed a similar analysis of eight Florida Gulf Coast fisheries, including both finfish and shellfish.

The Schaefer–Gordon model can be incorporated into an economic model to determine the optimum levels of harvest and stock. And this model can also be used to trace the effects of changes in environmental parameters on harvest, stock, and economic welfare. If an environmental quality parameter can be increased by public policy at some cost, optimum levels of investment in environmental quality can also be determined.

The standard approach is to specify a production function that makes the fishery industry's annual harvest, h, a function of economic inputs and the stock of fish to be caught, z. For simplicity, the economic inputs are aggregated into a measure of effort, e. A unit of effort can be called a "boat" and can be interpreted as an optimal combination of labor and capital. Assume that each unit of effort has a cost of p_e. For any chosen level of h, the production function can be solved for the required level of effort, given the size of the stock. And since the size of the stock at any point and time is the net result of past growth (which depends in part on q) and past harvest, the cost function can be written as

$$C = p_e \cdot e$$
$$= C(h, z, q)$$

Managing the fishery for economic objectives is a dynamic problem, since current harvests and costs depend on past harvest decisions and how they have affected the size of the stock. The economic objective is to maximize the present value of the net economic return from the fishery over time, subject to the biological constraint imposed by the growth function. If $p_h(h)$ is the inverse demand function for fish, the objective function is

$$\text{Max:}_{h_t} \int_0^\infty \left[\int_0^h p_h(h_t)dh - C(h_t, z_t, q) \right] e^{-r \cdot t} \cdot dt \qquad (14)$$

subject to

$$\frac{dz}{dt} = g(z_t, q) - h_t \qquad (15)$$

For any given initial conditions, this problem can be solved for the sequence of harvests that maximizes the objective function. The economic dimensions of the problem are easiest to see if we focus attention on the conditions for the long-run, steady-state equilibrium where growth and harvest are equal and the stock is of constant size. The first-order conditions for this optimum are

$$p_h = \partial C/\partial h + \rho \tag{16}$$

and
$$r = \frac{\partial q}{\partial z} - \frac{1}{\rho} \cdot \frac{\partial C}{\partial z} \tag{17}$$

where r is the interest rate and ρ is the shadow value of the stock, reflecting its effect on the rate of growth and the cost of harvest. Equation (16) says that the price of fish must equal the full marginal cost of their harvest, where this includes the opportunity cost of decreasing the stock, ρ. Equation (17) defines the optimum intertemporal trade-off. The interest rate is the opportunity cost of forgoing \$1 worth of harvest now. It must equal the benefit of forgoing harvest, which has two components. The first component is the contribution the additional stock makes to future growth. The second reflects the contribution that the extra stock makes to lowering harvest costs.[14]

VALUATION OF CHANGES IN Q UNDER THE SCHAEFER–GORDON MODEL

An increase in q shifts the cost function in equation (14) and changes the growth function of (15). This leads to new solution values for h and z in every period. The welfare value in each period is the increase in the net value term in brackets in equation (14), that is, the area between the marginal cost curves bounded by the demand curve. The stream of changes in that value must be discounted to its present value to obtain the total welfare value of the change in q.

Taking wetland acreage as a measure of q, Ellis and Fisher (1987) drew on the earlier work by Lynne, Conroy, and Prochaska (1981) to calculate the net welfare value of increases in wetlands in the blue crab fishery of the Gulf Coast of Florida. They abstracted from the dynamic, intertemporal dimension of the fishery management problem by focusing only on a single-period optimum and by assuming that the cost of harvest was independent of this stock. This is equiva-

---

[14]For a clear treatment of the optimization problem, see Fisher (1981). More detailed and rigorous treatments can be found in Clark (1976, 1985).

lent to asserting that $\rho = 0$, so that an optimum is defined by the price of harvest equalling its marginal cost.

The economic value of a change in environmental quality depends not only on the economic and biological parameters of the model, but also on the institutional arrangements for ownership or public management of the resource. The preceding discussion (and the Ellis–Fisher analysis) is based implicitly on the assumption of either private ownership with perfect competition or public regulation to achieve the economic optimum. But most fisheries resources are characterized by absence of private ownership and more or less open access to the resource. And where there is public regulation, it is seldom designed to achieve an economically efficient outcome. As a consequence, the net economic value of the resource will be lower; and the economic value of changes in q will also be affected by the ownership and management arrangements.

I will now present a simple static model to show the effect of open access on the welfare value of a change in the productivity of the fisheries resource such as might result from an improvement in water quality.[15] It is well known that under open access without regulation, competition drives rent to zero (Gordon, 1954; Scott, 1955). Let MC_x^0 in figure 9-9 be the marginal cost of harvesting fish at the original level of q. If the fishery is privately owned, or if it is optimally managed, the output and price will be given by the intersection of the marginal cost of harvest function and the demand curve at p_m^0, and x_m^0. An improvement in q would shift the marginal cost of harvest function outward to the right, resulting in a higher quantity and a lower equilibrium price. The new marginal cost curve is not shown in figure 9-9. The welfare measure would be the same as that presented above.

If there is open access to the fishery, the condition for equilibrium is that each fisherman be earning zero profits. This requires that price be equal to the average cost of harvesting fish. Compared to the efficient price and output under private ownership, more fishermen enter in pursuit of profit. The increase in fish caught decreases the stock, raising costs or decreasing the price of fish, or both. Entry continues until these forces eliminate the incentive for entry. At the initial level of q, the average cost of harvesting fish is AC_x^0, as shown in figure 9-9. The open-access equilibrium is at p_c^0 and x_c^0.

An improvement in q also shifts the average cost curve outward, resulting in a lower price and a higher quantity. But since price equals average cost both before and after the change in environmental quality, there is no change in producer surplus. The benefit consists en-

---

[15]See also McConnell and Strand (1989).

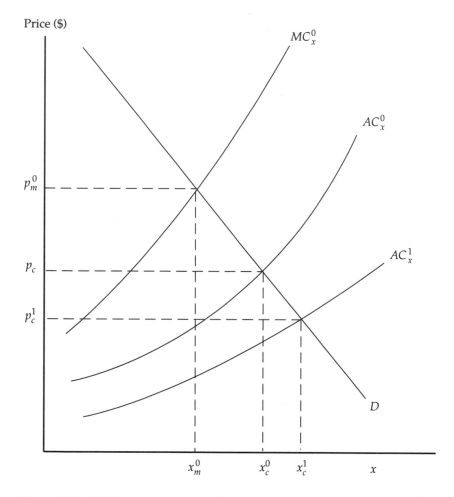

Figure 9-9. Welfare measurement for open-access resources: the case of a fishery

tirely of the increase in consumer surplus associated with the price decrease. The more elastic is the demand curve for fish, the smaller is the welfare gain associated with the environmental improvement.[16] In the limit, if this fishery is small relative to the market and the demand curve is perfectly elastic, there is no welfare gain. The physical improvement in productivity brought about by the higher water

---

[16]See Freeman (1991) for some illustrative calculations of the magnitude of these effects based on the Ellis–Fisher data.

quality is entirely dissipated by the uneconomic competition of fishermen for the potential increase in rents.

The Schaefer–Gordon model has been criticized as being too aggregated, and not being biologically realistic for many species. Also, one of its key predictions is not supported by observation for some species. The prediction is that if effort increases, eventually the stock must decrease, leading to a decline in growth and harvest and eventually to the biological and economic collapse of the fishery. Yet for several fisheries, plots of effort and harvest over time show increasing effort associated with almost no change in harvest.[17] This observed pattern of harvest and effort is entirely consistent with the Beverton–Holt biological model of the fishery.

The Beverton–Holt model combines a model of the growth of individual members of a species and the number of individuals in a single expression. The first element of the model is the recruitment of k individuals of the same age into a cohort at time $t = 0$. In the basic model, k is considered to be exogenous; but clearly it could be a function of environmental factors.[18]

The second element of the model is an expression for the population of the cohort at any point in time. Over the lifetime of the species, the total population changes because of natural mortality, m, and the harvest of fish, f. The latter is sometimes referred to as fishing mortality. Both of these forms of mortality can vary over time. Thus the population of the cohort at time t is given by

$$n_t = k \cdot e^{-(f_t + m_t)t} \tag{18}$$

The third element of the model is an expression giving the weight of the typical member of the cohort at any point in time. Each

---

[17]Townsend (1986) showed this to be the case for the North American lobster fishery originally analyzed by Bell (1972). McClelland (1991) has also shown this to be the case for several of the Florida Gulf Coast fisheries analyzed by Bell (1989).

[18]Consider, for example, the oyster. Recruitment could be defined as the attachment of larva to solid surfaces where shell development begins. The limiting factor in this stage of development may be the quantity of suitable surface available for attachment, not the number of eggs released and fertilized in the spawning stage.

fish grows over time, so its weight is given by

$$w_t = w_\infty [1 - e^{-\alpha(t)}]^\beta \qquad (19)$$

The total biomass of the cohort at time t is the product of equations (18) and (19).

The last element of the model shows the impact of fishing effort on the cohort. The total weight of fish harvested at any point in time is

$$h_t = w_t \cdot n_t \cdot f_t$$

where f_t depends at least in part on the level of effort.

Cost and revenue functions can be specified and the optimum time path of the exploitation of a cohort can be derived (see, for example, Clark, 1976, 1985). But since the number of recruits in one cohort is largely independent of the fishing mortality experienced by earlier cohorts, the intertemporal properties of the Beverton–Holt model are not nearly as interesting as those of the Schaefer–Gordon model.

What is of interest for our purposes is the number of places at which environmental quality parameters can enter into the biological model. For example, environmental variables can affect either or both the number of fish available for harvest at a point in time and their individual weights. Both recruitment and natural mortality may depend upon environmental quality. The individual growth rate, α, and the upper limit on size are also likely to depend on environmental quality. An interesting and potentially fruitful area for future research is the role of environmental variables in the Beverton–Holt model and the development of expressions for the economic value of these changes.

SUMMARY

When an environmental quality variable affects the production costs of firms, the welfare value of the change is measured by changes in the surpluses of producers and consumers. Measuring these values requires a model of the market for the output. In the simplest case of single-product firms in a competitive industry, developing such a model and deriving estimates is straightforward, given knowledge of how q affects the production or cost functions, or both. In this chapter I have shown how the simple model can be extended to deal with several kinds of complications, including the cases of multiproduct firms, vertically linked markets, monopoly power, and markets with price floors such as those for some agricultural products. And I have also shown how the models for optimum management of biological resources such as forests and fisheries can be adapted for the purposes of welfare measurement, and how environmental quality parameters can be incorporated into these models.

REFERENCES

Adams, Richard M., Thomas D. Crocker, and Richard W. Katz. 1984. "Assessing the Adequacy of Natural Science Information: A Bayesian Approach," *Review of Economics and Statistics* vol. 66, no. 4, pp. 568–575.

Adams, Richard M., Scott A. Hamilton, and Bruce McCarl. 1984. *The Economic Effects of Ozone on Agriculture*. Corvallis, Ore.: U.S. Environmental Protection Agency.

Adams, Richard M., and Bruce A. McCarl. 1985. "Assessing the Benefits of Alternative Ozone Standards on Agriculture: The Role of Response Information," *Journal of Environmental Economics and Management* vol. 12, no. 3, pp. 264–276.

Anderson, James E. 1976. "The Social Cost of Input Distortions: A Comment and a Generalization," *American Economic Review* vol. 66, no. 1, pp. 235–238.

Baumol, William J., John C. Panzer, and Robert D. Willig. 1982. *Contestable Markets and the Theory of Market Structure*. New York: Harcourt, Brace, Jovanovich.

Bell, Frederick W. 1972. "Technological Externalities and Common Property Resources: An Empirical Study of the U.S. Northern Lobster Fishery," *Journal of Political Economy* vol. 80, no. 1, pp. 148–158.

Bell, Frederick W. 1989. *Application of Wetland Evaluation Theory to Florida Fisheries*. Tallahassee, Fla.: The Florida Seagrant College.

Bowes, Michael D., and John V Krutilla. 1989. *Multiple-Use Management: The Economics of Public Forestlands*. Washington, D.C.: Resources for the Future.

Clark, Colin W. 1976. *Mathematical Bioeconomics: The Optimal Management of Renewable Resources*. New York: John Wiley.

Clark, Colin W. 1985. *Bioeconomic Modelling and Fisheries Management*. New York: John Wiley.

Ellis, Gregory M., and Anthony C. Fisher. 1987. "Valuing the Environment as Input," *Journal of Environmental Management* vol. 25, no. 2, pp. 149–156.

Fisher, Anthony C. 1981. *Resource and Environmental Economics*. New York: Cambridge University Press.

Freeman, A. Myrick, III. 1979. *The Benefits of Environmental Improvement: Theory and Practice.* Baltimore, Md.: The Johns Hopkins University Press for Resources for the Future.

Freeman, A. Myrick, III. 1982. *Air and Water Pollution Control: A Benefit-Cost Assessment.* New York: John Wiley.

Freeman, A. Myrick, III. 1991. "Valuing Environmental Resources Under Alternative Management Regimes," *Ecological Economics* vol. 3, no. 3, pp. 247–256.

Freeman, A. Myrick, III, and Winston Harrington. 1990. "Measuring Welfare Values of Productivity Changes," *Southern Economics Journal* vol. 56, no. 4, pp. 892–904.

Garcia, Phillip, Bruce L. Dixon, James W. Mjelde, and Richard M. Adams. 1986. "Measuring the Benefits of Environmental Change Using a Duality Approach: The Case of Ozone and Illinois Cash Grain Farms," *Journal of Environmental Economics and Management* vol. 13, no. 1, pp. 69–80.

Gordon, H. S. 1954. "The Economic Theory of a Common Property Resource: Fisheries," *Journal of Political Economy* vol. 62, no. 2, pp. 124–142.

Hartman, Richard C. 1976. "The Harvesting Decision When a Standing Forest Has Value," *Economic Inquiry* vol. 14, no. 1, pp. 52–58.

Heck, Walter W., Richard M. Adams, William W. Cure, Allen S. Heagle, Howard E. Heggestad, Robert J. Kohut, Lance W. Kress, John O. Rawlings, and O. Clifton Taylor. 1983. "A Reassessment of Crop Loss from Ozone," *Environmental Science and Technology* vol. 17, pp. 572A-581A.

Hyde, William F. 1980. *Timber Supply, Land Allocation and Economic Efficiency.* Washington, D.C.: Resources for the Future.

Just, Richard E., and Darrell L. Hueth. 1979. "Multimarket Welfare Measurement," *American Economic Review* vol. 69, no. 5, pp. 947–954.

Just, Richard E., Darrell L. Hueth, and Andrew Schmitz. 1982. *Applied Welfare Economics and Public Policy.* Englewood Cliffs, N.J.: Prentice-Hall, Inc.

Kopp, Raymond J., and Alan J. Krupnick. 1987. "Agricultural Policy and the Benefits of Ozone Control," *American Journal of Agricultural Economics* vol. 69, no. 5, pp. 956–962.

Kopp, Raymond J., William J. Vaughan, Michael Hazilla, and Richard T. Carson. 1985. "Implications of Environmental Policy for U.S. Agriculture: The Case of Ambient Ozone Standards," *Journal of Environmental Management* vol. 20, no. 4, pp. 321–331.

Lichtenberg, Erik, and David Zilberman. 1986. "The Welfare Economics of Price Supports in U.S. Agriculture," *American Economic Review* vol. 76, no. 5, pp. 1135–1141.

Lynne, Gary D., Patricia Conroy, and Frederick J. Prochaska. 1981. "Economic Valuation of Marsh Areas for Marine Production Processes," *Journal of Environmental Economics and Management* vol. 8, no. 2, pp. 175–186.

Madariaga, Bruce. 1988. "Ambient Air Quality Standards for U.S. Agriculture: The Correct Welfare Measure Revisited," *Journal of Environmental Management* vol. 27, no. 4, pp. 421–427.

Mathtech. 1982. *Benefits Analysis of Alternative Secondary National Ambient Air Quality Standards for Sulfur Dioxide and Total Suspended Particulates, Final Analysis*. Princeton, N.J.: Mathtech, Inc.

McClelland, John D. 1991. "On Selecting Models for Fishery Management," an honors paper for the Department of Economics, Bowdoin College, Brunswick, Me.

McConnell, Kenneth E., and Ivar E. Strand. 1989. "Benefits for Commercial Fisheries When Demand and Supply Depend on Water Quality," *Journal of Environmental Economics and Management* vol. 17, no. 3, pp. 284–292.

McGartland, Albert M. 1987. "Implications of Ambient Ozone Standards for U.S. Agriculture: A Comment and Some Further Evidence," *Journal of Environmental Management* vol. 24, no. 2, pp. 139–146.

Mjelde, James W., Richard M. Adams, Bruce L. Dixon, and Phillip Garcia. 1984. "Using Farmers' Actions to Measure Crop Loss Due to Air Pollution," *Journal of the Air Pollution Control Association* vol. 34, no. 4, pp. 360–364.

National Academy of Sciences. 1983. *Pest Control for Corn and Soybeans*. Washington, D.C.: National Academy of Sciences.

Russell, Clifford S., and William J. Vaughan. 1976. *Steel Production: Processes, Products, and Residuals*. Baltimore, Md.: The Johns Hopkins University Press for Resources for the Future.

Samuelson, Paul A. 1976. "Economics of Forestry in an Evolving Society," *Economic Inquiry* vol. 14, no. 4, pp. 466–492.

Schmalensee, Richard. 1976. "Another Look at the Social Valuation of Input Price Changes," *American Economic Review* vol. 66, no. 1, pp. 239–243.

Scott, A. D. 1955. "The Fishery: The Objectives of Sole Ownership," *Journal of Political Economy* vol. 63, no. 1, pp. 116–124.

Townsend, Ralph E. 1986. "A Critique of Models of the American Lobster Fishery," *Journal of Environmental Economics and Management* vol. 13, no. 3, pp. 277–291.

10

Valuing Longevity and Health

O ne of the basic services provided by the environment is the support of human life. Changes in the life-support capacity of the environment brought about by (for example) pollution of the air or water can lead to increases in the incidence of disease, impairment of daily activities and perhaps reduced life expectancy. Human alteration of the environment can affect health through a number of channels, including:

- organic compounds, which may contaminate aquifers used as sources of drinking water;
- poorly treated sewage or septic tank leachate, which may spread disease-causing bacteria and viruses into drinking water supplies and among shellfish bound for human consumption;
- elevated levels of ozone and sulfur dioxide air pollution, which can increase the frequency or severity of asthma attacks;
- emissions from manufacturing facilities, auto body repair and painting shops, and the like, which may include carcinogens;
- depletion of the stratospheric ozone layer, which can cause increased ground-level ultraviolet radiation and the incidence of skin cancer.

The purpose of this chapter is to describe and evaluate the currently available methods and techniques for estimating the monetary values of changes in human health that are associated with environmental changes.[1] Two links must be established in estimating these

---

[1]Parts of this chapter are based on a report prepared by Maureen L. Cropper and me for the U.S. Congressional Research Service (Cropper and Freeman, 1989) and on our contribution to the book by John Braden and Charles Kolstad (Cropper and Freeman, 1991). I am indebted to Cropper for

values. The first is the link between the environmental change and the change in health status. The second is that between the change in health status and its monetary equivalent, willingness to pay or willingness to accept compensation. There are two alternative strategies for valuing environmental change that affect human health. The first strategy is to develop a comprehensive model of individual behavior and choice in which environmental quality is one of the determining variables. Such models can provide a basis for measuring willingness to pay directly as a function of the environmental change. The second strategy is to deal with the two links separately. Economic values of changes in health status or health risk would be derived first. They would then be combined with independently derived predictions of health changes or risk changes as a function of environmental change. An example of this approach would be to use measures of the value of risk reduction derived from studies of wage rates and occupational risk, combined with micro- or macro-epidemiological studies of the relationship between air pollution and mortality rates, to arrive at an economic value of air quality improvement.

In the first major section of this chapter I discuss the types of health effects to be valued. In the following two sections I develop models and measurement techniques for valuing changes in mortality as measured by the probability of dying and for morbidity.

DEFINING AND MEASURING HEALTH EFFECTS

In this chapter I follow the conventional practice in distinguishing between mortality and morbidity effects. In the former case the primary end point of concern is death, while in the latter case the focus is on nonfatal illness or a set of symptoms. Mortality refers to a well-defined event—death—which for ceremonial and legal reasons almost always is noted and made part of an official record. The most basic measure of mortality is the crude mortality rate of the population. This is defined as the ratio of the total number of deaths from all causes during a period of time to the average number of persons in that population. The crude mortality rate can also be interpreted as the probability that an individual member of the population will die during a given period. For mortality, the measure of a change in health status is the change in the probability of dying for members of the population.

---

her contribution to this chapter, especially the portions dealing with models of morbidity benefits and the life-cycle models for valuing changes in mortality risk.

For purposes of valuation, it may be important to disaggregate changes in mortality rates by age, sex, socioeconomic status, and cause of death. This is because individuals' valuations of changes in their probability of dying are likely to depend in part on age and socioeconomic status and, perhaps, on cause of death. For example, people may place a higher value on reducing the risk of cancer death with its attendant pain and suffering than on an equal reduction in the risk of death from a plane crash. Age- and sex-specific mortality rates are readily available from official sources in the United States. But mortality rates are not typically calculated by any measure of socioeconomic status.

Morbidity is defined by the U.S. Public Health Service as ''a departure from a state of physical or mental well-being, resulting from disease or injury, of which the affected individual is aware'' (Peterson, 1975, p. 242). Morbidity can be classified in a variety of ways, among them duration of the condition (chronic or acute), degree of impairment of activity, or type of symptom. An episode of acute morbidity would last only a matter of days and would have a well-defined beginning and end. Chronic morbidity refers to cases of a longer-term illness of indefinite duration.

Published data are available for several categories of degrees of activity impairment. ''Restricted activity days'' are those on which a person is able to undertake some, but not all, normal activities; ''bed disability days'' are those in which a person is confined to bed, either at home or in an institution, for all or most of a day; ''work loss days'' are those on which a person is unable to engage in ordinary gainful employment (U.S. Department of Health, Education and Welfare, 1964).

These measures of morbidity reflect responses to ill health rather than the health condition itself. Whether a given clinical manifestation of ill health results in any restriction on activity, bed disability, or work loss depends upon a number of socioeconomic variables, such as employment and labor-force status, nonlabor sources of income, the presence of other income-earners in the household, and so forth.[2] This distinction between illness and the behavioral response to it may be important if, for example, restricted activity days are used in an epidemiological study of the effect of a pollutant on some measure of health.

Some studies also measure morbidity by ''symptom days,'' that is, by the occurrence of specific symptoms such as an asthma attack, a

---

[2]See Berkowitz and Johnson (1974), who show how other factors besides health affect disability status as reflected in labor force participation.

headache, a cough, throat irritation, or diarrhea. The choice of symptoms for defining and measuring morbidity has important implications for both the economic valuation of health effects and for biomedical research strategies into the effects of environmental quality on health. The economic perspective on health focuses attention on effects that people are aware of and wish to avoid, that is, effects that would reduce their utility. Yet much biomedical clinical research focuses on effects of questionable significance to individuals, and measures health effects that are difficult to relate to individual perceptions and behavior.

The question of significance is tied to a legal and policy issue that vexes the U.S. Environmental Protection Agency: What constitutes an adverse health effect? The Clean Air Act of 1970 calls for setting air quality standards so as to protect individuals from *adverse* health effects. The question of whether or not an effect is adverse arises (for example) when clinical studies reveal that exposure to an air pollutant under controlled conditions leads to detectable changes in organ structure or function without necessarily causing pain, impeding people's activities, or reducing life expectancy. Air chamber experiments with heavily exercising young adults, for instance, show that exposures to ozone at concentrations near the current primary air quality standard can lead to reductions in volume of forced expiration, which is a clinical measure of lung function (U.S. Environmental Protection Agency, 1987). Are these changes adverse?

From an economic perspective based on the willingness-to-pay definition of value, the answer depends on whether the changes are perceived by the individual and whether the individual reveals or expresses a willingness to pay to avoid the effect, or requires compensation to experience it. For example, the subjects in air chamber tests could be asked whether they could tell the difference between the high and low ozone concentrations and, if they could, whether they would require higher compensation to go through the higher exposure-level test again. If the answer to either question were no, and if the physical effect were reversible, then the economic benefits of avoiding the higher level of exposure and its attendant decrement in lung function would be zero. From an economic perspective, the effect would not be considered adverse. If the effect were irreversible and cumulative, then the higher concentration of the pollutant might cause an increase in the risk of chronic illness, which would be considered adverse. The methods developed in this chapter apply only to health effects that are adverse in this economic sense.

Measures of morbidity must also take into account the fact that, unlike mortality, morbidity is not a discrete event but a process involving time. Cases observed during a period of time may fall into one of four categories: (1) onset of morbidity occurs before the period and morbidity terminates by either recovery or death during the period; (2) onset of morbidity occurs before the period and morbidity terminates after the period; (3) onset of morbidity occurs during the period and morbidity terminates during the period; and (4) onset of morbidity occurs during the period and morbidity terminates after the period. The prevalence rate encompasses all four categories. It is defined as the total number of cases in the period as a percentage of the average number of persons in the population during the period. The incidence rate covers only the third and fourth categories. It is defined as the number of new cases during the period as a percentage of the average number of persons in the population. Incidence rate data would be more appropriate for investigating causal relationships between changes in exposure and changes in health status. Because willingness to pay to reduce morbidity is likely to depend on the length as well as the number of cases, the prevalence rate and measures incorporating data on duration would be more appropriate for analyzing the social costs of morbidity.

In addition to the mortality and morbidity effects discussed here, there is now increasing evidence that chemical contaminants can produced subtle physical and mental changes in people. For example, relatively low levels of lead in blood have been implicated in decrements to IQ in children, in low birth weight, and in lower body growth rates and, presumably, height at maturity (U.S. Environmental Protection Agency, 1986a, 1986b). To the extent that subtle effects such as low birth weight are associated with other adverse health effects and illness, they pose no special methodological problems from an economic perspective. The models and techniques discussed in this chapter can be used to estimate the monetary values of avoiding these adverse health effects. But more difficult problems arise in estimating monetary values for avoiding more subtle effects such as decrements in IQ and reduced growth rates. Some efforts have been made to estimate the costs of special education and lost productivity of individuals with impaired intellectual development (Mathtech, Inc., 1986). But as is established below, from a conceptual perspective these cost-of-illness and lost-wage measures are only partial and incomplete measures of the total willingness to pay to avoid adverse health effects. Surely there is a loss of utility or well-being associated with intellectual impairment; but basing estimates of the value of this loss on an individual's willingness to pay is problematic.

VALUING REDUCED MORTALITY RISKS

Because some forms of pollution may increase mortality, economists have had to confront the question of the economic value of life and life-saving. This is perhaps the most difficult and controversial aspect of valuing the health effects associated with environmental changes. To some, the idea of putting a price on human life is insensitive, crass, or even inhuman. But this point of view does not give sufficient weight to the fact that individuals in their day-to-day actions and governments in their decisions about social policy do in fact make trade-offs between changes in the probability of death and other goods that have monetary values. These trade-offs make it possible to infer the implicit prices that people are attaching to changes in the probabilities of their deaths.

The subject of this section is the economic theory of value as it is applied to the decisions made by individuals and governments concerning the trade-off between mortality risk and money. Here I present a brief description of the economic basis for assigning monetary values to certain kinds of life-saving activity. I also identify and briefly discuss some of the important economic and ethical issues that have been identified in the literature. I then turn to a more detailed discussion of models and methods for measuring willingness to pay for reduced mortality risk.

Willingness to Pay

In keeping with the assumption that individuals' preferences provide a valid basis for making judgments concerning changes in their economic welfare, reductions in the probability of death due to accident or illness should be valued according to what an individual is willing to pay to achieve the reductions or is willing to accept in compensation to forgo the reductions.[3] This presupposes that individuals treat longevity more or less like any other good rather than as a hierarchical value. Perhaps surprisingly, this seems to be a reasonable assumption, at least for small changes in mortality risk. Individuals in a variety of situations act as if their preference functions include life expectancy or the probability of survival as arguments. In their daily lives, they make a variety of choices that involve trading off changes

---

[3]Schelling (1968) appears to have been the first to propose applying willingness-to-pay concepts in this area. Other early contributors include Mishan (1971) and Jones-Lee (1974, 1976). Bailey (1980) provides a useful review of the earlier literature.

in the risk of death for other economic goods whose values can be measured in monetary terms. For example, some people travel to work in cars rather than by bus or by walking because of the increased convenience and lower travel time of cars, even though these people thus increase their risk of dying prematurely. Also, some people accept jobs with known higher risks of accidental death because the jobs pay higher wages.[4] In such cases, people must perceive themselves to have been made better off by the alternatives they have chosen; otherwise they would have chosen some other alternative. When what is being given up (or gained) can be measured in dollars, the individual's willingness to pay (or compensation required) is revealed by these choices. These choices are the basis of measures of the economic value of reductions in the risk of death.

As the ordinary, everyday nature of these examples should make clear, the economic question being dealt with here is not about how much an individual would be willing to pay to avoid his or her certain death or how much compensation that individual would require to accept that death. In this respect, the term "value of life" is an unfortunate phrase that does not reflect the true nature of the question at hand. Most people would be willing to pay their total wealth to avoid certain death; and there is probably no finite sum of money that could compensate an individual for the sure loss of life. Rather, the economic question is about how much the individual would be willing to pay to achieve a small reduction in the probability of death during a given period or how much compensation that individual would require to accept a small increase in that probability. This is an appropriate question to investigate because most environmental regulatory programs—even those aimed at fairly serious toxins—result in relatively small changes in individuals' mortality risks.

For this kind of situation, the theory of individual choice under uncertainty described in chapter 8 provides a useful analytical framework. Individuals do not know which of several alternative states of the world will exist at some specified date in the future, but they must make certain choices affecting their future utility before the future is revealed. Individuals are assumed to assign probabilities to alternative states of the world and to make their choices so as to maximize their expected utility. One aspect of uncertainty about the future is the date of one's death, or to put it differently, whether one will

---

[4]Evidence that risks to life do affect decisions in this way comes from a variety of empirical studies of behavior. See Peltzman (1975) for evidence about risk and driving. Evidence on the risk–wage trade-off is reviewed in chapter 12.

survive or succumb to some hazard during some time interval. Individuals can affect the probabilities of accidental death during present and future periods by the choices they make. The value of a reduction in risk to an individual is the amount of money that person would be willing to pay to achieve it, other things being equal.

An individual's willingness to pay for changes in the probability of his or her death can be translated into a more convenient figure for evaluating policies that reduce the risk of death, namely, the value of a *statistical life* or the value of a *statistical death avoided*. Suppose that there were a thousand people, each of whom has a probability of .004 of dying during the next year. Suppose a pollution control policy would reduce that probability to .003, a change of .001. Further suppose that each individual in that group expresses a willingness to pay $1,000 for this policy. Since the policy would affect all of the people equally, it is a form of collective good for the group. The total willingness to pay of the group is $1 million. If the policy is adopted, there will be on average one less death during the year. The total willingness to pay for the policy that results in one less death is $1 million. This is the value of statistical life.

As discussed thus far, the willingness-to-pay approach focuses on the individualistic dimensions of human behavior, that is, an individual's willingness to pay to increase his or her own life expectancy. But there is nothing in the logic of the willingness-to-pay approach to prevent consideration of the effects of kinship and friendship, that is, an individual's willingness to pay to reduce the probability of death of close relatives and friends.[5] Again, we should make it clear that the question is not the willingness to pay to prevent an imminent or highly probable death (that is, the ransom for the kidnapped child or the search for the lost hiker or boater) but the willingness to pay for a small reduction in the probability of death for the group of which the friend or relative is a part.

Ex Ante versus Ex Post Perspectives: The Ethical Issues

One way of characterizing the economic approach is by saying that it avoids the issue of valuing life, per se, by recognizing that what people actually "buy" and "sell" through their choices and trade-offs is not life versus death but small changes in the probability of dying. Another way of characterizing the economic approach is by saying that the economic value is derived by focusing on choices ex ante, that is, before

---

[5]See Mishan (1971) for further discussion of the external effects of the risk of death.

the uncertainty about the individual's death during a specified period of time is resolved. But at some point in time, the uncertainty is resolved. Each individual will know if he is to die now or live a while longer. From this ex post perspective, those who would die would be willing to pay their total wealth to change the outcome or would require an infinite compensation to accept it. Critics of the economic approach to valuation, such as Broome (1978), have argued that this difference in perspective can have no ethical or moral significance and that therefore the willingness-to-pay or compensation measures based on the ex ante perspective are morally unacceptable.

One defense of the economic perspective is based on the observation that people appear to be willing to make ex ante trade-offs involving risks of death. If people are rational and if their preferences are taken to be the basis of measures of economic value, then their willingness to consent to ex ante trade-offs must have some ethical significance. Furthermore, for many of the public policy issues where value of risk reduction information might be used, it will never be known ex post whose deaths were caused by failure to adopt a policy or whose lives were prolonged by a policy to reduce risks of death. For example, suppose a proposed air pollution control policy is predicted to reduce cancer mortality in the population by 100. If the policy is adopted, the identities of the 100 people whose lives were prolonged will never be known. If the policy is not adopted, it can never be known which 100 of the 400 thousand or so cancer deaths in any one year were "caused" by the failure to adopt the policy. Economists argue that it is consent and the veil of ignorance about who dies and who is saved that legitimize the ex ante perspective and its focus on the value of changes in risks rather than on the values of lives and deaths.[6]

An Alternative Concept of Value: Human Capital

There is an alternative to using willingness to pay for valuing lives or prolonging lives by public policies. It is based on the assumption that the value of an individual alive is what he or she produces, and that productivity is accurately measured by earnings from labor. Earnings before taxes are measured in order to reflect the government's, and therefore society's, interest in each individual's total productivity. With the death of the individual, that output is lost. This approach

---

[6]For further discussion of these and related issues, see Broome (1978, 1979, 1982, 1985, 1987), Buchanan and Faith (1979), Jones-Lee (1979, 1982, 1987), Williams (1979), Mishan (1981), Ulph (1982), Thaler (1982), and Linnerooth (1982).

has a long tradition; in fact, Landefeld and Seskin (1982) trace the idea back almost three hundred years. It has been the basis of some widely cited estimates of the benefits of air pollution control; see, for example, Lave and Seskin (1971, 1977).

The human capital approach is fundamentally at odds with the individualistic perspective of welfare economics and the theory of value. By in effect asking what the individual is worth to society, the human capital approach ignores the individual's own well-being, preferences, and willingness to pay. And it defines the social worth of the individual in a very narrow way, that is, as the individual's market productivity, thereby ignoring the value of that person's health and well-being to loved ones. Although I think that the human capital approach is inappropriate for valuing reductions in the risk of death, I will describe the approach in more detail and discuss its implementation, since human capital measures are often used in the literature.[7]

According to the human capital approach, the value of preventing the death of an individual who is presently of age t is the discounted present value of that individual's earnings over the remainder of his expected life:

$$\text{Value} = \sum_{i=1}^{T-t} \frac{\pi_{t+i} \cdot E_{t+i}}{(1 + r)^i} \qquad (1)$$

where π_{t+i} = probability of the individual surviving from age t to age $t + i$

E_{t+i} = expected earnings of the individual at age $t + i$

r = discount rate

T = age at retirement from the labor force

Several issues must be addressed concerning the implementation of the human capital approach. One question is: Should an individual's productivity be measured net of or inclusive of that person's own consumption? Netting out own consumption leaves a measure of the individual's worth as a producing asset to the rest of society. But this measure is the antithesis of the individualistic premise of conventional welfare economics.

---

[7]The human capital approach does make economic sense as the starting point for determining compensation for dependents in wrongful death settlements.

Another question concerns the role of nonmarket production in the measure of productivity and value. The omission of nonmarket productivity is particularly serious and troublesome in the case of housewives and nonworking parents. Some studies have attempted to correct for this omission by imputing earnings equal to the wages of domestic servants. Others have argued that the average earnings of employed females provide a measure of the opportunity cost of working at home for women. If women are rational, the value of home production must be equal to or greater than this opportunity cost. But even these adjustments do not capture the nonmarket productivity of other members of the household.

A third issue is what discount rate to use in calculating the present value. The human capital value of children and young adults will be particularly sensitive to the choice of a discount rate. Rates of 6 percent (Cooper and Rice, 1976) to 10 percent (Landefeld and Seskin, 1982) have been used in the literature. But these rates are quite high relative to estimates of the real rates of return on riskless financial assets discussed in chapter 7.

Some of the implications of the human capital approach are unsettling. Because of discounting and the time lag before children become productive participants in the economy, the human capital approach places a much lower value on saving children's lives compared with saving the lives of adults in their peak earnings years. Because of earnings differences by sex and race, the human capital approach values saving the lives of women and nonwhites less than it does saving the lives of adult white males. And the human capital approach assigns zero value to persons who are retired or totally disabled by illness or handicap. Human capital values for men of different ages, using alternative discount rates, are shown in table 10-1. According to these calculations, men in the age group 20 to 24 are "worth more" than males in other age groups. The table also shows the dramatic effect of discounting on the human capital value of children. The human capital values of women and nonwhite males would be lower than those shown in the table.

Even if it is accepted that the human capital approach is flawed in principle, one could ask if it might be a reasonable approximation to the value of statistical life based on willingness to pay. Both theoretical reasoning and empirical evidence suggest that human capital measures are a poor proxy for the desired willingness-to-pay measure of value for small changes in the risk of death.[8] While it is likely that an

---

[8]The theoretical reasoning is presented later in this chapter in the section

Table 10-1. Present Value of Future Earnings of Males by Selected Age Groups (1987 dollars)

Age group (years)	Real discount rate		
	2.5 percent	6 percent	10 percent
1 to 4	761,047	205,101	59,859
20 to 24	967,221	534,799	320,114
40 to 44	625,508	454,972	338,232
65 to 69	47,506	40,886	35,304

Note: Figures are based on the present value of both expected lifetime earnings and housekeeping services at 1977 wage rates and life expectancies and an annual increase in labor productivity of 1 percent.
Source: Landefeld and Seskin (1982).

individual's income and the consumption it allows are positively related to the utility that person derives from his or her own life, the human capital method does not reflect the probabilistic nature of death and death avoidance and individuals' differing attitudes toward risk. By definition an individual with no financial wealth could pay no more than the present value of his expected earnings stream to avoid certain death. But his statistical value of life based on willingness to pay for small probability changes could be several times his discounted earnings stream.

Individual Choice and Willingness to Pay

Accepting that the human capital approach does not provide a valid measure of reducing the risk of death, I turn now to describing some of the results generated by economic models of individual choice under uncertainty. The purposes of analyzing the models here include deriving hypotheses about the determinants of an individual's willingness to pay for a reduction in the risk of mortality, and examining the relationships between these predicted willingness-to-pay measures and other economic variables such as the individual's earnings and purchases of life insurance.

These models are based on the assumption that, as discussed earlier, individuals make choices that affect their risk of death so as to maximize the mathematical expectation of utility. This means that a key assumption of the models is that individuals know the relevant probabilities of dying and know how they are changed by the choices they

---

on Life-Cycle Models of Willingness to Pay. The empirical evidence comes from comparing the human capital values in table 10-1 with the values of statistical life obtained from hedonic wage studies; these are described in chapter 12.

make. Since individuals must choose alternatives before the uncertainty is resolved, expected utility is an ex ante concept. These models are used to derive the willingness to pay (WTP) for a reduction in the probability of death, defined as the maximum sum of money that can be taken from the individual ex ante without leading to a reduction in that person's expected utility. This is a compensating surplus measure of welfare change. Alternatively, these models could be used to derive the willingness-to-accept compensation for an increase in the probability of dying, defined as the sum of money that just compensates for the greater risk by increasing consumption sufficiently to equalize the expected utilities of the two alternatives.

Two aspects of the choice problem are of interest in this analysis: uncertainty and time. The uncertainty concerns the unknown timing of the individual's death and the effects of choices of occupation and consumption activity on the probabilities of surviving to any given date. The time aspect concerns the effects of choices made at any given time on the probabilities and utilities associated with future periods. This is of particular relevance for environmental problems, since many of the most important environmental policy issues are characterized by a substantial interval of time between exposure and the perceived effects on health. Most models of intertemporal choice have focused on identifying the optimal consumption stream given an income stream and opportunities for borrowing and lending. Our concern here is with choices made in the present that affect the probabilities of survival at future dates. The problem is to identify the marginal willingness to pay now for increases in the probabilities of survival during some future period.

Static Models

I begin with a simple one-period choice model to avoid the complications of intertemporal choice, identify the marginal willingness to pay for increases in survival probability, and then extend the model to the intertemporal case to see whether estimates of the value of statistical life based on intratemporal models can be applied to the intertemporal case.

Assume that an individual derives utility from the consumption of a composite good, X, with a price normalized to one. The initial endowment of X and the probability of surviving to enjoy its consumption, π, are both given to the individual. Let X^0 and π^0 represent this initial endowment. Expected utility is

$$E[u] = \pi^0 \cdot u(X^0) \tag{2}$$

Without loss of generality, I assume that the utility associated with death is zero. We can obtain an expression for the individual's marginal willingness to pay (*MWTP*) for a reduction in π^0 by taking the total differential of equation (2) and setting it equal to zero:

$$MWTP = \frac{u(X^0)}{\pi^0 \cdot \partial u / \partial X^0} \tag{3}$$

As equation (3) shows, one of the fundamental results of models of this sort is that if individuals are not free to adjust their survival probabilities in the market, their marginal willingness to pay for enhanced survival will depend on their initial survival situation. Other things being equal, the lower the initial survival probability is, the higher will be the marginal willingness to pay. Another result is that diminishing marginal utility of X implies a higher marginal willingness to pay for π with higher initial endowments of X, other things being equal.[9]

This model, while providing useful insights, does not suggest a method for estimating WTP other than by direct questioning. To use indirect market methods we must link WTP to voluntary risk-taking behavior, that is, π must be made endogenous. Now suppose that the individual has the opportunity to rearrange his consumption and survival position through exchange, for example by giving up some X in order to improve his chances of surviving to enjoy the remainder. Let p_π represent the price at which consumption can be exchanged for enhanced survival probability, or vice versa. For the simple model, I assume that there are no opportunities for insurance.[10] The individual chooses X and π so as to maximize expected utility, subject to the budget constraint

$$\text{Max: } E[u] = \pi \cdot u(X) + \lambda\{X^0 - X + p_\pi \cdot (\pi^0 - \pi)\}$$

The first-order conditions for a maximum of expected utility can be combined to obtain

$$\frac{u(X)}{\pi \cdot \partial u / \partial X} = p_\pi \tag{4}$$

---

[9] These propositions can be verified by taking the derivatives of equation (3) with respect to π and X. See, for example, Thaler and Gould (1982).

[10] Comprehensive models encompassing bequest motivation (utility derived from unconsumed X remaining at death) and insurance behavior can be developed. For examples, see Jones-Lee (1976), Conley (1976), Thaler and Rosen (1976), and Cropper and Sussman (1988).

This expression requires that the individual equate her marginal willingness to pay for enhanced survival (the left-hand side of the expression)[11] with the given price of enhanced survival.

The utility-maximizing individual may choose either to forgo consumption in order to enhance her survival probability or to take on increased risk (lower π) in order to enhance consumption opportunities. The actual choices depend on the initial endowment, X^0 and π^0, the price of π, and the individual's preferences. Whatever the final outcome, equation (4) allows us to infer the individual's marginal willingness to pay for enhanced survival, since this value will be equated to the observable price of changes in π.

If the risk in question is an environmental risk, it may not be feasible to have a market for reductions in π. But if there are other risks, all that is required is that the individual be able to affect the level of one of these risks. Suppose now that there are three sources of risk of death, one exogenous, one job-related, and one related to the level of consumption of a private good x_i, an element in the vector of market goods X. Assume that in addition to affecting the risk of death, this good conveys utility directly. The corresponding conditional probabilities of death are denoted ρ_e, ρ_j, and $\rho_x(x_i)$. Assuming that these causes of death are independent, the probability of surviving the current period is the product of the probabilities that the individual does not die from each of the three causes (Sussman, 1984); that is,

$$1 - \rho = (1 - \rho_e)(1 - \rho_j)(1 - \rho_x(x_i))$$

Suppose now that all types of jobs are alike in every respect except for the risk of accidental death and that more risky jobs have higher wage rates—in other words, the individual receives an annual wage:

$$M(\rho_j) \quad \text{where} \quad \partial M/\partial \rho_j > 0$$

In general, good x_i could either increase or decrease the risk of death. Some goods such as skydiving and cigarettes increase risk, while others such as smoke detectors decrease risk. Our interest is in risk-reducing goods, so that

$$\partial \rho_x/\partial x_i < 0$$

---

[11]This is similar to the expression for marginal willingness to pay for reductions in risk derived in chapter 8, since here utility and the marginal utility of consumption are zero in the event of death.

If I is exogenous income, then $M^* = I + M(\rho_j)$. Expected utility is given by

$$E(u) = (1 - \rho_e)(1 - \rho_j)[1 - \rho_x(x)] \cdot u[M^*(\rho_j) - p_x \cdot x_i, X] \quad (5)$$

By total differentiation of equation (5), willingness to pay for a marginal change in exogenous risk of death, $dI/d\rho_e$, is given by

$$w_{\rho_e} = (1 - \rho_j)(1 - \rho_x) \frac{u(\cdot)}{(1 - \rho) \dfrac{\partial u}{\partial M^*}} \quad (6)$$

This is the value of the utility lost if the individual dies ($u(\cdot)$), converted to dollars by dividing by the expected marginal utility of income ($(1 - \rho)\partial u/\partial M^*$) and multiplying by the probability that the individual does not die due to other causes ($(1 - \rho_j)(1 - \rho_x)$).

A question of interest is whether $MWTP$ can be estimated by observing risk-taking behavior in consumption or in the labor market. To answer this, assume that the individual chooses x_i and a job with its associated risk ρ_j, so as to maximize equation (5) and derive the first-order conditions for choice of job risk and consumption of x_i. They are:

$$\frac{\partial M}{\partial \rho_j} = (1 - \rho_e)(1 - \rho_x) \frac{u(\cdot)}{(1 - \rho) \dfrac{\partial u}{\partial M^*}} \quad (7)$$

and

$$-\frac{p_x}{\partial \rho_x / \partial x_i} = (1 - \rho_e)(1 - \rho_j) \frac{u(\cdot)}{(1 - \rho)\left[\dfrac{\partial u}{\partial M^*} - \dfrac{\partial u / \partial x_i}{p_x}\right]} \quad (8)$$

Equation (7) implies that the individual equates the marginal wage income forgone by moving to a safer job with the marginal benefit of a reduction in job risk. The latter is almost identical to the value of an exogenous risk change—equation (6)—except that the probability of not dying due to other causes is now $(1 - \rho_e)(1 - \rho_x)$ instead of $(1 - \rho_j)(1 - \rho_x)$. If

$$(1 - \rho_e) \approx (1 - \rho_j)$$

then the marginal price of risk reduction can be used as an estimate of the willingness to pay for a change in exogenous risk. This marginal price, which is an implicit price, can be estimated with a hedonic wage model; that method is described in chapter 12.

The first-order condition for the choice of x_i differs from the expression for marginal willingness to pay in part because it includes a term for the marginal utility of a dollar spent on x_i. This term is present because of the assumption that x_i conveys utility directly as well as through its effect on risk. Only if $\partial u / \partial x_i = 0$ can equation (8) be used as an approximation for marginal willingness to pay. Then, if $(1 - \rho_x) \simeq (1 - \rho_e)$, the left-hand side of (8), which is the marginal cost of reducing risk through purchasing x_i, can be used to approximate MWTP for an exogenous risk change.

Smoke detectors and seat belts are examples of goods whose primary purpose is to produce safety, that is, to reduce the risk of death for those who purchase them. Data on the purchase and utilization of these goods have been used to estimate the values of reducing risk of death (Dardis, 1980, and Blomquist, 1979). In order to use data on ''safety'' goods to estimate the value of risk reduction, one must also establish that the good's contribution to safety is known. What governs each individual's purchase decision is that person's perception of the risk-reduction capability of the good. In the absence of adequate information, individuals' perceptions might vary substantially and be difficult to observe. This would make the use of safety-good purchases for estimating values of risk reduction problematic.

The model presented above is based on the assumption that the safety good is divisible. But this assumption is often invalid. Some of the goods used in actual studies—for example, the smoke detector— are indivisible, that is, their purchase involves a 0–1 decision. The good is purchased if its marginal benefit is equal to or greater than its marginal cost. The equality of willingness to pay and price occurs only for the marginal purchaser of the good. Thus when Dardis (1980) divided the annual cost of operating a smoke detector by the corresponding increase in survival probability, she estimated MWTP for the marginal purchaser, not for the average purchaser.

In order to estimate an average value of MWTP we must have data on the cost of the safety good and on its effect in reducing risk of death for a cross section of individuals. If marginal cost and marginal risk reduction vary across individuals in the sample, we can estimate the average value of MWTP by using a logit or probit model of discrete choice, as described in chapter 4. Blomquist (1979) discusses this procedure in his study of seat belt use, but is unable to implement it due to lack of data. But if there is little variation in the price of the good across the sample, or if the proportion of the sample purchasing the good is very high, then little can be learned from applying the discrete choice model.

Safety could also be one of the characteristics of a differentiated product like an automobile. Different automobile models have measurable differences in accident rates (probability) and the severity of injury. If these differences are systematically related to the prices of different models of automobiles, then the hedonic price model described in chapters 4 and 8 can be applied to estimate individuals' willingness to pay for reductions in the risk of accident or death. For an example of this approach, see Atkinson and Halvorsen (1990).

Perhaps the most important conclusion to be drawn from this review of static models of individual choice and willingness to pay is that each person is likely to attach a different value to a small reduction in the probability of dying because of differences in underlying preferences, degree of risk aversion, wealth, the current level of risk exposure, age, the number of dependents, and, perhaps, the quality of life-years expected to be gained from the reduced risk. This conclusion must be kept in mind when interpreting and using the results of empirical estimates of willingness to pay that are based on averages of groups of perhaps quite heterogeneous people.

A second conclusion is that in the case of multiple risks of death where the individual can "purchase" reductions in some component of risk, the observed price or marginal cost of reducing that component of risk can be taken as a close approximation of the individual's willingness to pay for reductions in other components of risk, provided that the safety good or safer job does not also convey utility directly.

Introducing Time

The static models discussed above are relevant to questions such as transportation and occupational safety in which the individual's actions today affect the probabilities of dying today. But many of the important environmental and occupational health questions involve actions taken today whose effects on the probability of dying are realized only at some time, perhaps fifty years, in the future. How much would an individual be willing to pay now to control current pollution when the effects of improved health might be realized only twenty to fifty years in the future? The next step in our analysis is to develop a simple multiperiod model that allows us to investigate how willingness to pay now for a reduction in the probability of death is influenced by the time period to which the probability applies. To focus on the intertemporal aspect of the problem, assume only one cause of death, and one consumption activity.

As an extension of the model of intertemporal choice developed in chapter 7, assume that individuals maximize their expected lifetime utility u^*, where u^* is an additively separable function of the consumption stream:

$$E[u^*] = \sum_{t=1}^{T} \pi_t \cdot D^{t-1} \cdot u(X_t) \qquad (9)$$

where $D = 1/(1 + d)$

$d \equiv$ subjective rate of discount or own time preference

$\pi_t \equiv$ probability of surviving from period one through period t, that is, the probability of living for at least t years from now (year one), with

$$\pi_t = \prod_{i=1}^{t} \pi_i$$

$$= \prod_{i=1}^{t} (1 - \rho_i)$$

The term ρ_i is a conditional probability of dying during year i, that is, the probability of dying in year i, given that the individual has survived to the beginning of year i. Similarly, π_i is the probability of surviving the year given being alive at the beginning of the year. For now, assume that the stream of consumption X_t is given exogenously and cannot be altered by borrowing or lending.

The form of the intertemporal utility function deserves a brief comment. As explained in chapter 7, the assumption that individuals maximize discounted utilities does not necessarily impose any restrictions on the nature of individuals' preferences regarding present versus future consumption. The subjective rate of discount d could be positive, negative, or zero. A variety of types of preferences and behavior can be encompassed in this model, depending on the value of d.[12] From equation (9) the marginal rate of substitution (MRS)

---

[12]The marginal rate of substitution (MRS) between present and future consumption is

$$MRS_{X_j X_i} = D^{(j-i)} \frac{\partial u / \partial X_j}{\partial u / \partial X_i}$$

Since in equilibrium the MRS is equated with the market discount factor, $1/(1+r)^{j-i}$, borrowing and lending behavior depends on market interest rates, present and future income levels, and the rate at which the marginal utility of income is diminishing, as well as time preference.

between present consumption and present mortality reduction is

$$MRS_{\pi_1 X_1} = \frac{u(X_1) + \sum_{t=2}^{T} \frac{\pi_t}{\pi_1} D^{t-1} \cdot u(X_t)}{\pi_1 \cdot \partial u / \partial X_1} \tag{10}$$

The second term in the numerator shows the dependence of willingness to pay to reduce present risk on life expectancy and the associated stream of expected future utilities.[13]

Suppose now that the marginal rate of substitution between present mortality reduction and present consumption is known—from a study of the demand for occupational safety, for example. Can this marginal rate of substitution be used to estimate the demand for reductions in future mortality that might be obtained through an environmental or occupational health program? The marginal rate of substitution between a mortality reduction in future period t' and present consumption is

$$MRS_{\pi_{t'} X_1} = \frac{\sum_{t=t'}^{T} \frac{\pi_t}{\pi_{t'}} \cdot D^{t-1} \cdot u(X_t)}{\pi_1 \cdot \partial u / \partial X_1} \tag{11}$$

The two marginal rates of substitution might differ for any of three reasons. First, equation (10) could be greater than (11) because it includes a term for first-period utility, $u(X_1)$. Also, the more distant that period t' is from the present, the smaller is the stream of future periods being summed in (11). However, equation (11) could be greater than (10) if $\pi_{t'}$ is sufficiently smaller than π_1. Thus marginal rates of substitution estimated from one type of probability choice problem—for example, the single-period problem—cannot in general be used as predictors of the marginal rates of substitution for other types of probability choice problems with different intertemporal dimensions.

It should be noted that in this model the individual takes the intertemporal pattern of consumption opportunities as given. There is no borrowing or lending. Cropper and Sussman (1990) have extended this model by incorporating borrowing and lending at a riskless interest rate. This allows individuals to adjust their consumption

---

[13]Moore and Viscusi (1988) have empirically estimated a model relating wage differences to differences in the expectation of discounted life-years. Their results suggest that workers do discount future life-years.

streams over time. Measures of willingness to pay will be different when this opportunity is available to people. To see this, we turn now to a full life-cycle model of intertemporal choice.

Life-Cycle Models of Willingness to Pay

Several authors, including Usher (1973), Conley (1976), and Cropper and Sussman (1990), have used a life-cycle consumption/saving model with uncertain lifetime to analyze an individual's willingness to pay at age j for a change in the conditional probability of dying at age t, $t \geq j$. In this section, I follow the approach of Cropper and Sussman. This model can be used to examine:

1. the relationship between WTP for a change in current probability of death and age;

2. the relationship between the present value of expected lifetime earnings (the human capital measure) and WTP;

3. the relationship between WTP and the latency of the risk, that is, the interval between the exposure to the risk (a carcinogen, for example) and its manifestation (death due to cancer);

4. the relationship between the willingness to pay in advance to reduce a given risk and the willingness to pay at the time the risk is experienced (a discounting question).

I will show, among other things, that WTP will generally decrease with age; that under plausible circumstances, the present value of lifetime earnings will be less than WTP; that latency reduces WTP; and that the willingness to pay at year one for a reduction in risk in year t' is equal to the willingness to pay in year t' for that probability change discounted back to year one by a discount factor which in general will be different from the market interest rate.

THE BASIC LIFE-CYCLE MODEL

In the life-cycle model, an individual of any given age has a probability distribution over the date of his or her death. Let j denote the individual's current age (say 25); and let $\rho_{j,t}$ be the probability that the individual dies at age t (say at age 40, just before that person's 41st birthday)—that is, the person lives exactly $t - j$ more years. Since the $\{\rho_{j,t}\}$ constitutes a probability distribution, it must be true that

$$\rho_{j,t} \geq 0, t = j, j + 1, \ldots, T, \text{ and that}$$

$$\sum_{t=j}^{T} \rho_{j,t} = 1$$

where T is the maximum attainable age. The probability that the individual survives to his tth birthday, given that he is alive at age j, is $\pi_{j,t}$, which also is the probability that he dies at $t+1$ or later. Formally,

$$\pi_{j,t} = \sum_{s=t+1}^{T} \rho_{j,s}$$

Let δ_t be the probability of dying at age t conditional on being alive at the beginning of that year. Thus the conditional probability of surviving that year is $1 - \delta_t$. This term can also be derived from the survival probabilities

$$1 - \delta_t = \pi_{j,t+1}/\pi_{j,t}$$

Expected lifetime utility at age j is the sum of the utility of living exactly $t - j$ more years times the probability of doing so. As in the preceding section, assume that utility is additively separable. Also assume that there is no bequest motive. Then we can write expected lifetime utility at age j as:

$$V_j = E[u] = \sum_{t=j}^{T} \pi_{j,t} \cdot D^{t-j} \cdot u(X_t) \qquad (12)$$

The utility function for each period, $u(X_t)$, is assumed to be increasing in X_t, strictly concave, and bounded from below.

Two points about this formulation of the problem should be emphasized. First, the model is based on the assumption that the utility of living depends only on consumption and not on length of life per se. The concavity of the utility function implies that it is always desirable to spread a given amount of consumption over a longer time span. Thus lifetime utility is an increasing function of life expectancy. But this is only because of the effect on consumption, not because of the value of being alive per se. I return to this point below. Second, this model treats survival probabilities as exogenous to the individual. To keep things simple, I do not attempt to introduce opportunities for the individual to alter risk levels into this model; see Conley (1976) and Viscusi and Moore (1989) for examples of intertemporal models incorporating this additional element of choice.

The individual has to choose a time pattern of consumption, given initial wealth W_j, annual earnings M_t, $t=j$, ..., T, and capital market opportunities, so as to maximize expected lifetime utility (equation 12). Arthur (1981) and Shepard and Zeckhauser (1982,

1984) assume that the individual can save by purchasing actuarially fair annuities and borrowing via life-insured loans. If actuarially fair annuities are available, an individual who invests \$1 at the beginning of his jth year will receive $\$(1 + A_j)$ at the end of the year with probability $1 - \delta_j$ and nothing with probability δ_j. For the annuity to be fair, that is, to have an expected payout of $1 + r$ where r is the riskless rate of interest, there must be an annuity rate of interest a_j that satisfies

$$(1 + a_j)(1 - \delta_j) = 1 + r \tag{13}$$

If the individual borrows, she must cover the possibility that she might die before repaying the loan. Agreeing to pay $1 + a_j$ if she survives is equivalent to paying $1 + r$ on survival plus purchasing a life insurance policy in this amount at actuarially fair rates. So we can call a_j the actuarial rate of interest.

The individual's budget constraint can be expressed as the requirement that the present value of expected consumption equal initial wealth plus the present value of lifetime earnings, where discounting is done at the riskless rate, r:

$$\sum_{t=j}^{T} \pi_{j,t} \cdot (1 + r)^{j-t} \cdot X_t = W_j + \sum_{t=j}^{T} \pi_{j,t} \cdot (1 + r)^{j-t} \cdot M_t \tag{14}$$

Alternatively, making use of equation (13) and the fact that

$$\pi_{j,t} = \prod_{s=j}^{t=1} (1 - \delta_s)$$

the budget constraint can be expressed in terms of the actuarial rate of interest:

$$\sum_{t=j}^{T} \left[\prod_{s=j}^{t-1} (1 + a_s)^{-1} \right] \cdot X_t = W_j + \sum_{t=j}^{T} \left[\prod_{s=j}^{t-1} (1 + a_s)^{-1} \right] \cdot M_t \tag{14'}$$

The pattern of consumption over the life cycle is chosen so as to maximize equation (12) subject to (14) or (14'). Formally, it is the solution to the Lagrangian problem:

$$\underset{X_t}{\text{Max }} V_\lambda = \sum_{t=j}^{T} \pi_{j,t} \cdot (1 + d)^{j-t} \cdot u(X_t)$$

$$+ \lambda \left[W_j + \sum_{t=j}^{T} \pi_{j,t} \cdot (1 + r)^{j-t} \cdot (M_t - X_t) \right]$$

Now consider how a government health and safety regulation that reduces conditional probabilities of death affects lifetime utility. Assume that government regulation alters the probability that a person dies in any year only if that person is alive at the beginning of the year. Consider a regulation that reduces $\delta_{t'}$, the conditional probability of dying at age t', that is, the probability that the individual dies between his t'th and $t'+1$st birthdays, assuming he is alive on his t'th birthday. A program to increase the police force in a city in a single year reduces $\delta_{t'}$, for that year alone. A program that reduces an individual's exposure at age 30 to a carcinogen with a twenty-year latency period reduces the conditional probability of dying at all ages after 50 (δ_{50}, δ_{51}, δ_{52}, ...).

It should be emphasized that when the conditional probability of death is altered at age t', it affects the probability of surviving to ages $t'+1$ and beyond, $\pi_{j,t'+1}$, $\pi_{j,t'+2}$, ..., $\pi_{j,T}$, since, by repeated use of the definition of δ_t,[14]

$$\pi_{j,t'} = (1 - \delta_j)(1 - \delta_{j+1}) \ldots (1 - \delta_{t'-1})$$

In practice, the changes in $\delta_{t'}$ corresponding to some public project are likely to be small. For example, it has been estimated that the risk of dying of cancer due to all environmental causes is only 3.6×10^{-5} (Doll and Peto, 1981). For this reason the literature has focused on marginal changes in $\delta_{t'}$.

Formally, let $w_{j,t'}$ be the individual's marginal willingness to pay at age j for a change in $\delta_{t'}$. It is measured by the wealth that must be taken away from that person at age j to keep her expected utility constant, given the reduced risk of death, or:

$$w_{j,t'} \equiv \frac{dW_j}{d\delta_{t'}} = -\frac{dV_j/\delta_{t'}}{dV_j/dW_j}$$

The envelope theorem implies that

$$\frac{dV_j/d\delta_{t'}}{dV_j/dW_j} = \frac{\partial V_\lambda/\partial \delta_{t'}}{\partial V_\lambda/\partial W_j}$$

---

[14]A change in $\delta_{t'}$ also affects life expectancy, which is the sum of survival probabilities from the current age onward,

$$\sum_{t=j}^{T} \pi_{j,t}$$

Note, however, that there are many changes in $\{\delta_t\}$ that result in equivalent changes in life expectancy.

Thus, using equation (12), we can express marginal willingness to pay as

$$
\begin{aligned}
w_{j,t'} = (1 - \delta_{t'})^{-1} \sum_{t=t'+1}^{T} \pi_{j,t} \cdot \Big[& (1 + d)^{j-t} \cdot u(X_t) \cdot \lambda^{-1} \\
& + (1 + r)^{j-t} \cdot (M_t - X_t) \Big]
\end{aligned} \tag{15}
$$

Willingness to pay at age j for a reduction in the conditional probability of death at age t' equals the gain in expected utility from year t' onward, converted to dollars by dividing by the marginal utility of income in year j, λ, and discounted at the individual's own rate of time preference. Added to this is the effect of a change in $\delta_{t'}$ on the budget constraint. A reduction in $\delta_{t'}$ makes the individual wealthier by increasing the present value of her expected lifetime earnings from age $t' + 1$ onward. But an increase in survival probabilities also decreases the consumption that the person can afford in each of the years $t' + 1$ through T from a given earnings stream. Thus her willingness to pay is reduced by the change in the present value of the consumption stream.

It should be noted that $w_{j,t'}$ is the rate at which the individual is willing to trade wealth for a unit change in risk. To compute the dollar value of a small change in risk, equation (15) must be multiplied by the magnitude of the risk change. For example, if $w_{j,t'} = \$2 \times 10^6$ but the change in risk is only 10^{-6}, willingness to pay for the risk change is $2.

There are a number of points to be made about willingness to pay for reductions in the risk of death on the basis of equation (15). First, for a policy that affects the conditional probabilities for dying over a number of years, the total marginal willingness to pay is the sum of the willingness to pay for the changes in each of the $\delta_{t'}$.

Second, a key assumption of models of this type is that expected lifetime utility depends only on expected lifetime consumption, as Linnerooth (1979) has noted. So what is being calculated is the willingness to pay for the opportunity to continue consumption. As Bergstrom (1982) has pointed out, if the intertemporal objective function is derived from preferences among alternative lotteries, it should include a term that values survival per se. If this term is an increasing function of the $\{\pi_{j,t}\}$, any willingness-to-pay measure derived from equation (12) must be regarded as a lower bound to true willingness to pay.[15]

---

[15]This condition was first noted in a static context by Conley (1976) and Cook (1978).

Third, even granting the assumption that utility depends only on consumption, as long as the individual's average utility of consumption exceeds his marginal utility, willingness to pay exceeds human capital, and human capital must be interpreted as a lower bound to willingness to pay. By use of the first-order conditions for utility maximization, the term in brackets in equation (15) can be written as

$$(1 + r)^{j-t} \left(\frac{u(X_t)}{\partial u/\partial(X_t)} - X_t + M_t \right)$$ (16)

This implies that if $u(X_t) \cdot (\partial u/\partial X_t)^{-1} - X_t > 0$ for all t, then each year's contribution to $w_{j,t'}$ as given by equation (16) exceeds that year's contribution to the present value of lifetime earnings, $(1 + r)^{j-t} \cdot M_t$, and willingness to pay must exceed the present discounted value of lifetime earnings. As noted by Conley (1976) and Cook (1978), the condition that $u(X_t) \cdot (\partial u/\partial X_t)^{-1} - X_t > 0$ implies that the average utility of consumption exceeds its marginal utility, a condition that holds for all increasing, concave utility functions, provided consumption exceeds a subsistence level.

Blomquist (1981), expanding on Linnerooth's analysis, reviewed the results of nine empirical estimates of willingness to pay and showed that the implied value of statistical life was typically much larger than the expected lifetime earnings of the members of the sample population used in each study. His analysis lends strong support to the assertion that values of life based on lifetime earnings are a poor proxy for the willingness to pay for reduced risk of mortality.

Fourth, equation (15) implies that as t' (the age at which risk of death changes) increases, $w_{j,t'}$ must decline.[16] This implies that the value of reducing a person's current probability of dying must always be greater than the value of reducing exposure at age j to a carcinogen with a latency period of $t' - j$ years. In the second case fewer expected life-years are saved. Furthermore, the longer the latency period, the smaller is willingness to pay, other things being held constant. This means that willingness-to-pay measures based on behavior toward contemporaneous risk (for example, from studies of wages and occupational accident mortality) will not be good proxies for willingness to pay to reduce latent risks.

Cropper and Sussman (1990) have shown that there is an alternative approach to dealing with the latency problem. Suppose we have obtained a measure of $w_{t',t'}$ for a group at age t by examining, for

---

[16]Strictly speaking, this holds only if equation (16) is positive; however, this will be the case as long as the individual is above subsistence.

example, the group members' trade-offs between wages and con-temporaneous job risks. Cropper and Sussman have shown that $w_{j,t'}$ ($j < t'$) can be calculated from the following expression:

$$w_{j,t'} = \prod_{s=j}^{t'} (1 + a_s)^{-1} \cdot w_{t',t'} \tag{17}$$

The proof involves the following steps:
(1) Substitute the first-order conditions into equation (15) to obtain:

$$w_{j,t'} = (1 - \delta_{t'})^{-1} \sum_{t=t'+1}^{T} \pi_{j,t} \cdot (1 + r)^{j-t} \cdot \left[\frac{u(X_t)}{\partial u/\partial X_t} + M_t - X_t \right]$$

(2) In a similar manner,

$$w_{j+1,t'} = (1 - \delta_{t'})^{-1} \sum_{t=t'+1}^{T} \pi_{j+1,t} \cdot (1 + r)^{j+1-t} \cdot \left[\frac{u(X_t)}{\partial u/\partial X_t} + M_t - X_t \right]$$

(3) Into this expression, substitute

$$\pi_{j+1,t} = \frac{\pi_{j,t}}{(1 - \delta_j)} \text{ for all } t$$

(4) Divide the second expression by the first to obtain

$$\frac{w_{j+1,t'}}{w_{j,t'}} = \frac{(1 + r)}{(1 - \delta_j)}$$

(5) Use equation (13) and rearrange to obtain

$$w_{j+1,t'} = (1 + a_j) w_{j,t}$$

Repeating this procedure for all pairs of years between j and t' and rearranging terms results in equation (17). The willingness to pay in advance for a reduction in risk in t' is the contemporaneous willing-ness to pay discounted back by a factor derived from the actuarial interest rates over the interval. Recall that these actuarial interest rates are a combination of the riskless interest rate and conditional probabilities of death. As such, they vary over time for the individual and across individuals, depending on age and other factors.

The fifth point to be made is that it is interesting to see how $w_{j,j}$ varies over the life cycle. Shepard and Zeckhauser (1982, 1984) have used an expression like equation (15) to examine this question. If consumption were constant for all t, as would be the case if the riskless rate of interest were equal to the subjective rate of time preference, $w_{j,j}$ would decline monotonically with age. Younger persons would always have a higher WTP to reduce current risk of death than older persons because there would be more years of consumption and utility in the summation. If, however, consumption increases over some portion of the life cycle, $w_{j,j}$ may also increase with age up to some point, and then decline. If, for example, the individual cannot be a net borrower but can lend at the riskless rate of interest, her consumption is likely to be constrained by income at the beginning of her life. This will cause the present value of the utility of consumption and, hence, $w_{j,j}$, to increase up to some point, and then decline.[17]

Finally, in the special case in which the utility function exhibits constant elasticity of marginal utility, that is, $u(X_t) = X_t^\beta$, $0 < \beta < 1$, we can examine the effect of changes in risk preferences on willingness to pay. In the special case where $d = r$, equation (15) becomes

$$w_{j,t'} = (1 - \delta_{t'})^{-1} \sum_{t=t'+1}^{T} \pi_{j,t} \cdot (1 + r)^{j-t} \cdot [M_t + X_t \cdot (1 - \beta)/\beta] \qquad (15')$$

which is a decreasing function of β. Since $1 - \beta$ is the relative risk-aversion index for the constant elasticity utility function, equation (15') indicates that more risk-averse individuals (those with smaller values of β) have higher willingnesses to pay to reduce risk of death.

EXTENSIONS OF THE BASIC LIFE-CYCLE MODEL

Arthur (1981) has embedded the life-cycle model for an individual into a one-sector general equilibrium model with population growth. He derives an expression for the value to society of a change in an individual's conditional probability of death. Society's WTP for a change in δ_t differs from the individual's WTP due to reproductive effects: if risk of death is reduced during the reproductive years, there is a gain due to the productivity of the increased population. Arthur also notes that the consumption cost of additional life-years reflected in equation (14) may not be borne by the individual, depending on institutional arrangements; however, it is borne by society.

---

[17]The expression for $w_{j,t'}$ in the no-net-borrowing case is identical to equation (15) except that the term $(1 + r)^{j-t} \cdot (M_t - X_t)$ does not appear inside the brackets.

Recognition that each individual is typically part of a family has led to efforts to model individual choice and willingness to pay in a family context (Cropper and Sussman, 1988). Cropper and Sussman assume that an individual's expected utility depends on the size of the bequest that the person would leave heirs in the event of his death and that the utility and size of the bequest depend on the number of heirs and their degree of economic dependence on the individual's earnings. The individual is also assumed to gain utility while alive from the consumption of his dependents, as well as from their existence. A bequest motive, other things being equal, reduces willingness to pay for a reduction in risk of death because one can provide for one's heirs by saving or by purchasing life insurance. On the other hand, the fact that a married person may get more utility while alive than a single person should raise his willingness to pay above that of the single person. A study of willingness to pay by Jones-Lee (1986) suggests that the latter effect dominates (see also Jones-Lee, Hammerton, and Philips, 1985).

An individual's purchases of life insurance clearly reflect some aspect of the value of that individual's staying alive. But since a life insurance policy is paid to the beneficiaries (usually the dependents) in the event of death, it is the value to the dependents of the individual's earning capacity that is being insured. This is not the same thing as the individual's willingness to pay for his own enhanced survival probability.

Summary

In this section I have described the use of the willingness to pay to reduce the risk of death as the basis for valuing policies that save lives; and I have presented models for inferring this willingness to pay from individuals' choices. These models suggest that individuals' valuations are likely to vary across individuals according to income levels, age, background risk levels, and preferences toward risk-taking. The most common application of these models has been to the wage-risk trade-off revealed in the job market. Since these wage-risk studies are an application of the hedonic price technique, discussion of them is put off until chapter 12.

VALUING REDUCED MORBIDITY

A number of approaches to determining the monetary value of reduced morbidity have been utilized in the literature on the economics of health and safety. These approaches can be broadly categorized as being based on individual preferences (willingness to pay or required compensation) or on resource or opportunity costs. The resource or opportunity cost approach is an example of the damage function method described in chapter 2. It seeks to identify the real costs of illness in the form of lost productivity and output and the increase in resources devoted to medical care. Costs per illness or per day are multiplied by the number of cases or days sick to determine an aggregate value. As I will show formally in this section, the resource-opportunity cost approach to valuation can be misleading because it fails to capture the variety of behavioral responses to illness and the threat of illness.

To see this in a simple way, consider the case of an individual who experiences one less day of asthma attacks because of an improvement in air quality. The benefit to that person might include avoiding the lost wages associated with being unable to work one day and the reduction in costs for medicine and treatment. But in addition, the individual avoids the discomfort associated with the attack itself. The first two components are captured by the resource-opportunity cost approach. Only a comprehensive willingness-to-pay measure would capture the discomfort component as well.

Now consider an individual who experiences no asthma attacks at present levels of air pollution because he spends money operating an air purifier and stays at home indoors on high-pollution days in order to prevent the attacks that would be associated with exposure to the outdoor air. If air quality is improved, this individual benefits from being able to reduce the monetary expenditures and the lost wages and opportunities for leisure activities that are associated with these defensive activities. This individual will benefit from reduced air pollution even though there is no observed reduction in the actual incidence of asthma attacks related to air pollution.

To summarize, air pollution that affects human health can reduce people's well-being through four channels: the medical expenses associated with treating disease induced by air pollution (including the opportunity cost of time spent in obtaining such treatment); the lost wages resulting from the inability to work; the defensive or averting expenditures and activities associated with attempts to prevent disease induced by air pollution (including the opportunity cost of time); and the disutility associated with the symptoms and lost opportunities for leisure activities caused by the illness. Improving environmental quality can yield benefits to individuals by reducing some or all of these adverse affects. Portions of the first three of these effects have readily identifiable monetary counterparts, but effects of the fourth kind may not. A truly comprehensive benefit measure should be capable of capturing all of these relevant effects. Measures based solely on decreases in medical costs or lost wages are not comprehensive because they omit major categories of beneficial effects.

Although individual willingness to pay is the correct starting point for analyzing health-related values, there is one important respect in which society's valuation of changes in health might diverge from that of the affected individual. Society has developed several mechanisms for shifting some of the costs of illness away from the individual who is ill and onto society at large. These mechanisms include medical insurance, which spreads the costs of treatment among all policyholders, and sick leave policies, which shift at least part of the cost of lost work days onto the employer and ultimately onto the consumers of the employer's products. An individual's expressed willingness to pay to avoid illness would not reflect those components of the costs of her illness that are borne by or shifted to others. But the value to society of avoiding her illness includes these components. Empirical measures of the value of reducing illness must take account of these mechanisms for shifting costs.

In this section, I develop a general model of individual choice that captures the principal ways by which individuals can affect their health status, and use this model to derive measures of the value of improvements in health and decreases in pollution that affect health. I also describe several extensions of the basic model and consider the problems that arise when this approach to valuation is applied to reductions in the incidence of chronic disease.

A Basic Model of Health Production and Choice

Most of the formal models used for deriving the value of reduced morbidity use some variant of the health production function first developed by Grossman (1972). Cropper (1981) introduced a pollution variable into the health production function. Harrington and Portney (1987) extended the model to examine explicitly the relationships among willingness to pay for a reduction in pollution, reductions in the cost of illness, and changes in defensive expenditures. In this section I develop an expanded version of the Harrington and Portney model.

The health production function relates exogenous variables (including environmental variables such as air pollution) and choice variables (such as preventive medicine and treatment costs) to some measure of health status. It is assumed that individuals know their health production function, choose the output level optimally, and choose inputs so as to minimize the cost of production of any level of health. Of course, these are strong assumptions.

As originally formulated by Grossman (1972), the health production function was dynamic, allowing for "investments" in "health capital" that yielded benefits in the form of reduced illness over several time periods. The simple model presented here will abstract from this intertemporal dimension of the problem. Intertemporal models are briefly discussed in the section on Multiple Symptoms below.

Let health in any time period be measured by the number of days sick. This is a simplification in that it makes no distinction between one episode of illness of two days' duration and two separate illnesses of one day each, and in that differences in the severity of illness are ignored. Some of the implications of richer specifications of the health variable are discussed in the section on Health Capital below.

Assume that one of the determinants of health status is the level of exposure to or dose of some environmental contaminant. Dose is represented by the scalar variable d, which depends on the concentration of pollution c, and the extent of an averting activity a, undertaken to avoid or reduce exposure to pollution. Examples of averting activities include filtering tap water before drinking and staying indoors on days of high air pollutant levels. If the contaminant is an air pollutant, c could be interpreted as the number of days during which some measure of air pollution exceeds a stated standard, the mean value of the pollutant averaged over the relevant time period, or the highest value recorded for the pollutant during that period. Since a change in pollutant emissions is likely to change all of these measures in a predictable way, the choice of a measure for d should be based on whatever is the best predictor of changes in health status. Similar questions arise for other forms of environmental contamination, such as chemicals in drinking water and pesticide residues in food.

Assume also that there is a set of mitigating activities and treatments that can be undertaken to reduce the health impact of any given exposure to pollution. Examples of mitigating activities include taking antihistamines and visiting a doctor to obtain relief from a sinus headache. In this model, assume that the level of mitigating activities can be chosen by the individual so as to maximize utility. This is an extension of the Harrington and Portney model (1987) in which medical and treatment costs were assumed to be determined by the number of days spent sick rather than chosen in an optimizing framework.

The health production function for an individual can be written as

$$s = s(d, b)$$
$$d = d(c, a)$$

and by substitution $\qquad\qquad s = s(c, a, b)$

with $\qquad\qquad\qquad\qquad\qquad \partial s / \partial c > 0$

$$\partial s / \partial b, \; \partial s / \partial a < 0$$

and where $\quad$ $s =$ days sick

$d =$ realized exposure to pollution

$c =$ pollution

$b =$ medical treatments and mitigating activities

$a =$ averting and avoidance activities

The health production function can be estimated from cross-section data on illness, pollution, and averting and mitigating activities. It would also be necessary to control for other determinants of health status such as physical and socioeconomic characteristics of individuals—for example, age, sex, use of tobacco, income, and education.[18]

The individual derives utility from the consumption of a numeraire good, X, normalized with a price of one, and leisure, f. Illness causes disutility; so:

$$u = u(X, f, s)$$

with $\quad$ $\partial u / \partial X, \; \partial u / \partial f > 0$

$$\partial u / \partial s < 0$$

The individual chooses X, f, a, and b so as to maximize utility subject

---

[18]Perhaps the most difficult part of this model specification to handle properly in empirical estimation is the averting and mitigating activities. The national surveys typically used in micro-epidemiological studies (for example, the Health Interviews Survey used in Portney and Mullahy, 1986) do not include data on mitigating and averting activities.

to the budget constraint of:[19]

$$I + p_w(T - f - s) = X + p_a \cdot a + p_b \cdot b$$

where I = nonlabor income

        p_w = the wage rate

        T = total time available

        p_a = the price of averting activities

        p_b = the price of mitigating activities

The first-order conditions for a maximum include:

$$\partial u / \partial X = \lambda$$

$$\partial u / \partial f = \lambda \cdot p_w$$

and

$$\lambda \cdot \frac{p_b}{\partial s / \partial b} = \frac{\partial u}{\partial s} - \lambda \cdot p_w = \lambda \cdot \frac{p_a}{\partial s / \partial a} \tag{18}$$

where λ is the Lagrangian multiplier and can be interpreted as the marginal utility of income.

THE MARGINAL VALUE OF REDUCED POLLUTION

This model of choice can be used to derive an observable measure of the individual's marginal willingness to pay to reduce pollution. An individual's willingness to pay for a small reduction in ambient pollution is the largest amount of money that can be taken away from that person without reducing her utility. If pollution enters the utility function directly because, for example, of aesthetic disamenities associated with pollution, then there are additional benefits that are not associated with health. In the health production model, in which pollution affects utility only through health, willingness to pay is the reduction in the cost of achieving the optimal level of health made possible by the decrease in pollution. For example, if a reduction in ozone levels from .16 to .11 parts per million (ppm) reduces the number of days of respiratory symptoms from 6 to 4, and if an expenditure on b of $20 has the same effect, then (all else being equal) the individual should be willing to pay no more than $20 for the ozone reduction.

---

[19]Where there are time costs associated with the averting or mitigating activities, they should be incorporated into the full income budget constraint; for an example, see Gerking and Stanley (1986).

Formally, marginal willingness to pay for a reduction in pollution (w_c) is given by the reduction in sick time associated with the reduction in pollution times the marginal cost of reducing sick time. The latter is given by the cost of an additional mitigating input divided by the reduction in sick time that input produces, or, alternatively, by the cost of averting behavior divided by the reduction in sick time that averting behavior produces. This can be verified by first totally differentiating the indirect utility function, $v(I, p_w, p_a, p_b, c)$, and solving for

$$w_c = \frac{dI}{dc} = - \frac{\partial v/\partial c}{\partial v/\partial I} = - \frac{\partial v/\partial c}{\lambda}$$

$$\frac{\partial v}{\partial c} = \frac{\partial u}{\partial s} \cdot \frac{\partial s}{\partial c} - \lambda \cdot p_w \cdot \frac{\partial s}{\partial c}$$

$$= \left(\frac{\partial u}{\partial s} - \lambda \cdot p_w \right) \frac{\partial s}{\partial c}$$

Assuming that mitigating and averting activities are optimally chosen, then the first-order condition (18) can be substituted into this expression to obtain

$$w_c = - p_b \cdot \frac{\partial s/\partial c}{\partial s/\partial b} = p_b \cdot \frac{\partial b}{\partial c}$$

$$= - p_a \cdot \frac{\partial s/\partial c}{\partial s/\partial a} = p_a \cdot \frac{\partial a}{\partial c} \tag{19}$$

The right-hand terms follow from application of the implicit function rule.[20]

---

[20]Most of these results can also be derived from the expenditure function, as follows:

$$e \equiv \min [X + p_a \cdot a + p_b \cdot b + p_w \cdot f + p_w \cdot s$$
$$+ \mu [u^0 - u(X, f, s)]$$

where $s = s(c, a, b)$.
The first-order conditions include

$$\frac{p_a}{\partial s/\partial a} = \mu \cdot \frac{\partial u}{\partial s} - p_w$$

So $$w_c = - \frac{\partial e}{\partial c} = \left(\mu \cdot \frac{\partial u}{\partial s} - p_w \right) \cdot \frac{\partial s}{\partial c}$$

$$= p_a \cdot \frac{\partial s/\partial c}{\partial s/\partial a}$$

The other expressions in the text can be found by similar manipulations.

There are several things to note about these expressions for marginal willingness to pay. First, the ratios

$$\frac{\partial s/\partial c}{\partial s/\partial b} \quad \text{and} \quad \frac{\partial s/\partial c}{\partial s/\partial a}$$

can be interpreted as marginal rates of technical substitution (MRTS) between pollution and the other variable in producing a constant level of sickness. Marginal willingness to pay can be expressed in terms of any of the MRTS between pollution and any other input in the production of health, since to minimize the cost of producing health, the values of marginal products of all inputs must be equal at the margin. Second, all of the measures are functions of observable variables that can be calculated given knowledge of the health production function. And third, as shown by the right-hand terms, marginal willingness to pay can be calculated from the reductions in expenditures on either mitigating or averting behavior that are required to attain the original health status, holding all else constant. This will not in general be equal to the observed reduction in mitigating or averting behavior associated with the reduction in pollution.

In order to compute equation (19), it is necessary to estimate a production function for the health outcome of interest and evaluate the numerator and denominator of the equation at current levels of all inputs. In practice this has proven to be a difficult task.[21] In addition to having data on the relevant health outcome and on ambient pollution levels, one must identify averting and mitigating behaviors and measure their costs. In practice the most effective method of reducing exposure, given ambient pollution levels, is to spend time indoors. Although the amount of time spent indoors could be measured, determining its cost would be difficult. Devices that reduce indoor pollution concentrations (air conditioners, air filters) have costs that can be measured; but they produce other services, such as reducing indoor temperature, so that it is inappropriate to allocate all of these costs to pollution avoidance. The implications of this possibility are discussed in the section on "Impure" Models below.

Because of the difficulties of implementing these measures, it is useful to consider an alternative expression that shows the relationship between the observable cost of illness and marginal willingness to pay. The first step in deriving this expression is to obtain the demand functions for a and b: $a^*(I, p_w, p_a, p_b, c)$ and $b^*(I, p_w, p_a, p_b, c)$.

---

[21]For one effort to implement this basic model, see Gerking and Stanley (1986).

These functions give the optimal quantities of a and b as functions of income, prices, and pollution. The second step is to take the total derivative of the health production function:

$$\frac{ds}{dc} = \frac{\partial s}{\partial a} \cdot \frac{\partial a^*}{\partial c} + \frac{\partial s}{\partial b} \cdot \frac{\partial b^*}{\partial c} + \frac{\partial s}{\partial c}$$

This gives the impact of a change in pollution on illness after taking account of the optimal adjustments of a and b to the pollution change. This expression can be rearranged as follows:

$$\frac{\partial s}{\partial c} = \frac{ds}{dc} - \frac{\partial s}{\partial a} \cdot \frac{\partial a^*}{\partial c} - \frac{\partial s}{\partial b} \cdot \frac{\partial b^*}{\partial c}$$

and multiplied by the first-order conditions of equation (18)

$$-\frac{p_a}{\partial s / \partial a} = p_w - \frac{\partial u / \partial s}{\lambda} \tag{20}$$

to obtain

$$-p_a \frac{\partial s / \partial c}{\partial s / \partial a} = \left(p_w - \frac{\partial u / \partial s}{\lambda} \right) \cdot \frac{ds}{dc} - \left(p_w - \frac{\partial u / \partial s}{\lambda} \right) \cdot \frac{\partial s}{\partial a} \cdot \frac{\partial a^*}{\partial c}$$

$$- \left(p_w - \frac{\partial u / \partial s}{\lambda} \right) \cdot \frac{\partial s}{\partial b} \cdot \frac{\partial b^*}{\partial c}$$

or, after rearranging:

$$w_c = p_w \cdot \frac{ds}{dc} + p_b \cdot \frac{\partial b^*}{\partial c} + p_a \cdot \frac{\partial a^*}{\partial c} - \frac{\partial u / \partial s}{\lambda} \cdot \frac{ds}{dc} \tag{21}$$

This expression says that MWTP is the sum of the observable reductions in the cost of illness and averting activities and the monetary equivalent of the disutility of illness. The change in the cost of illness consists of the economic value of reductions in sick time and mitigating expenditures. The term $p_w \cdot ds/dc$ includes both actual lost wages and lost leisure time valued at the wage rate.

To compute ds/dc we need not estimate a health production function, but can instead estimate a dose-response function—a reduced-form relationship between illness and ambient pollution controlling for other variables that affect health status. In the health production framework a dose-response function is obtained by substituting the demand functions for b and a into the health production function. Full implementation of equation (21) as a measure of value therefore requires estimation of these demand functions.

As a practical matter, the first three terms in equation (21) can be approximated after the fact by using the observed changes in illness and averting and mitigating expenditures. In this way, equation (21) can be used to derive a lower bound to individual WTP. Since the last term in the equation is negative ($\partial u/\partial s < 0$), the first two terms—the value of lost time plus the change in averting and mitigating expenditures—give a lower bound to WTP. In the health literature, the term "cost of illness" typically refers only to the social cost of lost earnings plus recuperative (mitigating) medical expenditures associated with illness. This term therefore ignores two components of the social cost of illness—the social value of averting expenditures and the cost of lost leisure time that results from illness.

VALUING NONMARGINAL CHANGES IN POLLUTION

In the preceding subsection I defined a measure of value for a small change in health status. This value measure is likely to be a function of health status itself. Specifically, for reasons analogous to the standard assumption of diminishing marginal utility, the marginal willingness to pay for further decreases in morbidity is likely to decrease as health status increases. In principle, this functional dependence of marginal willingness to pay on health status should be taken into account whenever large (that is, nonmarginal) changes in morbidity are being valued. The proper way to do this is straightforward in principle, but may be hard to implement in practice.

If the function relating marginal willingness to pay to health status is known, the value of the nonmarginal change is simply the integral of this function over the relevant range. Typically, however, what empirical methods produce is a point estimate of marginal willingness to pay. If it can be assumed that marginal willingness to pay is (approximately) constant over the relevant range, the value of the nonmarginal change can be calculated by multiplying the marginal willingness to pay times the change in health status. The larger is the change in morbidity and the more rapidly marginal willingness to pay decreases with increases in health status, the larger is the error in using this simple approach. Alternatively, one could compute a lower bound to MWTP. As shown in chapter 4, the change in averting and mitigating expenditures, holding illness constant, constitutes a lower bound to willingness to pay for pollution improvement.

Extensions of the Health Production Model

The basic model of choice and value described in the preceding section involves a considerable amount of simplification in order to obtain some insights about the relationship between observed behavior and economic value. In this section I extend the model in several directions toward greater realism to see to what extent the conclusions about values are preserved.

MULTIPLE SYMPTOMS

The model can be generalized to the case of many symptoms and various forms of averting and mitigating behavior. Consider the special case in which each mitigating activity enters the separate health production function for one symptom, $s_i(c, a, b_i)$. Following procedures similar to those of the basic model, it can be shown that

$$w_c = \sum p_i \cdot \frac{\partial s_i/\partial c}{\partial s_i/\partial b_i} \tag{22}$$

where p_i is the price of b_i. Estimation of w_c in this case involves estimating production functions for all symptoms. Similarly, it can be shown that

$$w_c = p_a \cdot \frac{\partial s_i/\partial c}{\partial s_i/\partial a}$$

for any symptom.[22]

_____

[22]These conclusions parallel those of Dickie and Gerking (1991). And they hold whether all or only some symptoms involve restrictions on activities or work. Let $s_1[d(c, a), b_1]$ and $s_2[d(c, a), b_2]$ represent the health production function. Suppose that only s_2 restricts work. The maximization problem is

$$\text{Max } u(X, f, s_1, s_2) - \lambda[I + p_w(T - f - s_2) - X - p_a \cdot a$$
$$- p_1 \cdot b_1 - p_2 \cdot b_2]$$

The first-order conditions for b_1 and b_2 are

$$\frac{\partial u}{\partial s_1} = \lambda \frac{p_1}{\partial s_1/\partial b_1}$$

Unfortunately, simple expressions such as these cannot be obtained when mitigating activities affect more than one symptom. Suppose that there are two symptoms, s_1 and s_2, neither of which restricts activities, and two mitigating activities, b_1 and b_2, with health production functions:

$$s_1 = s_1 (c, a, b_1, b_2)$$

and

$$s_2 = s_2 (c, a, b_1, b_2)$$

For example, b_1 could be a multisymptom cold remedy. The first-order conditions are:

$$\frac{\partial u / \partial s_1}{\lambda} \cdot \frac{\partial s_1}{\partial b_1} + \frac{\partial u / \partial s_2}{\lambda} \cdot \frac{\partial s_2}{\partial b_2} - p_1 = 0$$

$$\frac{\partial u / \partial s_1}{\lambda} \cdot \frac{\partial s_1}{\partial b_2} + \frac{\partial u / \partial s_2}{\lambda} \cdot \frac{\partial s_2}{\partial b_2} - p_2 = 0$$

$$\frac{\partial u / \partial s_1}{\lambda} \cdot \frac{\partial s_1}{a} + \frac{\partial u / \partial s_2}{\lambda} \cdot \frac{\partial s_2}{\partial b_2} - p_2 = 0$$

and

$$\left(\frac{\partial u}{\partial s_2} - \lambda \cdot p_w \right) = \lambda \cdot \frac{p_2}{\partial s_2 / \partial b_2}$$

therefore

$$w_c = - \frac{\partial v / \partial c}{\lambda} = - \frac{\partial u}{\partial s_1} \cdot \frac{\partial s_1}{\partial d} \cdot \frac{\partial d}{\partial c} + \left(\frac{\partial u}{\partial s_2} - \lambda \cdot p_w \right) \frac{\partial s_2}{\partial d} \cdot \frac{\partial d}{\partial c}$$

$$= - p_1 \cdot \frac{\frac{\partial s_1}{\partial d} \cdot \frac{\partial d}{\partial c}}{\partial s_1 / \partial b_1} - p_2 \cdot \frac{\frac{\partial s_2}{\partial d} \cdot \frac{\partial d}{\partial c}}{\partial s_2 / \partial b_2}$$

The first-order condition for a can be written as

$$\frac{\partial d}{\partial c} \left[\frac{\partial u}{\partial s_1} \cdot \frac{\partial s_1}{\partial d} + \left(\frac{\partial u}{\partial s_2} - \lambda \cdot r \right) \frac{\partial s_2}{\partial d} \right] = \lambda \cdot p_a \frac{\partial d / \partial c}{\partial d / \partial a}$$

which can be substituted into the expression for w_e to obtain

$$w_c = - p_a \cdot \frac{\partial d / \partial c}{\partial d / \partial a}$$

and since

$$\frac{\partial s_i}{\partial d} \cdot \frac{\partial d}{\partial c} = \frac{\partial s_i}{\partial c}$$

then

$$w_c = - p_a \cdot \frac{\partial s_i / \partial c}{\partial s_i / \partial a}$$

As Dickie and Gerking (1991) show, as long as the derivatives of the health production function are known and the number of mitigating and averting activities (three here) equals or exceeds the number of symptoms (two), this system of equations can be used to solve for the unobserved utility terms, for example,

$$\frac{\partial u / \partial s_1}{\lambda}$$

The solution can be used to substitute into the following expression to obtain a measure of willingness to pay:

$$w_c = -\frac{\partial v / \partial c}{\lambda} = -\frac{\partial u / \partial s_1}{\lambda} \cdot \frac{\partial s_1}{\partial c} - \frac{\partial u / \partial s_2}{\lambda} \cdot \frac{\partial s_2}{\partial c}$$

But if the number of symptoms exceeds the number of mitigating and averting activities, then the expression for MWTP will necessarily include some unobservable marginal utility terms.

SHIFTING THE COST OF ILLNESS

The measures derived above give each individual's willingness to pay to avoid pollution-induced illness. But where social mechanisms have been developed for spreading some of the costs of illness, the social value of reducing pollution may exceed the aggregate of the individuals' willingness to pay. For example, where employers grant a certain number of days of paid sick leave, the individual does not incur any financial cost for the loss of a work day. Thus if a truthful person were asked a contingent valuation question about the value of reducing pollution, the response would not include the lost wages component of the first term on the right of equation (21). Nevertheless, society loses that person's output for that day.

The burden of this lost output is borne initially by the employer, but ultimately shows up as some combination of higher prices for outputs and lower equilibrium wages. A nonmarginal reduction in pollution that resulted in fewer lost work days would reduce the costs of paid sick leave plans. In principle this would result in changes in general equilibrium wages and prices.

Similarly, medical insurance has the effect of shifting the marginal cost of treatment away from the sick individual. A 100 percent coverage plan reduces the price of medical treatment to zero. Thus marginal willingness-to-pay measures based on the health production function would have to focus on changes in averting expenditures. Again, changes in the cost of treatment covered by insurance should be added to the aggregate of individuals' willingness to pay to obtain the social benefit of reduced pollution-induced illness.

As Harrington and Portney (1987) point out, the introduction of sick leave and medical insurance plans could alter individuals' choices of mitigating and averting activities and days of illness by changing the prices of these activities. If the effect of these plans on prices has been taken into account in estimating the models of individual behavior, then the measures of social values described here will be correct.

HEALTH CAPITAL AND THE HEALTH PRODUCTION FUNCTION

The basic model outlined above ignores any dynamic aspects of health and illness. In that model, averting or mitigating behavior and pollution in any one period of time affect only illness in that time period. It would be more in the spirit of Grossman's (1972) original treatment of the health production function to formulate the choice problem as one of optimizing investment in health capital over time. Health capital can be thought of as a measure of health status that depreciates over time unless it is maintained or augmented with some kind of investment. To formalize this, let H_t be the stock of health capital at t. H_t can be augmented by investment of time, y_t, in activities such as exercising and by the purchase of "health" goods, z_t. Health capital is also subject to depreciation at the rate of g percent per year. The health stock changes over time according to

$$dH_t/dt = I_t(z_t, y_t) - g \cdot H_t$$

The rate of depreciation itself can be made a function of age and other characteristics.

One approach to deriving willingness to pay is to assume that health capital is an argument in the symptom production functions. Then, measures of willingness to pay similar to those of the basic model can be derived. Cropper (1977, 1981) used a different approach. She made the rate of depreciation a function of pollution, among other things. She also assumed that illness was not a direct argument in the utility function. She derived a willingness-to-pay measure that involved two terms. The first term is the value of the reduction in sick time. The second is the change in the rate of investment in health capital caused by the change in pollution. The sign of this term is ambiguous, since reducing pollution both lowers the price of an investment in health capital and increases the desired quantity of health capital. This second term is analogous to the expressions involving averting and mitigating activities. Once again, the conclusion is that true willingness to pay is different from changes in the cost of illness because of the induced changes in other activities and expenditures that affect health.

"IMPURE" MODELS

The models outlined so far have been "pure" in the sense that pollution affects only sick days and averting behavior serves only to reduce exposure to pollution. But if either pollution or averting activities also affect utility directly, then the model becomes "impure" in the sense that it is no longer possible to derive expressions for marginal willingness to pay that are functions only of observable data. Rather, the expressions for MWTP become "contaminated" with unobservable marginal utility or marginal disutility terms, or both.

Suppose that in addition to causing illness, pollution causes disutility directly—for example, by impairing the view from one's house. Then the expression for willingness to pay in this modified version of the basic model is

$$w_c = \frac{\partial v/\partial c}{\lambda} = -\frac{(\partial u/\partial s - \lambda \cdot p_w)\, \partial s/\partial c + \partial u/\partial c}{\lambda}$$

The first-order condition can be used to eliminate $(\partial u/\partial s)/\lambda$; but the unobservable $(\partial u/\partial c)/\lambda$ remains. Observations on averting and mitigating behavior are not sufficient to measure MWTP.[23] Similarly, suppose a generates utility in addition to reducing exposure. For example, a could be air conditioning or water filters that reduce unpleasant tastes in tap water. Then the first-order condition for a becomes

$$\left(\frac{\partial u}{\partial s} - \lambda \cdot p_w\right)\frac{\lambda \cdot p_a}{\partial s/\partial a} - \frac{\partial u/\partial a}{\partial s/\partial a}$$

When this is substituted into the indirect utility function, the unobservable $(\partial u/\partial a)/\lambda$ remains.

Chronic Morbidity

We turn now to some of the problems that arise when we try to apply these models of choice and value to the case of chronic morbidity. The

---

[23]If pollution varies across housing sites, then the hedonic housing price model provides a way around this problem. See chapter 11.

first problem is how to quantify and measure chronic morbidity. Chronic morbidity is more like a state of being than like an illness where incidence and duration are the primary characteristics of concern. It may be more useful to model chronic morbidity in terms of state-dependent preferences; for example,

$$u = u(X, s)$$

$$s = 0 \text{ for health}$$

$$s_1 - 1 \text{ for chronic disease}$$

The second problem is that it may be important to distinguish between the willingness to pay to avoid the chronic state and the willingness to pay to avoid symptoms associated with the disease. For example, for individuals with chronic heart disease, exposure to carbon monoxide air pollution can lead to an increase in the incidence of angina. An individual's willingness to pay for reducing the number of angina attacks given that he has the chronic disease can be estimated by using one of the models for acute morbidity. But on the question of willingness to pay to avoid the chronically ill state, we must choose between an ex ante and an ex post perspective. We have already discussed some of these issues in the context of mortality risks.

Suppose that air pollution increases the risk of chronic lung disease. The ex ante perspective asks for a healthy individual's willingness to pay to reduce the risk of incurring chronic lung disease at some time in the future.[24] The ex post perspective would be applicable to individuals already suffering from chronic lung disease who would be asked their willingness to pay to be restored to a healthy state. As discussed in chapter 8, the ex ante and ex post perspectives might give quite different values to the same chronic health effect, even leaving aside the likelihood that an individual who actually experiences a chronic health effect might learn something about it that would alter his willingness to pay. Which perspective is appropriate would depend, at least in part, on the nature of the policy question being asked. If a potential cure for the chronic condition is available, the ex post perspective yields the correct answer to the

---

[24]This is essentially the approach used by Viscusi, Magat, and Huber (1991). In a variation on the contingent ranking method, they asked individuals to choose between different bundles of chronic bronchitis and other attributes to estimate their risk–risk and risk–dollar marginal rates of substitution.

valuation question. If the policy in question is one that would reduce exposure to, say, air pollution leading to reduced risk of chronic disease, then the ex ante perspective is appropriate.

Measurement

In order to implement measures based on the health production function, such as equation (18), we would require the following data for a cross section of individuals over some time period:

    1. frequency, duration, and severity of pollution-related symptoms;

    2. ambient pollution levels to which the individual is exposed;

    3. actions the individual takes to avoid or mitigate the effects of air pollution;

    4. costs of avoidance and mitigating activities;

    5. other variables affecting health outcomes (age, general health status, presence of chronic conditions, and so forth).

These data would be used to estimate health production and input demand functions, which, in turn, would be used to calculate the expression for marginal willingness to pay. The same data could be used to implement a cost-of-illness measure. Items 1, 2, 4, and 5 would be used to estimate a dose-response function. The predicted effect of a change in pollution on illness would be multiplied by the value of lost time plus the observed change in medical and avoidance expenditures in order to calculate the cost of illness.

There is an important difference, however, between the data requirements of the averting behavior approach and the cost-of-illness approach. To infer MWTP using data on avoidance and mitigating behavior we should, ideally, know each individual's beliefs about the efficacy of these behaviors, since it is these beliefs that motivate observed behavior. Individuals must believe that air pollution affects illness in order for them to take actions to reduce their exposure. Rowe and Chestnut (1985) report that half of the asthmatics they surveyed believed that air pollution aggravated their asthma. The spirit of the cost-of-illness approach is to measure observable costs associated with actual changes in illness induced by air pollution. In practice, it is the objective effect of avoidance and mitigating behavior that is measured under the averting behavior approach, with the tacit assumption that individuals correctly perceive the effects of their actions.

An alternative approach to the direct observational method based on the health production function is to ask some form of hypothetical question. For example, people could be asked directly what they would be willing to pay to reduce pollution. Alternatively, people could be asked to value a reduction in the number of symptom days; then the value per symptom day could be multiplied by the predicted change in the number of symptom days based on an estimate of the dose-response function. In theory these responses could be collected without obtaining any of the information in the five data categories listed above. Collecting information on the age, health status, and income of respondents is necessary if values from hypothetical responses from a given study are to be used in other contexts.

If willingness to pay for morbidity benefits varies with health status, age, and income, pollution control policies may have very different benefits depending on the characteristics of the target population. To value the morbidity benefits of different policies we must therefore know the distribution of key variables in the population affected by the policy, and how WTP (or the cost of illness) varies with these characteristics.

In principle, each of the valuation techniques presented above is capable of describing how the value of reduced morbidity varies with the characteristics of the respondent. The averting behavior models suggest that WTP depends on any variables that affect the marginal product of pollution, mitigating activities, or avoidance activities. In practice these variables would include health status (whether or not the respondent has a chronic respiratory condition), age, and perhaps education. WTP will also vary with factors such as earned income or education that affect the cost of averting activities. The effect of these variables on WTP could be calculated from the health production function or from equations describing the unit cost of averting activities.

When WTP is estimated by the contingent valuation method, data are typically gathered on variables that would enter the health production function or affect the level of averting (mitigating) behavior undertaken. WTP responses can then be regressed on these variables. A similar approach can be taken with the cost of illness, provided that cost-of-illness data are gathered as part of a survey that also elicits socioeconomic and demographic data.

What are the prospects for measuring WTP using the averting behavior approach? It is usually difficult to measure the cost of activities that reduce exposure to pollution or that prevent symptoms from occurring altogether. The mere presence of an air conditioner in a home or in a car is an imperfect measure of reduced exposure to air pollution. The cost of averting behavior is inherently difficult to measure. For example, to determine the cost of spending leisure time indoors rather than outdoors, we cannot rely on observed prices but must question individuals directly. Also measuring the cost of averting behavior is complicated by the fact that many avoidance activities produce joint products.

Another possibility is to use mitigating behavior, in the form of medication purchased, to measure WTP. Medication has an advantage in that it is a pure averting good with a cost that is easily measurable. If we can find a product that is appropriate for each symptom of interest and that affects only that symptom, then we can sum the willingness-to-pay estimates produced by equation (21) across symptoms.

SUMMARY

In this chapter I have described the economic framework for valuing reductions in illness and risk of death on the basis of what an individual is willing to pay. The aggregate benefits of a reduction in the risk of death and incidence of illness is the sum of what each of the affected people is willing to pay to reduce his or her own risks of death and illness, plus the sum of what everyone in society is willing to pay over and above this amount to reduce the risks of death and illness for any of the other exposed people. This second component of willingness to pay might arise because of altruistic feelings, especially toward one's children or loved ones, or because members of society are forced to bear the costs of illness and premature deaths of others—through health insurance and paid sick leave, for example.

To assert that an individual has a WTP for a reduction in risk of death or illness implies that the individual can perceive and is aware of changes in his health status. It does not require that the individual know that the reduction is attributable to a specific environmental change. If the value of reduced risk or symptom days is known, policy analysts can calculate benefits if they can predict the magnitude of the reduction caused by pollution control. Assumptions about individuals' knowledge do play important roles in the empirical estimation of WTP in some circumstances, however. For example, if WTP for risk reductions is to be inferred from wage-risk premia, then it must be assumed that individuals know the relative risk levels of different jobs. Also, if WTP for reduced illness is to be inferred from an individual's averting or mitigating behavior in response to changes in pollution, then it must be assumed that the individual knows the relationship between pollution and her illness experience.

In valuing risks to life, two forms of the indirect approach have been widely used. One is based on compensating wage differentials received by workers in risky occupations; the other is based on the cost of goods that increase safety, such as seat belts and smoke detectors. Safety goods are typically 0–1 activities whose benefits exceed their costs for most persons. Thus, unless the data permit estimation of the parameters of a discrete choice model, basing risk valuations on these activities is likely to understate the value of a risk reduction. On the other hand, if the safety good also conveys utility directly, then value of risk reduction alone will be overestimated. The compensating wage approach may also understate the value of a risk reduction because persons who are willing to be paid to accept increased risk (such as structural iron workers) may have lower values for risk reduction than the average person.

A serious shortcoming of studies of job-risk premia and use of safety equipment is that they value only voluntarily assumed risks of accidental death (for example, risk of death on the job, or risk of dying in an auto accident). By contrast, environmental risks are largely involuntary, and may lead to a painful illness (such as cancer) before death occurs. These considerations suggest that individuals might be willing to pay more to reduce environmental risks than to reduce risk of death in an auto accident. On the other hand, to the extent that environmental risks may not occur until after a long latency period, fewer years of life will be lost than are risked from death in an auto accident during the current year. This suggests that willingness to pay estimates for a reduction in current risk of death may overstate willingness to pay to reduce environmental risks.

One advantage of the direct (survey) approach is that the risks being valued can be tailored to circumstances relevant to environmental health policy. For example, people can be asked their willingness to pay to reduce exposure to a pollutant that increases their risk of dying of cancer in twenty years. However, as the risk being valued becomes smaller in magnitude and more distant in time, individuals may have trouble understanding what it is they are valuing.

It may not be appropriate to use the value of a reduction in risk of accidental death during the coming year to value the reduction in risk of death associated with exposure to an environmental carcinogen. Exposure to a carcinogen may not increase risk of death until twenty years hence, which suggests that the value of reducing risk of death today overstates the value of reducing exposure to the carcinogen. On the other hand, death due to cancer may be feared more than accidental death. It is therefore unclear what the relationship between the two risk valuations should be.

We have also reviewed models for inferring this value of reduced morbidity from observable data. The value of a reduction in symptoms can be inferred either from activities that reduce exposure to air pollution (running an air purifier, staying indoors) or that mitigate the severity or duration of symptoms (taking medication, visiting a doctor).

The principal advantage of these indirect approaches is that they are based on observed behavior; however, to implement them requires a considerable amount of data. We must be able to measure and estimate the relationships among the health outcome of interest, the amount of mitigating or averting behavior undertaken, and its price. In addition, we must either know how effective the individual thought his or her averting behavior was, or we must know its actual effectiveness and assume that the individual correctly perceived this effectiveness.

REFERENCES

Arthur, W. B. 1981. "The Economics of Risks to Life," *American Economic Review* vol. 71, no. 1, pp. 54–64.

Atkinson, Scott E., and Robert Halvorsen. 1990. "Evaluation of Risks to Life: Evidence from the Market for Automobiles," *Review of Economics and Statistics* vol. 72, no. 1, pp. 133–136.

Bailey, Martin J. 1980. *Reducing Risks to Life: Measurement of the Benefits.* Washington, D.C.: American Enterprise Institute.

Bergstrom, Theodore C. 1982. "When Is a Man's Life Worth More Than His Human Capital?" in Michael W. Jones-Lee, ed., *The Value of Life and Safety.* Amsterdam, The Netherlands: North-Holland.

Berkowitz, Monroe, and William G. Johnson. 1974. "Health and Labor Force Participation," *Journal of Human Resources* vol. 19, no. 1, pp. 117–128.

Blomquist, Glenn. 1979. "Value of Life Savings: Implications of Consumption Activity," *Journal of Political Economy* vol. 87, no. 3, pp. 540–558.

Blomquist, Glenn. 1981. "The Value of Human Life: An Empirical Perspective," *Economic Inquiry* vol. 19, no. 1, pp. 157–164.

Broome, John. 1978. "Trying to Value a Life," *Journal of Public Economics* vol. 9, no. 1, pp. 91–100.

Broome, John. 1979. "Trying to Value a Life: Reply," *Journal of Public Economics* vol. 12, no. 2, pp. 259–262.

Broome, John. 1982. "Uncertainty in Welfare Economics, and the Value of Life," in Michael W. Jones-Lee, ed., *The Value of Life and Safety.* Amsterdam, The Netherlands: North-Holland.

Broome, John. 1985. "The Economic Value of Life," *Economica* vol. 52, no. 207, pp. 281–294.

Broome, John. 1987. "The Economic Value of Life: A Reply," *Economica* vol. 54, no. 215, p. 401.

Buchanan, James M., and Roger L. Faith. 1979. "Trying Again to Value a Life," *Journal of Public Economics* vol. 12, no. 2, pp. 245–248.

Conley, B. C. 1976. "The Value of Human Life in the Demand for Safety," *American Economic Review* vol. 66, no. 1, pp. 45–55.

Cook, Philip J. 1978. "The Value of Human Life in the Demand for Safety: Comment," *American Economic Review* vol. 68, no. 4, pp. 710–711.

Cooper, B. S., and Dorothy P. Rice. 1976. "The Economic Cost of Illness Revisited," *Social Security Bulletin* vol. 39, no. 2, pp. 21–36.

Cropper, Maureen L. 1977. "Health, Investment in Health, and Occupational Choice," *Journal of Political Economy* vol. 85, no. 6, pp. 1273–1294.

Cropper, Maureen L. 1981. "Measuring the Benefits from Reduced Morbidity," *American Economic Review* vol. 71, no. 2, pp. 235–240.

Cropper, Maureen L., and A. Myrick Freeman III. 1989. "Estimating Individuals' Values for Health Benefits," in Congressional Research Service, U.S. Library of Congress, *Health Benefits of Air Pollution Control: A Discussion.* Washington, D.C.: Library of Congress.

Cropper, Maureen L., and A. Myrick Freeman III. 1991. "Valuing Environmental Health Effects," in John B. Braden and Charles D. Kolstad, eds., *Measuring the Demand for Environmental Quality.* Amsterdam, The Netherlands: Elsevier.

Cropper, Maureen L., and Frances G. Sussman. 1988. "Families and the Economics of Risks to Life," *American Economic Review* vol. 78, no. 1, pp. 255–260.

Cropper, Maureen L., and Frances G. Sussman. 1990. "Valuing Future Risks to Life," *Journal of Environmental Economics and Management* vol. 20, no. 2, pp. 160–174.

Dardis, Rachel. 1980. "The Value of a Life: New Evidence from the Marketplace," *American Economic Review* vol. 70, no. 5, pp. 1077–1082.

Dickie, Mark, and Shelby Gerking. 1991. "Valuing Reduced Morbidity: A Household Production Approach," *Southern Economic Journal* vol. 57, no. 3, pp. 690–702.

Doll, Richard, and Richard Peto. 1981. "The Causes of Cancer: Quantitative Estimates of Avoidable Risks of Cancer in the United States Today," *Journal of the National Cancer Institute* vol. 66, pp. 1193–1308.

Gerking, Shelby, and Linda R. Stanley. 1986. "An Economic Analysis of Air Pollution and Health: The Case of St. Louis," *Review of Economics and Statistics* vol. 68, no. 1, pp. 115–121.

Grossman, Michael. 1972. "On the Concept of Health Capital and the Demand for Health," *Journal of Political Economy* vol. 80, no. 2, pp. 223–255.

Harrington, Winston, and Paul R. Portney. 1987. "Valuing the Benefits of Health and Safety Regulations," *Journal of Urban Economics* vol. 22, no. 1, pp. 101–112.

Jones-Lee, Michael W. 1974. "The Value of Changes in the Probability of Death or Injury," *Journal of Political Economy* vol. 99, no. 4, pp. 835–849.

Jones-Lee, Michael W. 1976. *The Value of Life: An Economic Analysis.* Chicago: University of Chicago Press.

Jones-Lee, Michael W. 1979. "Trying to Value a Life: Why Broome Does Not Sweep Clean," *Journal of Public Economics*, vol. 12, no. 2, pp. 249–256.

Jones-Lee, Michael W. 1982. *The Value of Life and Safety.* Amsterdam, The Netherlands: North-Holland.

Jones-Lee, Michael W. 1986. Personal communication to M. L. Cropper, November 1986.

Jones-Lee, Michael W. 1987. "The Economic Value of Life: A Comment," *Economica* vol. 54, no. 215, pp. 397–400.

Jones-Lee, Michael W., M. Hammerton, and P. R. Philips. 1985. "The Value of Safety: Results of a National Sample Survey," *Economic Journal* vol. 95, no. 377, pp. 49–72.

Landefeld, J. Steven, and Eugene P. Seskin. 1982. "The Economic Value of Life: Linking Theory to Practice," *American Journal of Public Health* vol. 72, no. 6, pp. 555–566.

Lave, Lester B., and Eugene P. Seskin. 1971. "Air Pollution and Human Health," *Science* vol. 169 (August 21), pp. 723–731.

Lave, Lester B., and Eugene P. Seskin. 1977. *Air Pollution and Human Health.* Baltimore, Md.: Johns Hopkins University Press for Resources for the Future.

Linnerooth, Joanne. 1979. "The Value of Human Life: A Review of the Models," *Economic Inquiry* vol. 27, no. 1, pp. 52–74.

Linnerooth, Joanne. 1982. "Murdering Statistical Lives...?" in Michael W. Jones-Lee, ed., *The Value of Life and Safety.* Amsterdam, The Netherlands: North-Holland.

Mathtech, Inc. 1986. "Methodology for Valuing Health Risks of Ambient Lead Exposure: Draft Report." Princeton, N.J.: Mathtech, Inc.

Mishan, Ezra J. 1971. "Evaluation of Life and Limb: A Theoretical Approach," *Journal of Political Economy* vol. 79, no. 4, pp. 687–705.

Mishan, Ezra J. 1981. "The Value of Trying to Value a Life," *Journal of Public Economics* vol. 15, no. 2, pp. 133–137.

Moore, Michael J., and W. Kip Viscusi. 1988. "The Quantity-Adjusted Value of Life," *Economic Inquiry* vol. 26, no. 3, pp. 369–388.

Peltzman, Sam. 1975. "The Effects of Automobile Safety Regulation," *Journal of Political Economy* vol. 83, no. 4, pp. 677–725.

Peterson, William. 1975. *Population*, 3d ed. New York: Macmillan.

Portney, Paul R., and John Mullahy. 1986. "Urban Air Quality and Acute Respiratory Illness," *Journal of Urban Economics* vol. 20, no. 1, pp. 21–38.

Rowe, Robert D., and Lauraine G. Chestnut. 1985. *Oxidants and Asthmatics in Los Angeles: A Benefits Analysis.* Report No. EPA-230–07–85–010. Washington, D.C.: U. S. Environmental Protection Agency.

Schelling, Thomas C. 1968. "The Life You Save May Be Your Own," in Samuel B. Chase, Jr., ed., *Problems in Public Expenditure Analysis.* Washington, D.C.: Brookings Institution.

Shepard, Donald S., and Richard J. Zeckhauser. 1982. "Life Cycle Consumption and Willingness to Pay for Increased Survival," in Michael W. Jones-Lee, ed., *The Value of Life and Safety.* Amsterdam, The Netherlands: North-Holland.

Shepard, Donald S., and Richard J. Zeckhauser. 1984. "Survival vs. Consumption," *Management Science* vol. 30, pp. 423–439.

Sussman, Frances G. 1984. "A Note on the Willingness to Pay Approach to the Valuation of Longevity," *Journal of Environmental Economics and Management* vol. 11, no. 1, pp. 84–89.

Thaler, Richard. 1982. "Precommitment and the Value of a Life," in Michael W. Jones-Lee, ed., *The Value of Life and Safety.* Amsterdam, The Netherlands: North-Holland.

Thaler, Richard G., and William Gould. 1982. "Public Policy Toward Lifesaving: Should Consumer Preferences Rule?" *Journal of Policy Analysis and Management* vol. 1, no. 2, pp. 223–242.

Thaler, Richard G., and Sherwin Rosen. 1976. "The Value of Life Savings," in Nester Terleckyj, ed., *Household Production and Consumption.* New York: Columbia University Press.

Ulph, Alistair. 1982. "The Role of Ex Ante and Ex Post Decisions in the Valuation of Life," *Journal of Public Economics* vol. 18, no. 2, pp. 265–276.

U.S. Department of Health, Education and Welfare, National Center for Health Statistics. 1964. *Health Survey Procedure, Vital and Health Statistics Programs and Collection Procedures,* Series 1, no. 2. Washington, D.C.

U.S. Environmental Protection Agency. 1986a. *Air Quality Criteria for Lead.* Research Triangle Park, N.C.: Environmental Criteria and Assessment Office.

U.S. Environmental Protection Agency. 1986b. "Review of the National Ambient Air Quality Standards for Lead: Assessment of the Scientific and Technical Information," OAQPS Draft Staff Paper. Research Triangle Park, N.C.: Office of Air Quality Planning and Standards.

U.S. Environmental Protection Agency. 1987. *Summary of Selected New Information on Effects of Ozone on Health and Welfare.* Corvallis, Ore.: U.S. Environmental Protection Agency.

Usher, Dan. 1973. "An Imputation to the Measure of Economic Growth for Changes in Life Expectancy," in Milton Moss, ed., *The Measurement of Economic and Social Performance.* Vol. 38 of *Studies in Income and Wealth.* New York: National Bureau of Economic Research.

Viscusi, W. Kip, Wesley A. Magat, and Joel Huber. 1991. "Pricing Health Risks: Survey Assessments of Risk-Risk and Risk-Dollar Tradeoffs," *Journal of Environmental Economics and Management* vol. 21, no. 1, pp. 32–51.

Viscusi, W. Kip, and Michael J. Moore. 1989. "Rates of Time Preference and Valuation of the Duration of Life," *Journal of Public Economics* vol. 38, no. 3, pp. 297–317.

Williams, Alan. 1979. "A Note on 'Trying to Value a Life'," *Journal of Public Economics* vol. 12, no. 2, pp. 257–258.

11

Property Value Models

E conomists were documenting the relationship between the prices of housing units and quantities of environmental amenities even before this relationship had been recognized as an application of the newly developed theory of hedonic prices.[1] The latter half of the 1970s and the decade of the 1980s saw an explosion of both theoretical and empirical studies of the monetary values of nonmarket amenities and disamenities based on hedonic price theory.[2] Now it is well accepted that housing price differentials do reflect differences in the quantities of various characteristics of housing and that these differentials have significance for applied welfare analysis. The two questions that have been explored most intensively in this literature are the proper specification and estimation of the model relating housing prices to amenities, and the development of measures of welfare change that make the best use of the available data and that are consistent with the underlying economic theory. The purpose of this chapter is to review the current state of knowledge concerning these two questions.

After a brief review of the evolution of economic thinking about property prices and environmental amenities, I present a more detailed exposition of the cross-section hedonic property value model. This presentation includes discussions of: problems in estimating the hedonic price function; approaches to recovering

---

[1]One of the principal early contributors to the development of the hedonic price theory as applied to differentiated consumer products such as automobiles was Griliches (1971). But the concept can be traced back at least to Court (1941).

[2]Smith and Huang (1991) recently conducted a meta-analysis of hedonic studies of air pollution and housing prices. They reported finding 37 studies and more than 160 separate estimates of the effect of some air quality variable on housing prices.

information on preferences and the demands for characteristics from the hedonic price function; and the measurement of welfare change. Also included are discussions of the relationships among property prices, rents, and taxes and how these relationships affect the measurement of welfare changes, the relationship between the hedonic model and the averting behavior model, and the application of the hedonic model to markets for agricultural and commercial land. I then describe two alternative approaches to interpreting property price information: the repeat sales model and discrete choice models. All of these models have their foundations in individual choice theory and a model of the equilibrating processes of the housing market.

HISTORICAL BACKGROUND

The theory of rents holds that the equilibrium price for a parcel of land will be the present value of the stream of rents produced by the land. Economic theory has long recognized that the productivity of land differs across sites. These productivity differentials will yield differential rents to land and therefore differential land values. Where land is a producer's good, competition and free entry are sufficient to ensure that productivity differentials are fully reflected in the land rent structure. For any property where the land rent is less than the productivity, the activity occupying that land must be earning a profit. Some potential entrant will be willing to bid above the going rent in order to occupy that site and reap the rewards of a superior productivity. It is this competition that bids up land rents and eliminates the profit. Rent differentials will be equal to productivity differentials. And since the price at which a unit of land sells in the market is the present value of the stream of future rents, productivity differentials will also be reflected in land prices.

Some environmental characteristics such as air or water quality may affect the productivity of land either as a producer's good or a consumer's good. Where this is so, the structure of land rents and prices will reflect these environmentally determined productivity differentials. These results from classical rent theory aroused considerable interest among economists about the possibility of using data on land rent or land value for residential properties to measure the benefits to households brought about by improvements in environmental characteristics such as air or water quality. Ridker (1967) was the first economist to attempt to use residential property value data as the basis for estimating the benefits of changes in measures of environmental quality such as air pollution. He reasoned as follows:

> If the land market were to work perfectly, the price of a plot of land would equal the sum of the present discounted streams of benefits and costs derivable from it. If some of its costs rise (e.g., if additional maintenance and cleaning costs are required) or if some of its benefits fall (e.g., if one cannot see the mountains from the terrace) the property will be discounted in the market to reflect people's evaluation of these changes. Since air pollution is specific to locations and the supply of location is fixed, there is less likelihood that the negative effects of pollution can be significantly shifted on to other markets. We should therefore expect to find the majority of effects reflected in this market, and we can measure them by observing associated changes in property values. (p. 25)

The last sentence of this passage raises three questions. The first is whether environmental variables such as air pollution do systematically affect land prices. Assuming an affirmative answer to this question, the second is whether knowledge of this relationship is sufficient to predict changes in land prices when, say, air pollution levels change. And the third question is whether changes in land prices accurately measure the underlying welfare changes.

Ridker (1967) and Ridker and Henning (1967) provided the first empirical evidence that air pollution affects property values by regressing median census tract property values on a measure of sulfate air pollution. They then went on to assert positive answers to the second and third questions. Specifically, they argued that the coefficient on the air pollution variable in the regression equation could be used to predict the change in the price of any residence conditioned on a change in its air pollution level, and that the sum of all such changes could be taken as a measure of the benefit of improving air quality in an urban area (Ridker, 1967, pp. 136-137, and Ridker and Henning, 1967, p. 254).

This work stimulated a by-now large literature on the proper theoretical interpretation of the observed air pollution–property value relationship. Early contributions included those by Freeman (1971, 1974a, and 1974b), and Anderson and Crocker (1972). Subsequent efforts to provide a sound theoretical basis for interpreting the air pollution–property value relationship have taken one of two paths. The first has been the development of models of the urban land market to determine whether and under what circumstances changes in aggregate land values accurately measure the benefits associated with environmental improvements. Early efforts in this direction included those by Strotz (1968), Lind (1973), Pines and Weiss (1976), and Polinsky and Shavell (1976). A recent contribution has been Kanemoto (1988). Although some of these models can be given an interpretation in the context of hedonic price theory, they do not lend themselves to empirical application, so they are not covered in this book.[3]

The second path drew on hedonic price theory to interpret the derivative of the cross-section regression equation with respect to air pollution as a marginal implicit price and therefore a marginal value for the air quality improvement (see Freeman, 1974b, and Rosen, 1974). In this chapter I describe this approach to welfare measurement, along with more recent developments such as the repeat sales hedonic model and discrete choice models of housing choice. My primary emphasis is on model specification and interpretation rather than econometric estimation. My objectives are to provide an overview of the methods of welfare measurement based on property values and to identify the major conceptual issues.

THE CROSS-SECTION HEDONIC PROPERTY VALUE MODEL

The hedonic price theory first introduced in chapter 4 provides the basis for deriving welfare measures from observed differences in prices of houses. In this section I first review the basic theory of hedonic prices in the context of the housing market. I then discuss a number of topics involving the estimation of the model, its interpretation, and its application.

The Basic Model

Assume that each individual's utility is a function of that person's consumption of a composite commodity X, a vector of location-specific environmental amenities Q, a vector of structural characteristics of the house the person occupies (such as size, number of rooms, age, and type of construction) denoted by S, and a vector of characteristics of the neighborhood in which the house is located (such as quality of local schools, accessibility to parks, stores, and work place, and crime rates) denoted by N.

---

[3]I discussed some of these models in my earlier book (Freeman, 1979). For further discussion of this branch of the literature, see Bartik and Smith (1987) and Palmquist (1991).

Any large area has in it a wide variety of sizes and types of housing with different locational, neighborhood, and environmental characteristics. An important assumption of the hedonic technique is that the urban area as a whole can be treated as a single market for housing services. Individuals must have information on all alternatives and must be free to choose a house anywhere in the urban market. It is as if the urban area were one huge supermarket offering a wide selection of varieties. Of course, individuals cannot move their shopping carts through this supermarket. Rather, their selections of residential locations fix for them the whole bundle of housing services. It is much as if shoppers were forced to make their choices from an array of already filled shopping carts. Individuals can increase the quantity of any characteristic by finding an alternative location alike in all other respects but offering more of the desired characteristic.

Since our interest is in the values of characteristics to buyers of houses, there is no need to model formally the supply side of this market. Let us simply assume that the housing market is in equilibrium, that is, that all individuals have made their utility maximizing residential choices given the prices of alternative housing locations, and that these prices just clear the market given the existing stock of housing and its characteristics. Under these assumptions, the price of the ith residential location can be taken to be a function of the structural, neighborhood, and environmental characteristics of that location. In other words,

$$P_{h_i} = P_h(S_i, N_i, Q_i) \tag{1}$$

As explained in chapter 4, this relationship can be linear in a characteristic if repackaging of that characteristic is possible. But in general this need not be the case. Two living rooms with six-foot ceilings are not equal to one living room with a twelve-foot ceiling. Where repackaging is not possible, equation (1) will be nonlinear.

To model the problem more formally, consider an individual who occupies house i. Her utility is given by

$$u = u(X, Q_i, S_i, N_i) \tag{2}$$

Assume that preferences are weakly separable in housing and its characteristics. This assumption makes the demands for characteristics independent of the prices of other goods, a convenient property for empirical work. The individual maximizes $u(\cdot)$ subject to the budget constraint:[4]

$$M - P_{h_i} - X = 0 \tag{3}$$

The first-order condition for the choice of environmental amenity q_j is

$$\frac{\partial u / \partial q_j}{\partial u / \partial x} = \partial P_{h_i} / \partial q_j$$

Assume now that the hedonic price function $P_h(\cdot)$ has been estimated for an urban area. Its partial derivative with respect to any of its arguments, for example q_j, gives the implicit marginal price of that characteristic, that is, the additional amount that must be paid by any household to move to a bundle with a higher level of that characteristic, other things being equal. If this function is nonlinear, the marginal implicit price of a characteristic is not constant, but depends on its level and perhaps the levels of other characteristics as well. If the individual is assumed to be a price taker in the housing market, that person can be viewed as facing an array of implicit marginal price schedules for various characteristics. An individual maximizes utility by simultaneously moving along each marginal price schedule until she reaches a point where her marginal willingness to pay for an additional unit of that characteristic just equals the marginal implicit price of that characteristic. If an individual is in equilibrium, the marginal implicit prices associated with the housing bundle actually chosen must be equal to the corresponding marginal willingnesses to pay for those characteristics.

Panel A in figure 11-1 shows the partial relationship between P_{h_i} and q_j as estimated from equation (1). Panel B shows the marginal implicit price of q_j, $\partial P_{h_i} / \partial q_j$. It also shows the marginal willingness-to-pay curves for two individuals, k and m, who have chosen utility maximizing bundles of housing characteristics; these are labelled $b_k(q_j)$ and $b_m(q_j)$. These curves show each individual's marginal willingness to pay for changes in the characteristic, holding utility constant at the level achieved by maximizing equation (2) subject to (3).

---

[4]For the urban form in which most jobs are located at the city center, distance from the center, d, is one important characteristic of a house. The budget constraint becomes

$$M - P_{h_i} - t(d_i) - X = 0$$

where $t(d_i)$ is the cost of commuting to the job.

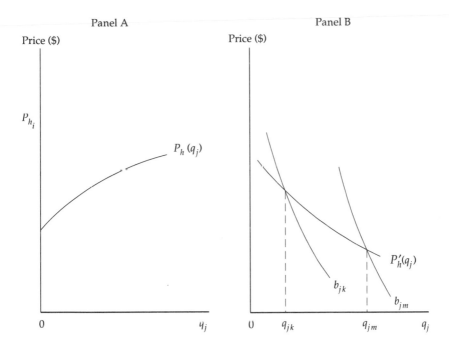

Figure 11-1. The hedonic price function, marginal implicit price function, and individual willingnesses to pay

Let this level be u^*. Both individuals in figure 11-1 have chosen locations where their marginal willingnesses to pay for q_j are equated with its marginal implicit price.

The hedonic price function shown in figure 11-1 is concave from below, meaning that the marginal implicit price function is downward-sloping. Although in principle the hedonic price function could take any form (concave, convex, or linear), there are a priori reasons for expecting the hedonic price function to be concave from below for an amenity like clean air. In that case, q_j is an inverse function of the concentrations of pollutants in the atmosphere. As pollutant concentrations approach zero, q_j approaches some upper limit. If the hedonic price function were concave upward, this would imply that as pollutant concentrations approach zero, the marginal implicit price of q_j approaches its maximum. This in turn would mean that those who have obtained the highest available q_j would also have the highest marginal willingness to pay for additional q_j. If the hedonic price function were concave from below, those individuals with high marginal willingness-to-pay curves who already experience high levels of q_j would have a low willingness to pay at the margin for additional q_j.

The analysis described here results in a measure of the price of and the marginal willingness to pay for q_j but does not directly reveal the marginal willingness-to-pay function. The second stage of the hedonic technique is to combine the quantity and implicit price information in an effort to identify the marginal willingness-to-pay function for q_j. The individual's demand price or willingness to pay for q_j is a function of the level of q_j. Since there may be substitute and complementary relationships among characteristics, willingness to pay for q_j may also depend on the levels (or marginal implicit prices) of other characteristics. It is convenient to assume that the utility function is weakly separable in housing so that prices of other goods can be omitted in the specification of the marginal willingness-to-pay function. Also, it is convenient to assume that each individual purchases only one housing bundle. If more than one were purchased, it would be necessary that the bundles be identical or that the hedonic price function be linear in all characteristics. This is because there can be only one marginal implicit price recorded for each individual for each characteristic.

Given these assumptions, for the individual who chooses to locate at i we have:

$$b_{ij} = b_{ij}(q_j, Q_i^*, S_i, N_i, u^*) \tag{4}$$

where Q_i^* is all amenities except q_j. The task of identifying this function from observed choices is addressed later in this section. If (4) can be identified, it can be used to estimate the welfare change of an individual associated with changes q_j, assuming that other things are held equal. Specifically, if the quantities of other characteristics and amenities do not change, the welfare change can be found by integrating b_{ij} over the relevant range of the change in q_j. However, a change in the quantity of one characteristic can result in changes in the quantities of other characteristics the individual chooses and in changes in the hedonic price function itself. The task of welfare measurement when individuals can fully adjust to the new supply of amenities and characteristics is discussed below, in the section on Measuring Welfare Changes.

Estimating the Hedonic Price Function

THE DEPENDENT VARIABLE

In specifying the hedonic price function, the first question we must ask is whether the dependent variable to be explained should be a pure site or land value or the full price of the house and land together. Since the environmental amenities of interest are location-specific but not a part of the structure, the values of the environmental amenities should be reflected in the price of land alone. However, at least in the United States land is not usually traded separately from the structures placed upon it, so the observed prices reflect the values of both the land and its structural improvements. This causes no problems at the theoretical level, but it does require that the hedonic price equation adequately control for structural characteristics.

The second question concerns the source of data on housing prices. Data on actual market transactions are preferable. For rental housing there is a regular monthly "market transaction," from which fairly accurate data on housing rents could be gathered. However, the majority of residential housing is owner-occupied, and only a very small percentage of the total owner-occupied housing stock is exchanged through the market each year.

The earliest hedonic property value studies used data from the U.S. Census of Population and Housing (see, for example, Ridker and Henning, 1967; Anderson and Crocker, 1971; Polinsky and Rubinfeld, 1977; and Harrison and Rubinfeld, 1978). The census asks each owner to estimate the value of his or her property. The census also gathers other data on structural characteristics, as well as socio-economic data on occupants. These data are aggregated by census tracts and reported as means or medians. Although the census tract observations represent a convenient source of data for property value studies, there are two kinds of problems with them.

The first concerns the degree of accuracy of estimates of values by individual owners. Nelson (1978a) was able to compare median owners' estimates with median professional assessments by census tracts for Washington, D.C. He found that owners' estimates were systematically higher by between 3 and 6 percent, while zero-order correlation coefficients were approximately 0.9 or better. He concluded that as long as errors in owners' estimates are random, statistical estimates of price functions will be unbiased.

The second problem with census tract data is the loss of detail and reduced ability to control for relevant housing and location characteristics, both because of the limited number of variables reported and because of aggregation of individual data by census tract. Census tract boundaries are chosen in an effort to construct relatively homogeneous units in terms of housing and socioeconomic characteristics. If within-tract variation is relatively small compared with the variation among tracts, then relatively little is lost by aggregating a given set of observations to census tract units before undertaking the statistical analysis. But even within generally homogeneous communities there may be substantial variation in relevant characteristics, such as number of rooms. The effect of these variables on individual property values would be masked by aggregation.

An alternative source of property value data would be professional appraisals of individual properties for taxation or other purposes. Some jurisdictions have developed computer-based systems of appraisals, made for tax purposes, which include data not only on appraised values but also on a variety of structural and site characteristics. As these systems are developed and extended, they can provide a valuable data source for further property value studies. However, the appraisals must be used with caution.

The most preferred source of data is systematically collected information on actual sales prices of individual dwellings, along with relevant characteristics. Fortunately, in many parts of the country these data are collected by multiple-listing services and tax assessing agencies. The first study to exploit this source of data was by Crocker (1970). In recent years, virtually all published hedonic property value studies have used micro data on individual transaction prices.

The presumed superiority of individual transaction data over nonmarket estimates provided by experts is based on the assumption that the housing market is in an equilibrium in which all opportunities for possible gains from further trade at the revealed set of prices have been exhausted. This is a heroic assumption. Buyers and sellers often operate with substantial ignorance about the true willingness-to-pay and willingness-to-accept offers of other potential buyers and sellers. Sellers typically state an asking price that effectively truncates potential offers at that price. Sellers must choose to accept or reject offers more or less at the time they are received, without knowing when (or even if) a higher offer might come along. And buyers lack information on possible prior bids made by others for a given property (Horowitz, 1986). Horowitz developed an alternative model of the bidding and acceptance strategies of buyers and sellers, and estimated both a standard hedonic model and his alternative model with the same data set. He found that the statistical performance of his bidding model was substantially superior to the standard model in predicting sales prices. However, since he used principal components in his estimation procedure rather than actual attributes, it is not possible to analyze the impact of his alternative modeling strategy on marginal implicit prices for environmental attributes.

EXPLANATORY VARIABLES

In choosing the appropriate explanatory variables, the first question to be addressed is the way in which environmental amenities and location characteristics enter the hedonic price function. The typical practice has been to enter a simple scalar measure of an amenity—for example, parts per million (ppm) of an air pollutant, or distance to a park. But Parsons (1990) has shown that this practice is not consistent with a restriction imposed on the hedonic price function by profit maximizing behavior on the supply side of the housing market. The implication of profit maximization is that the effect of the environmental amenity can only be captured without bias by weighting the amenity by the area of the lot on which the house sits. The restriction is that if an area of land of given q_j is developed and sold in two or more different lot sizes, the prices of the lots must be such that the return per acre is independent of the sizes of the lots. For example, if two one-acre lots sell for \$X apiece, one two-acre lot must sell for \$2X. This means that the premium on lots with higher levels of q_j must be twice as high for the two-acre lot compared with the one-acre lots. The higher premium on the larger lot is necessary to compensate the landowner for the forgone opportunity to capture two premiums with the smaller lots.[5]

---

[5]Parsons' (1990) specification of the amenity variable is algebraically equivalent to the earlier model of Barton Smith (1978) when the hedonic price function is linear. Smith measured a premium on the sales prices of lots and found the premium to be negatively associated with air pollution levels. The premium was defined as

$$\text{PREM}_i = (P_{a_i} - P_{a*}) \cdot s_i$$

where P_{a_i} = unit price of land of quality q_i,
P_{a*} = unit price of land of minimum quality, and
s_i = area of lots with q_i.

Rearranging and assuming no structures and no other amenities, we have

$$P_{h_i} = s_i \cdot P_{a_i}$$
$$= s_i \cdot P_{a*} + \text{PREM}_i$$

Specifying a linear hedonic price function with a weighted amenity variable gives

$$P_{h_i} = s_i \cdot c_1 + s_i \cdot c_2 \cdot q_{ji}$$

So c_1 can be interpreted as the unit value of land with the minimum value q_{ji} and $c_2 \cdot q_j$ is the unit premium for the amenity. Notice also that in the Parsons specification, the marginal implicit prices are

$$\text{MIP}_s = c_1 + c_2 \cdot q$$
and
$$\text{MIP}_q = c_2 \cdot s$$

Although the argument is correct in principle, there is some question about its relevance in practice. As Parsons points out, once lots are developed, the cost of changing the size of lots on which houses sit may be too high to force amenity premiums to take the weighted form in the secondhand market for houses. If existing houses on large lots have the same characteristics as new houses on the same lot size, and new and old houses are considered to be very close substitutes, arbitrage between new and secondhand houses could maintain the weighted amenity price. But the substantial transactions costs in the housing market make it difficult to place much weight on arguments based on arbitrage.

The levels of some environmental amenities are fixed by location, while the levels of others, especially those related to air quality, vary over time with changes in emissions and meteorological conditions. With time-varying amenities, there is the question of how best to represent the level of the amenity in the regression equation. The typical practice in air pollution–property value studies has been to use the annual mean as a summary statistic. But Murdoch and Thayer (1988) have shown that in the case of visibility, using more information on the probability distribution of visual range improves the statistical performance of the hedonic price function.

A number of other conceptual and practical issues must be resolved in the course of selecting a set of explanatory variables for a hedonic price function. These include: Which measures of environmental quality should be used to characterize environmental amenities? Is it possible to separate the effects of different amenities on property values when measures of the amenities are correlated? What objective data best capture "neighborhood" characteristics? Does the spatial scale of the socioeconomic data often used in these studies correspond to peoples' perceptions of these characteristics? (For example, is there sufficient homogeneity within census tracts so that census tract means or medians adequately measure "neighborhood"?) And is there sufficiently close correspondence between peoples' perceptions of amenity levels (which presumably govern the choices reflected in property prices) and the objective measures of amenity levels that are available to the researcher?

Since the objective of the hedonic analysis is to determine the effect of one amenity on property values, other things being equal, a key issue is the control for structural, neighborhood, and other environmental variables. The issue is made more difficult by the likelihood of multicollinearity among housing characteristics. This raises the troublesome question of the trade-off between increasing bias through the omission of variables that are correlated with the variable of concern and increasing the variance or imprecision of coefficient estimates when collinear variables are included. Theory does not provide any hard-and-fast answers to this question. Recent work by Atkinson and Crocker (1987) and Graves and coauthors (1988) suggests the value of approaching this question systematically, using Bayesian principles. These authors have also examined the effects of errors in the measurement of other explanatory variables on the estimates of the coefficients on the environmental variables of concern.

FUNCTIONAL FORM

Functional forms for the hedonic price function that have been proposed or used in the literature include the linear, the quadratic, the log–log, the semi-log, the inverse semi-log, the exponential, and the Box–Cox transformation. The first step in choosing a functional form is to see what theory can tell us. According to theory, a hedonic price function is an equilibrium relationship derived from the interaction of individuals' preferences and suppliers' cost or profit functions. The only obvious general restriction on the form of the hedonic price function is that its first derivative with respect to an environmental characteristic be positive (negative) if the characteristic is a good (bad). As discussed above, it is plausible that the second derivative of the function be negative. But this property cannot be deduced from the general features of the hedonic model. Beyond that, there is not much that can be derived from first principles.

Rosen (1974) and Epple (1987) have shown that it is possible to solve for the hedonic price function analytically after making specific assumptions about the form of individual utility functions and the distribution of suppliers' characteristics (Rosen) or the exogenous supply of housing characteristics (Epple). These analytical solutions are only possible for a very limited set of assumed forms of preferences and supply. For example, Rosen assumed that individuals utility functions were linear. This is not an attractive assumption, especially if the ultimate objective of the analysis is to measure welfare values for changes in supplies of environmental characteristics.

Early researchers tried alternative functional forms for the hedonic price function (typically the semi-log, inverse semi-log, and log–linear) and selected one on the basis of goodness-of-fit criteria. Goodman (1978) was one of the first to experiment with a flexible functional form. He employed a Box–Cox transformation of the dependent variable:

$$P_h{}^{[\lambda]} = \frac{P_h^{\lambda} - 1}{\lambda}$$

For $\lambda = 1$, this is a simple linear function. As λ approaches zero, this becomes the semi-log form. Some authors have found estimates of λ that were significantly different from both zero and one, indicating that this more complicated form fit the data better than either the linear or semi-log forms.

Transforming only the dependent variable still produces a very limited range of possibilities. Halvorsen and Pollakowski (1981) proposed that what they called a quadratic Box–Cox functional form be estimated. It would have the form

$$P_h{}^{[\lambda]} = a_o + \sum_{i=1}^{n} a_i \cdot z_i{}^{[\beta]} + 1/2 \sum_{i=1}^{n} \sum_{i=1}^{n} c_{ij} \cdot z_i{}^{[\beta]} \cdot z_j{}^{[\beta]} \qquad (5)$$

for all i, j and where i and j index characteristics, and:

$$P_h{}^{[\lambda]} = (P_h^{\lambda} - 1)\,\lambda \text{ for } \lambda \neq 0,$$
$$P_h{}^{[\lambda]} = \ln P_h \text{ for } \lambda = 0,$$
$$z_i{}^{[\beta]} = (z_i^{\beta} - 1)\,/\beta \text{ for } \beta \neq 0, \text{ and}$$
$$z_i{}^{[\beta]} = \ln z_i \text{ for } \beta = 0$$

Cassel and Mendelsohn (1985) pointed out that for welfare analysis, it is not the goodness of fit of the hedonic price function that matters; rather, it is the estimate of the marginal implicit price of the environmental attribute. In the regression equation for housing price, the environmental variable is likely to have relatively little influence in determining the estimated magnitude of β. But the estimate of β would have a major impact on the estimated marginal implicit price of the environmental characteristic.

The most general flexible form would be an extension of the quadratic Box–Cox of equation (5) to:

$$P_h{}^{[\lambda]} = a_o + \sum_{i=1}^{n} a_i \cdot z_i{}^{[\beta_i]} + 1/2 \sum_{i=1}^{n} \sum_{i=1}^{n} c_{ij} \cdot z_i{}^{[\beta_i]} \cdot z_j{}^{[\beta_j]} \qquad (6)$$

for all i, j. This functional form allows for different transformations for each independent variable. But it is not feasible to estimate this form for any realistic number of characteristics. A compromise, proposed by Palmquist (1991), would be to set $\beta_i = \beta_j$ for all $i = 1, \ldots, n - 1$, where n indexes the environmental attribute of interest. This would be responsive to the point raised by Cassel and Mendelsohn. Allowing for a separate transformation of the environmental amenity should give better results.

One question about functional form is whether the form chosen allows the marginal implicit price of the environmental characteristic to depend on the levels of the other attributes of houses. Of the commonly used functional forms, only the log and the Box–Cox transformation make the implicit prices of characteristics depend on the levels of other characteristics; the other forms impose independence. But this is a question that should be answered by the data, not by assumption.

A recent study by Cropper, Deck, and McConnell (1988) sheds some light on the question of functional form. These authors simulated the performance of a housing market using real data on buyer and housing characteristics drawn from the Baltimore, Maryland, area. After specifying the functional form and parameters of individuals' utility functions and the distribution of characteristics that reflect taste differences across individuals, they solved an assignment problem that produces a housing market equilibrium with each house being sold to the individual with the highest willingness to pay for its bundle of characteristics. With knowledge of the utility function parameters, it was then possible to calculate the true marginal implicit price for each individual and for the mean across all individuals. Cropper, Deck, and McConnell used the equilibrium prices to estimate six alternative functional forms for the hedonic price function. They were then able to compare the mean true marginal bids with the bids calculated from each hedonic price function. They found that when all of the housing characteristics were included in the hedonic price function, the linear and quadratic versions of the Box–Cox transformation provided the most accurate estimates of marginal implicit prices. However, when they experimented with various forms of misspecified hedonic price functions (by omitting variables or using proxy variables), they found that the linear version of the Box–Cox transformation was consistently superior in generating marginal implicit prices. This was apparently because with the quadratic version of the function there are more coefficients to estimate for each characteristic; and omitting a variable leads to bias in more of the coefficients. Since there is no reason to expect these biases to be offsetting, the net result is a larger bias in the estimated marginal implicit prices (which are functions of all of the relevant coefficients).

IS THE MARKET IN EQUILIBRIUM?

Interpreting the marginal implicit prices as measures of households' marginal willingnesses to pay requires the assumption that each household is in equilibrium with respect to a given vector of housing prices and that the vector of housing prices is the one that just clears the market for a given stock of housing. In other words, it is the price vector that makes all participants in the market in aggregate just willing to hold the existing stock of housing. For these two aspects of equilibrium to be fully achieved, we require first that households have full information on all housing prices and attributes and that their transactions and moving costs be zero; and second, that the price vector adjust instantaneously to changes in either demand or supply. The market for housing can be viewed as a stock-flow model where the flow (change in stock) is a function of prices, but the prices at any point in time are determined only by the stock at that point in time.

This idealized model is clearly not an accurate representation of real-world housing markets. But in evaluating the strength of this criticism of the hedonic price model, we must focus on several distinct issues. The first concerns the accuracy of the price data itself. Where the data are based on assessments, appraisals, or self-reporting, they may not correspond to actual market prices. The errors in measuring the dependent variable will tend to obscure any underlying relationship between true property value measures and environmental amenities. But estimates of the relationship will not be biased unless the errors themselves are correlated with other variables in the model.

A second issue concerns the speed of adjustment of the market to changed conditions of supply and demand. If adjustment is not complete, observed marginal implicit prices will not accurately measure household marginal willingnesses to pay. A major question is whether imperfect adjustment will lead to systematic biases in estimates of willingness to pay.

Consider households' imperfect adjustment to changing prices. First, an increase in housing prices need not affect the marginal implicit prices of attributes, in which case no adjustment of the attribute bundles is necessary. Even if marginal implicit prices change, households will not move unless the potential utility gain to returning to full equilibrium exceeds the information costs, transactions costs, and moving costs associated with the change. These costs help to define a band within which observed marginal implicit prices can diverge from household marginal willingnesses to pay for housing attributes.

If housing prices change so that the marginal implicit price schedule for an attribute moves consistently in one direction, households will consistently lag in their adjustment to that change. And the marginal willingnesses to pay will be overstated or understated according to whether the marginal implicit price is rising or falling.[6]

A third issue concerns expectations about future environmental amenity levels. Market prices for long-lived assets such as housing reflect the discounted present value of the stream of expected future services from that asset. A change in expectations about future environmental amenity levels can affect housing prices and marginal implicit prices independently of the present level of these amenities. For example, if there are widespread expectations of an improvement in air quality, and the market adjusts reasonably quickly to these expectations, the price differential between presently dirty houses and clean houses should decrease. Correlating these prices with existing levels of air pollution would lead to an underestimate of the marginal implicit price of air quality.

To summarize, divergences from full equilibrium of the housing market in many circumstances will only introduce random errors into the estimates of marginal willingnesses to pay. However, where market forces are moving continuously in one direction, or are expected to move in one direction, incomplete market adjustment or full adjustment to changing expectations, or both, can introduce biases in both directions. We should be much more cautious about utilizing the cross-section hedonic price model in those cities and at those points in time during which market forces and environmental quality levels are changing rapidly (granted that "rapidly" is an imprecise term). However, it is also possible in these circumstances to determine the direction of bias. Thus, estimates of marginal willingness to pay derived from such studies can be labeled as an upper bound or a lower bound on the basis of that analysis.

LIMITED RANGE OF MODELS OF HOUSES

In order to interpret observed marginal implicit prices as marginal willingnesses to pay, it is necessary to assume that there is a sufficiently wide variety of housing models available such that every household is in equilibrium. In other words, individuals are not forced to corner solutions to their maximization problems. The hedonic price function

---

[6]In their meta-analysis of hedonic property value studies, Smith and Huang (1991) found a negative time trend for the implicit price of an air pollutant (total suspended particulates), other things being equal.

defines an opportunity locus across attribute space. A household chooses a housing model such that its indifference surface is tangent to the given opportunity locus, provided that a model with that precise set of attributes is available. If the optimum model is not available, the household must pick the nearby housing model that gives the highest utility level; but then the first-order conditions for utility maximization are not satisfied as equalities.[7]

The hedonic model is based on an assumption that the implicit price function is differentiable and continuous. But this is an artifact of the statistical and mathematical technique. If this assumption is not satisfied in practice, two sorts of problems can arise. The first problem is that the statistically fitted hedonic price function is a better approximation only when the number of housing units is large and there is more continuous variation in characteristics among units. A small number of distinctly different types of housing units might be better analyzed with one of the discrete choice models described later in this chapter. The second type of problem arises if there are no units available with particular combinations of attributes. If there are substantial gaps in the opportunity locus, some households will not be able to satisfy the first-order conditions as equalities. This could be a problem for certain subsets of the urban population.

Harrison and Rubinfeld (1978) have uncovered an anomaly that might be explained by the presence of gaps in the range of available models of houses. They estimated both an implicit price function for housing in the Boston Standard Metropolitan Statistical Area and marginal willingness to pay as a function of nitrogen oxides levels (q_j) and incomes. Figure 11-2 shows the marginal implicit price function for (q_j) and the estimated marginal willingness-to-pay functions for high-income (b_{jm}) and low-income (b_{jk}) households. In examining the distributional implications of a postulated change in nitrogen oxides levels, Harrison and Rubinfeld found that some high-income households experienced a large benefit due to their initial location in a high nitrogen oxide area. This finding is contrary to the implications of the simple model, for two reasons. First, according to the simple model, high-income households should locate in low-pollution areas and thus have low willingnesses to pay at the margin for air quality improvements. Second, the postulated change in air pollution generated lower improvements in air quality in those areas with higher initial air quality levels.

One possible explanation for the anomaly is that some high-income households wished to consume bundles of housing with low

---

[7] Mäler (1977) discusses this set of issues on pp. 361–362.

Price ($)

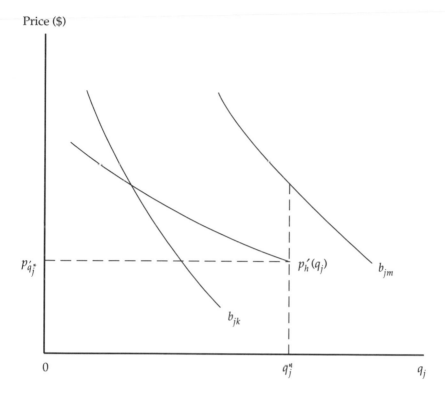

Figure 11-2.   The effect of a gap in the range of available models

nitrogen oxides and high levels of some other attribute—say the cultural amenities of the urban environment—but that there were no housing units available that supplied these two attributes jointly in the appropriate combination. If that was the case, the marginal implicit price function would not exist to the right of, say, q_j^* in figure 11-2. The high-income households would settle at q_j^* because that was the best they could do. When this occurs, the observed marginal implicit price ($P_{q_j^*}$) can no longer be taken as representing the equilibrium marginal willingness to pay for high-income households.

How important is this problem in practice? As the example indicates, one cannot rule it out on a priori grounds. In fact, the problem is almost certain to exist for some subgroups in some urban areas. But we need not conclude that aggregate estimates are so unreliable as to be of no use just because the aggregate model does not perfectly replicate the situation of every component of the aggregate. This is a problem to which empirical researchers must be sensitive. Examination of the disaggregated behavior of the model such as that carried out by Harrison and Rubinfeld could be helpful in identifying the existence of such problems and judging their seriousness.

MARKET SEGMENTATION

Straszheim (1974) was the first to raise the question of market segmentation in the context of estimating hedonic price functions for housing within an urban area. He argued that the urban housing market really consisted of a series of separate, compartmentalized markets with different hedonic price functions in each. As evidence in support of the segmentation hypothesis, Straszheim showed that estimating separate hedonic price functions for different geographic areas of the San Francisco Bay area reduced the sum of squared errors for the sample as a whole.

For different hedonic price functions to exist in an urban area, two conditions must be met. The first is that either the structure of demand, the structure of supply, or both, must be different across segments. Either buyers in separate submarkets must have different structures of demands, or the structure of characteristics of the housing stocks must be different. The second condition is that purchasers in one market segment must not participate significantly in other market segments. In other words, there must be some barrier to mobility of buyers among market segments that prevents arbitrage from occurring in response to differences in marginal implicit prices. Such barriers could be due to geography, discrimination, lack of information, or a desire for ethnically homogeneous neighborhoods. Even with buyer immobility, if demand and supply structures are the same they will produce similar structures of hedonic prices. And perfect mobility and information on the part of buyers will eliminate differences in the implicit prices for any characteristic across market segments.

If market segmentation does exist, the hedonic price function estimated for the urban area as a whole will provide faulty estimates of the implicit prices facing subsets of buyers in different market segments. Thus estimates of benefits and estimates of demand functions based on faulty price data will also be faulty. If market segmentation does exist, separate hedonic price functions must be estimated for each segment; and benefit and demand functions must be separately estimated for each segment with a different set of implicit prices.

It is not clear how significant the problem of market segmentation is for air pollution–property value studies within single urban areas—although there are enough positive results in the literature to suggest that it is not a problem that can be dismissed out of hand. Some authors have found evidence supporting the hypothesis that segmentation leads to different hedonic price functions, but others have not found evidence of segmentation in their data.[8]

It should be noted that the existence of market segmentation does not render the hedonic price technique invalid; rather, it makes application of the technique more difficult. If the appropriate basis for segmentation can be identified, it is conceptually possible to estimate separate implicit price functions for each submarket. Although these functions would be different across markets, they each would accurately reflect the outcome of the market processes in each submarket. Thus the functions could be used to estimate equilibrium marginal willingnesses to pay.

Estimating Characteristics Demands: The Identification Problem

The attractiveness of the hedonic price model for applied welfare analysis lies in the potential for using estimates of individuals' marginal implicit prices for a characteristic to recover information on the underlying structure of preferences, specifically, the marginal willingness-to-pay function.

Rosen (1974) had argued that marginal willingness-to-pay and marginal supply price functions could be estimated from the information contained in the hedonic price function in the following manner:

> . . . compute a set of implicit *marginal* prices . . . for each buyer and seller evaluated at the amounts of characteristics . . . actually bought or sold, as the case may be. Finally, use estimated marginal prices . . . as endogenous variables in the second-stage simultaneous estimation of [the marginal willingness-to-pay and supply price functions]. Estimation of marginal prices plays the same role here as do direct observations on prices in the standard theory and converts the second-stage estimation into a garden variety identification problem. (p. 50)

This suggestion has been the source of a large literature for more than a decade. Since the emphasis in this book is on models and basic

---

[8]Those finding differences in hedonic price functions include Straszheim (1974), Schnare and Struyk (1976), Harrison and Rubinfeld (1978), Goodman (1978), and Michaels and Smith (1990). But see also Nelson (1978b), Ball and Kirwan (1977), and Sonstelie and Portney (1980).

economic method rather than on econometric issues, I provide here only an overview of the sources of the identification problem and alternative approaches to solving it.[9]

The difficulties in identifying the marginal willingness-to-pay function from hedonic price data come in two forms. The first arises from the fact that the source of data for the dependent variable in the marginal willingness-to-pay function is not direct observation of b_{kj}; rather, it is the calculation of the marginal implicit price $\partial P_{hi}/\partial q_j$ from the estimated hedonic price function. But this variable is itself computed as the function of the same characteristics that are explanatory variables in the marginal willingness-to-pay function. Brown and Rosen (1982) and Mendelsohn (1987) show that at least in some cases this procedure leads to parameter estimates for the marginal willingness-to-pay function that are identical to the estimated coefficients in the hedonic price function. As Brown and Rosen put it,

> Contrary to Rosen's original statement, we claim that marginal attribute prices constructed as above will not necessarily play the same role in estimation that direct observation on prices would play if they were available. Because such constructed prices are created only from observed sample quantities, any new information that they may provide (i.e., any information beyond that already provided directly by observed sample quantities) can only come from a priori restrictions placed on the functional form of the price function $P_h(\cdot)$. In the absence of such additional restrictions, second-stage "structural" estimation of the sort suggested by Rosen may only reproduce the information already provided by the first-stage estimation of the $P_h(\cdot)$ function. (1982, p. 176; notation changed by the author.)

In other words, since the second-stage estimation procedure utilizes no additional data beyond that already contained in the hedonic price function, it can do no more than reproduce the coefficients estimated from the hedonic price function.

The second source of difficulty lies in the fact that both the quantity of the characteristic and its marginal implicit price are endo-

---

[9]Readers interested in more technical discussion, especially from an econometric perspective, should consult Brown and Rosen (1982), Epple (1987), Bartik (1987), and McConnell and Phipps (1987). Additional contributors to this literature include Quigley (1982), Murray (1983), Mendelsohn (1984, 1985), and Horowitz (1987). Bartik and Smith (1987) and Palmquist (1991) also provide useful reviews.

genous in the hedonic price model. Unlike the standard market model in which an individual faces an exogenously determined price and chooses a quantity, and unlike a quantity-rationed market in which an individual faces an exogenously determined quantity and reveals a marginal willingness to pay, the individual chooses both a point on the hedonic price schedule *and* its associated quantity. The choice of that point simultaneously determines the marginal willingness to pay and the quantity of the characteristic. As Palmquist (1991) put it, "the other marginal prices [on the individual's marginal willingness-to-pay function] are only observed for other individuals with other socio-economic characteristics and provide no information on the original consumer's bid for different quantities of the characteristic" (p. 96). This makes it very difficult to separate out the effects of demand-shifters from the price–quantity relationship itself.

The nature of the problem can be illustrated with figure 11-3. Consider two identical individuals with uncompensated bid functions of the form

$$b^* = b^* (q_j , M)$$

where M is income, a demand-shifter.[10] If the two individuals choose different levels of q_j—for example, q_{j1} and q_{j2} — it must be because of differences in income. This means that the demand-shifter, which is unobserved in the data of figure 11-3, is correlated with the observed choices of q_j.

Turning to the supply side of the market, if characteristic supplies are not exogenous, the two equilibrium points must be on different marginal supply price curves. If the supplier trait that shifts the marginal supply price curves is used as an instrument in the estimation of the marginal willingness-to-pay function, the supplier trait will be correlated with the unobserved demand-shifter. Again, the estimate of the marginal willingness-to-pay function will be biased.

One approach to identification is to find truly exogenous variables to be used as instruments. This appears to be a difficult task, however; for some of the suggested possibilities and their problems, see Mendelsohn (1984, 1985), Bartik (1987), Bartik and Smith (1987) and Palmquist (1991). Another approach is to impose sufficient structure on the problem by assumption to assure that the conditions for identification of the marginal willingness-to-pay function are met; an early example is provided by Quigley (1982). He assumed a

---

[10]This example is due to Bartik (1987, pp. 84–85).

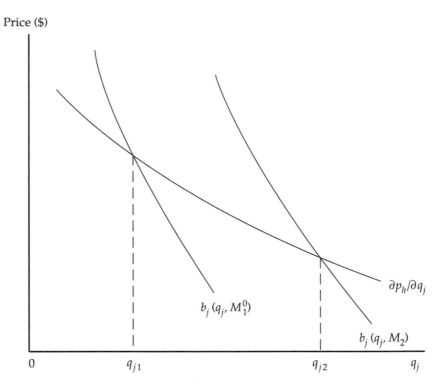

Price ($)

$b_j (q_j, M_1^0)$

$\partial p_h / \partial q_j$

$b_j (q_j, M_2)$

0 q_{j1} q_{j2} q_j

Figure 11-3.   The identification problem when income and the quantity of the characteristic are correlated

functional form for preferences that included homotheticity as a property. By specifying the relationship between income and demand, this assumption made it possible to separate the effects of income and quantity change on the marginal willingness to pay for characteristics.

Another possibility is to assume that there is a characteristic in the marginal implicit price function that is not an argument in the marginal willingness-to-pay function for another characteristic. This makes it possible to use the omitted characteristic as an instrument. Recently many authors have been critical of this approach, since the results are only as good as the assumptions imposed to obtain them.[11] Unfortunately, such assumptions are not testable.

----

[11]See Mendelsohn (1987), Bartik and Smith (1987), Horowitz (1987), and Palmquist (1984).

Probably the most reliable approach to solving the identification problem is to find cases where the marginal implicit prices of characteristics vary independently of the other demand-shift variables. Specifically, this means finding cases where individuals with the same preferences, income, and other traits face different marginal implicit prices. This can only occur if similar individuals must choose in markets with different hedonic price functions, which in turn implies either segmented markets within a city or observations taken from several different housing markets (as, for example, in different cities).

The first step in implementing this approach is to estimate separate hedonic price functions for each housing market, using the same specification. The second step is to compute the marginal implicit price faced by each individual from the hedonic price function in that market. Then the computed marginal implicit prices can be regressed on the observed quantities of the characteristics and the exogenous demand-shifters to obtain the uncompensated bid function. Assuming sufficient independent variation across markets and assuming that there are no unobserved differences in preferences across individuals, this approach will lead to reliable and properly identified bid functions.[12]

Measuring Welfare Changes

It has been established that in an equilibrium of the housing market utility-maximizing individuals equate their marginal willingness to pay for housing characteristics with the marginal implicit prices of these characteristics and that in some circumstances it may be possible to estimate inverse demand or bid functions on the basis of this information. We now ask the question how can we use the information on prices and preferences that can be extracted from the hedonic housing market to calculate measures of welfare change for changes in environmental amenities? The basic concepts of welfare measurement at the level of the individual are straightforward, and have already been introduced in chapter 4. But welfare measurement in the aggregate may be more difficult because of induced price changes and the possibility that individuals will adjust their consumption positions in response to changes in prices and quantities. Also, in princi-

---

[12]For examples of this approach, see Palmquist (1984) and Bartik (1987). See Ohsfeldt and Smith (1985, 1988) for an investigation of the relationship between the variability of marginal implicit prices and the reliability of the parameter estimates of bid functions.

ple, it is necessary to consider changes in the supply side of the hedonic property market.

In this sub-section I first define the basic welfare measure for marginal and nonmarginal changes in a characteristic or environmental amenity holding other things constant, in particular, individuals' choices of housing bundles. I rule out by assumption people moving to different housing sites with different bundles of characteristics in response to the change in the environmental characteristic or any induced changes in the hedonic price function. This measure looks only at benefits to purchasers of housing bundles. I then consider a fully general measure of welfare change that includes possible changes in profits on the supply side of the hedonic market as well as the consequences of individuals' adjustments on the demand side of the market. I go on to describe some models that can be applied as special cases of the more general definition of welfare change, or that can provide lower or upper bound approximations of the correct measure, at least in certain circumstances.

In this section, I must distinguish between the task of defining a conceptually correct welfare measure and identifying what must be known in order to implement such a measure. Some of the welfare measures identified in this section require knowledge of the marginal willingness-to-pay function for each individual. But even when the identification problems discussed above can be solved, the models generally provide only estimates of the uncompensated bid function. I also include here some discussion of how to obtain the necessary marginal willingness-to-pay functions from data on inverse demands.

BASIC PRINCIPLES

Since a change in an environmental amenity in an urban area is nonexcludable and nondepletable, it is, in effect, a public good. The marginal value of the change, then, is simply the sum of the marginal willingnesses to pay of each of the affected individuals evaluated at the existing housing equilibrium. In other words, for the amenity q_j:

$$w_q = \sum_{i=1}^{n} b_{ij} = \sum_{i=1}^{n} \frac{\partial P_h}{\partial q_i} \tag{7}$$

where w_q is the aggregate marginal welfare change. Although most proposed environmental policy changes are nonmarginal in magnitude, the ease of calculating equation (7) may make it useful for indicating whether some improvement is desirable, by comparing this measure with an estimate of the marginal cost of the improvement.

Similarly, in a partial equilibrium setting where there are no changes in other prices or characteristics, the welfare value of a change in an environmental amenity, say from q_j^0 to q_j^1, is given by

$$W_q = \sum_{i=1}^{n} \int_{q_j^0}^{q_j^1} b_{ij}(q_{ij}, P, u_j^*) \, dq_j \qquad (8)$$

where W_q is the aggregate benefit. Notice that this measure requires knowledge of the marginal willingness-to-pay functions of individuals. If the uncompensated bid functions from the second stage of the hedonic price estimation are used, the welfare gain (loss) will be overestimated (underestimated) for compensating measures of welfare change. The magnitude of any errors introduced will depend upon the factors discussed in chapter 3.

There is a method for calculating exact welfare measures for non-marginal changes in a characteristic, holding all other things constant. It is based on an adaptation of Hausman's technique for exact welfare measurement for price changes (see chapter 3); this adaptation is due to Horowitz (1984). Suppose that an individual's bid function

$$b_j^* = b_j^*(q_j, Q^*, M - P_h)$$

has been identified. In equilibrium,

$$\frac{\partial u / \partial q_j}{\partial u / \partial M} = b_j^*(\cdot) \qquad (9)$$

and

$$b_j^* - \partial P_h / \partial q_j$$

The left-hand side of equation (9) is the slope of the indifference curve between the numeraire, M, and q_j. So, in equilibrium

$$\frac{dM}{dq_j} = b_j^*(\cdot) - \partial P_h / \partial q_j$$

This expression can be solved for

$$M = f(q_j, Q^*, C)$$

where C is a constant of integration. The benefit of an increase in q_j is

$$W_q = f(q_j^0, Q^*, C) - f(q_j^1, Q^*, C)$$

Horowitz (1984) presents a simple example where the inverse demand function is derived from a Cobb–Douglas utility function.

In principle at least, measures of welfare change based on equations (7) or (8) can be generalized to incorporate the effects of general equilibrium adjustments. At any location, the value of a nonmarginal change can be taken to be the integral of the values of a series of infinitesimal changes in the amenity (Bartik and Smith, 1987, p. 1223). The value of each small change is taken to be the willingness to pay of the occupant of that site at that point in the sequence of changes. The measures for all sites together is the sum of the values for each site. It is given by

$$W_q = \sum_{i=1}^{n} \int_{q_j^1}^{q_j^1} \frac{\partial P_h\ (q_j,\, Z)}{\partial q_j}\ dq_j \tag{10}$$

where Z is the vector of all other site characteristics, which are held constant by assumption.

In principle this measure allows individuals to relocate in response to changes in the quantity and price of the amenity, since it in effect sums individuals' marginal values as the amenity changes at each site. This is important, because a major limitation of some of the measures to be described below is their inability to account for individual relocation decisions. Furthermore, this measure does not require knowledge of either the marginal willingness-to-pay or the bid function. It relies on the fact that at each point in the sequence of changes, each individual's marginal bid is revealed by the marginal implicit price of the characteristic.

However, the hedonic price function is likely to shift as a consequence of the change in the amenity level. Therefore, in order to implement this measure, it is necessary to know how the hedonic price function and the marginal implicit prices at each location change as the levels of the amenities at each location change along the path of integration. As a practical matter, this is a major limitation of the measure.

A GENERAL, COMPREHENSIVE MEASURE OF BENEFITS

Suppose there are increases in several environmental amenities in an urban area. These increases need not be uniform across the area.

Specifically, let us consider the case where the vector Q increases from Q^0 to Q^1. We will first look at the immediate welfare effects of this change and then examine how the welfare gains are magnified and redistributed by the rational adjustments of individuals and ensuing price changes in the hedonic market.[13]

First, assuming that individuals cannot move to new locations and that the hedonic price function does not change, the benefit to individuals is given by an expanded version of equation (8):

$$W_q = \sum_{i=1}^{n} \int_{Q^1}^{Q^1} b_i(Q, P, u_i^*)\, dQ \tag{11}$$

where each individual's welfare gain is computed from a path-independent line integral over the changes in the individual elements in Q and where $b_i(\cdot)$ is the vector of individual marginal willingness-to-pay functions for the characteristics.

Now, at the existing hedonic price function, some people may wish to choose different bundles of characteristics. If they do change, it must be because they perceive themselves to be better off after the adjustment. This welfare gain is in addition to that given by equation (11). Furthermore, the effort to adjust to different characteristics bundles is likely to affect the hedonic price function, unless the number of people wishing to do so is quite small relative to the market. And it is possible that the suppliers of housing will respond to changes in the hedonic price function by offering different bundles of housing characteristics, S. This could have further repercussions on the hedonic price function, and it will increase the profits of housing suppliers.

When all of these repercussions have worked themselves out, the aggregate benefit to individuals can be defined in terms of each individual's total willingness to pay for a housing unit with given characteristics, holding utility constant. Let this total willingness to pay be given by

$$B_i(Q_i^{j*}, S_i^{j*}, u^*)$$

where Q_i^{j*} and S_i^{j*}, ($j = 0, 1$) indicate the vectors of environmental and other characteristics actually chosen by the individual in the original and new equilibrium. Each individual's total benefit is the increase in total willingness to pay for the characteristics actually chosen, holding utility constant, minus any increase in actual expen-

---

[13]This section is based on the analysis by Bartik (1988).

diture on housing. Summing across all individuals, we obtain

$$W_q = \sum_{i=1}^{n} \left[B_i(Q_i^{1*}, S_i^{1*}, u^*) - B_i(Q_i^{0*}, S_i^{0*}, u^*) \right]$$

$$- \sum_{i=1}^{n} \left[P_h^1 (Q_i^{1*}, S_i^{1*}) - P_h^0 (Q_i^{0*}, S_i^{0*}) \right] \tag{12}$$

Turning to the supply side of the market, there are m ($k = 1, ..., k,$ ..., m) housing producers. In aggregate, they realize a change in aggregate profits given by the increase in expenditures on housing net of any change in their costs. This is given by

$$\Delta \text{Prof} = \sum_{k=1} \left[P_h^1 (Q_i^{1*}, S_i^{1*}) - P_h^0 (Q_i^{0*}, S_i^{0*}) \right]$$

$$- \sum_{k=1}^{p} \left[C_k^1(Q^1, S_1^*) - C_k^0(Q_i^{0*}, S_0^*) \right] \tag{13}$$

where $C_k(\cdot)$ is the cost function for producers.

The welfare change for society as a whole is the sum of equations (12) and (13). Notice that one component of this sum is simply a transfer of revenue from buyers to sellers, so it nets out. The total welfare change is the sum of the increase in total willingness to pay of individuals minus any cost increase on the part of producers of housing. Full implementation of this welfare measure would require enormous amounts of information. But it should be noted that this measure reduces to equation (11) if the hedonic price function does not change and if the change in environmental amenities does not affect the costs of supplying housing amenities for producers.

Even if this set of conditions is not satisfied, equation (8) can be interpreted as a lower bound on the true measure of benefits. This can be seen by decomposing the true benefit measure into a three-step sequence of changes and adjustments. Consider first the change in amenity levels without any adjustment on the part of individuals or suppliers. The welfare change associated with this step is given by equation (8) plus any reduction in the costs of supplying existing houses at the affected locations. Second, suppose hypothetically that the hedonic price function is shifted to its new equilibrium position but that no adjustments to the new price function by individuals or suppliers are permitted. At this stage, although some individuals and suppliers may gain while others lose, on net all of the price changes sum to zero. At this stage, there is no net change in welfare.

Finally, allow individuals and suppliers to respond to the new hedonic price function. Any adjustments that take place at this stage must represent welfare improvements for those responding. The total welfare change is the sum of equation (8), any costs reduction to suppliers, and the benefits of adjusting to the price change. The latter two components are either zero or positive. Thus, equation (8) represents a lower bound on the true measure of benefits. And the smaller the adjustment to the changes in the hedonic price function is, the smaller is the error involved in using (8).

A SPECIAL CASE

If the hedonic price function does not shift, then exact welfare measurement may be a relatively easy task. One situation in which the hedonic price function could be assumed to be constant is when the number of sites at which there is a change in the amenity level is small relative to the total urban market. If this is the case, and if individuals can move without cost from one site to another in response to the change in environmental amenity levels, then exact welfare measurement is straightforward. The hedonic price function can be used to predict the changes in the prices of affected properties. Benefits are exactly measured by the increase in the values of the affected properties. And knowledge of the marginal bid functions is not required.

Consider the case of an improvement from q_j^0 to q_j^1 at just one site, as shown in figure 11-4. Assume that moving costs for occupants who choose to relocate are zero (the impact of positive moving costs are discussed in the next sub-section). The change in the amenity level results in an increase in the price of this house from P_h^0 to P_h^1. The owner of the property is made better off by this increase in wealth. Even though the occupant of the property experiences the increase in amenity level, he or she is made worse off because of the increase in the cost of occupying the property. The occupant is shifted from point a on the curve B_j^0 to point b on the curve B_j^1. But with costless moving, the occupant can relocate to his or her original equilibrium position. So the net welfare change is the increase in wealth to the owner. If the owner and occupant are the same person, the result is still the same. This individual might choose to move to a property with an amenity level somewhat greater than q_j^0 because of a wealth effect. But the increase in wealth fully captures the benefit of the amenity improvement to this individual.

Price ($)

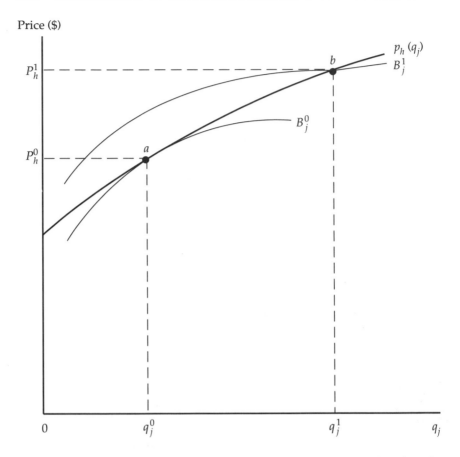

Figure 11-4.   The benefit of an amenity improvement at one site when the hedonic price function is unchanged

APPROXIMATIONS

Several lines of reasoning can be used to arrive at reasonable approximations or upper or lower bounds on the true measure of benefits. First consider the case where the number of affected sites is small so that the hedonic price function does not change, but where moving costs are positive. This makes the analysis a bit more complicated. But it can be shown that the predicted increase in prices of affected houses is an upper bound on the net benefits of the increase in the amenity. Those renter occupants who choose to move in response to the amenity change are made worse off by the amount of moving costs. Thus, moving costs must be deducted from the increase in property prices in order to obtain the measure of net benefits. If moving costs are so high as to preclude moving as a response to the change in amenity levels, some people are left worse off by the amenity improvement. But the increase in property prices is still an upper bound on the total benefit.

As long as the cost of moving is independent of the price of housing and the amenity level, moving costs can be represented as a parallel upward shift in the hedonic price function. Figure 11-5 reproduces the case shown in the preceding figure but adds moving costs. If moving costs c are equal to or greater than the vertical distance $a–b'$, then the individual will be indifferent between staying at the original property with amenity level q_j^1 at point b and adjusting to a new equilibrium at point a'. If moving costs exceed this amount, the individual would prefer to be compensated by the vertical distance $b–c$ to stay at the original property rather than move. This maximum required compensation is thus an upper bound on the error involved in using the increase in property prices as a measure of benefits.[14]

Recall that the measures described here rely on there being only a few affected houses so that the hedonic price function does not change. But suppose that the number of properties affected by the change in amenity levels is sufficiently large so as to cause a shift in the hedonic price function. Under this condition, there is an intuitive argument that using the original hedonic price function to predict the changes in property prices will result in an overestimate of the true benefits for the amenity change (see Bartik and Smith, 1987, p. 1223). As was discussed above, if it is possible to predict how the hedonic price function and the marginal implicit price schedule shift, then equation (10) can be used directly to compute benefits. But suppose that it is not known exactly how the marginal implicit price function will change. If equation (10) is used with the original function, then the result will be in error. The sign of the error depends on the direction of the shift of marginal implicit price schedule for the amenity. It seems plausible to expect that as the amenity level increases, the marginal implicit price schedule will actually shift downward. Since the benefit calculation does not take account of this shift,

---

[14]Palmquist (1992) has performed a similar analysis for reductions in amenity levels and required compensation as the welfare measure. Owners lose an amount equal to their reduction in property price. Renters require additional compensation for moving costs. So the reduction in property prices as predicted by the hedonic price function is a lower bound on the measure of loss.

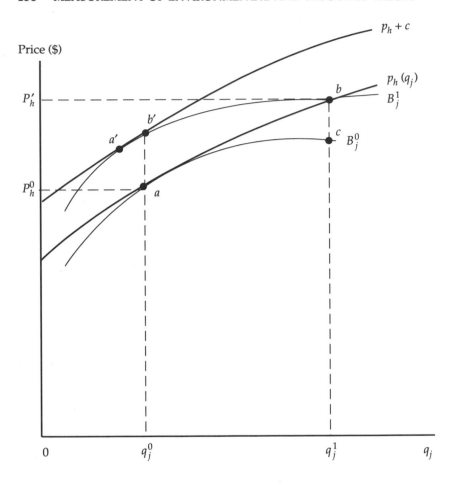

Figure 11-5. The benefit of an amenity improvement with moving costs

the result will be an overestimate of the true benefits of the amenity change.[15]

Another, more ad hoc, approach to calculating welfare changes is to fall back on some assumption as to the shape of this function through the original equilibrium point. Three alternative assumptions can be used to establish bounds on the true measure. Under the first,

---

[15]For a formal analysis supporting this intuition, see Bartik (1988). Kanemoto (1988) establishes a similar result. Strictly speaking, his analysis is not based on the hedonic price model; rather, Kanemoto assumes two different classes of land and examines the relationship between price differences and benefits.

it is assumed that the marginal willingness to pay for each individual is constant, that is, that the marginal willingness-to-pay function is a horizontal line through the known point. In this case, each individual's benefit for the postulated improvement in the amenity is approximated by the product of the (assumed) constant marginal willingness to pay and the change in the amenity. The aggregate benefit is obtained by summing over all individuals. This assumption leads to an estimate of aggregate benefits that is likely to be biased upward.

A second convenient assumption would be that each individual's marginal willingness-to-pay curve decreases linearly from its observed point to the point of the highest attainable level of the amenity. Marginal willingness to pay would be zero at this point. It is not clear whether this approximation would lead to an overestimate or an underestimate of true benefits. The third assumption would be that all individuals' marginal willingness-to-pay functions are identical. Then, as discussed in chapter 4, the marginal implicit price curve is identified as the marginal willingness-to-pay curve for the representative individual.

Rents, Taxes, and Property Prices

In the development and exposition of the theoretical model, I have ignored the temporal dimension of housing prices and how welfare measures based on property prices might be converted to the annualized form usually used in welfare evaluation. It is typically the market price of a property that is observed. Inferences about the streams of rents and of benefits are drawn by converting observed present values to annual streams. The institutions of income and real property taxation affect the way in which the market capitalizes rents (and changes and differentials in rents) into market prices for properties. These effects must be properly understood if the process of retracing these steps to infer rents from property value observations is to be successful.

In the simplest case of a stream in perpetuity and with no taxes, the conversion of property value to rent is given by

$$R = P_h \cdot r$$

where r is the appropriate discount rate. I proceed by examining first the effects of the two forms of taxation separately, then their combined effects. The conclusion is that unless the analyst takes account of the effects of taxation, the use of property value data is likely to lead to an overestimation of benefits.

Ad valorem taxation of property can be viewed as a device for capturing some of the rent of land for the government. Since taxation affects the net return to the property owner, it should affect the market value of property as an asset.[16] An individual would purchase a property as an asset only if its market price, P, is equal to or less than the discounted present value of the rental stream net of property taxation. Market forces would establish the following relationship between property values and rents:

$$P_h = (R - t \cdot P_h) / r \tag{14}$$

where t is the ad valorem tax rate.

If property values are known, the rental stream they represent can be computed by rearranging equation (14)

$$R = P_h \cdot (r + t) \tag{15}$$

Assume that the property value–amenity relationship, $P_h(q_j)$, has been estimated. The marginal benefit of a change in q_j at a site is

$$w_{q_j} = \partial R / \partial q_j$$
$$= (r + t) \, \partial P_h / \partial g_j \tag{16}$$

The present value of this stream of benefits is

$$\frac{w_{q_j}}{r} = \left(1 + \frac{t}{r}\right) \partial P_h / \partial q_j \tag{17}$$

In other words, when the hedonic price function is defined in terms of property value, ignoring the effect of property taxation on the capitalization of rents can lead to the underestimation of benefits. The term (t/r) is a measure of the percentage error resulting from omitting the tax term in the calculation of benefits. For an interest rate of 10 percent and an effective tax rate of 11–20 mils per dollar of market value (1–2 percent), the error is between 11 and 20 percent.

But this is not the whole story. The income tax code treats the imputed rental income of homeowners differently than it does the rental income of landlords. The absence of a tax liability for imputed rent further complicates the task of inferring annual rents and bene-

---

[16]The effects of property taxation on benefit estimation were first discussed by Niskanen and Hanke (1977).

fits from observations of (capitalized) market prices for housing assets. This is because the market will place different values on two assets with the same rental stream if one is subject to income taxation while the other is not.

Assume two perpetual assets indexed as a and b with equal annual returns of R per year. The return to the first asset is taxable at the rate g percent, while the return to the second asset incurs no tax liability. If r represents the market rate of return on assets with taxable returns, the two assets will be priced so as to equalize the after-tax rate of return:

$$P_a = R_a/r$$

and
$$P_b = \frac{1}{(1-g)} \frac{R_a}{r} > P_a$$

If P_b is observed, the tax-free rental stream can be computed from

$$R_b = r(1-g) P_b$$

Taking account of this adjustment and using equations (16) and (17), the marginal annual benefits of amenity changes to homeowners and their present value are

$$w_{q_i} = r(1-g)\, \partial P_b/\partial q_i \tag{18}$$

$$\frac{w_{q_i}}{r} = (1-g)\, \partial P_b/\partial q_i \tag{19}$$

Ignoring the effects of income taxation leads to an overestimation of benefits.[17]

The tax code confers an additional benefit on homeowners by allowing them to deduct property tax payments in calculating taxable income. This lowers the real cost of the property tax by g percent.[18]

---

[17] The discount factor $r(1-g)$ is analogous to the municipal bond rate, and it arises for the same reason. However, where the marginal tax rate is itself a function of income, g varies across individuals; and (18) and (19) must be computed separately for each individual.

[18] The code also permits the deduction of mortgage interest. But since the interest cost (either out-of-pocket or opportunity cost) of holding an asset does not enter into the calculation of the present value of its stream of returns, except as shown below, this feature of the code does not affect the capitalization equations.

Combining all these effects, ad valorem taxation, deductibility, and exemption of imputed rental income, we have

$$P_b = [R_b - (1 - g) \cdot t \cdot P_b] \, r \cdot (1 - g) \tag{20}$$

Solving for R_b gives

$$R_b = (r + t) \, (1 - g) \, P_b$$

and marginal benefits are calculated by

$$w_{q_j} = (r + t) \, (1 - g) \, \partial P_b / \partial q_j \tag{21}$$

and
$$\frac{w_{q_j}}{r} = \left(1 + \frac{t}{r} \right) (1 - g) \partial P_b / \partial q_j \tag{22}$$

The effects of ignoring taxation in calculating benefits depend on the magnitudes of g and t/r. The higher the marginal income tax rate, the more likely benefits would be overstated if taxes were ignored. For an example, suppose the marginal income tax rate is 30 percent, the opportunity cost of capital is 10 percent, and the property tax rate is 2 percent. Then the terms in parentheses come to 0.84. Ignoring tax effects would lead to an overstatement of benefits by almost 20 percent.

An alternative approach to dealing with taxation and discounting is to base the hedonic equation on measures of user cost (called gross rent by Sonstelie and Portney, 1980). This variable captures the full cost of owning (and using) an asset such as a house. User cost would include property taxes and the opportunity cost of capital plus any change in the market price of the asset over the interval, say a year. It would be calculated as follows:

$$u = (r + t + m) \, P_h$$

where m is the percentage rate of change in market value.

The user cost approach differs from that outlined above in two respects. The first is the inclusion of the change in market value over time. P_h could be changing because of physical depreciation of the house, general price inflation, changes in the price of housing relative to other goods, and changes in the variables determining P_h. Only the latter changes have relevance for benefit estimation, and they would be captured by modified versions of equations (21) and (22), which generalized from the assumption of constant streams in perpetuity. However, the depreciation term might be useful in empirical work as an approximation of expected changes in these variables, provided that it were adjusted to net out general price level effects.

The second difference arises in considering the effects of some provisions of the income tax code on user cost. For one thing, the tax exemption for imputed rent does not affect the user cost of holding a house, since user cost is an opportunity cost. For another, the tax deductibility of mortgage interest does affect user cost, but it does not affect the market capitalization of streams of benefits. If user cost is used to compute benefits, the net result of these two effects is to overstate benefits in comparison with equations (21) and (22).

The Hedonic Model with Averting Behavior

The use of the hedonic property value model for welfare measurement described in this section requires the assumption that preferences are separable in housing characteristics and those environmental amenities specific to housing locations. This is probably not always a reasonable assumption. If the assumption is not valid, it is necessary to explore the nature of the interdependence between housing characteristics and other arguments in the individual's utility function and how the welfare measures derived from hedonic theory are related to those derived from other models of behavior. In this section I explore some implications of one kind of interdependence between a location-specific amenity, say air quality, and other arguments of an individual's utility function.

Suppose that air quality as measured by q produces an amenity such as a better view of the distant mountains. Suppose that q also reduces the incidence of illness and disease related to air pollution. I examine here the relationship between the marginal value of improved health and the marginal implicit price of q as revealed from a hedonic property value analysis. (Models for valuing reduced morbidity have been explored in chapter 10.)

As in chapter 10, assume that the number of days spent sick (by whatever measure) is inversely related both to q and to the quantity of an averting activity, a. This relationship is $s = s(q, a)$. The averting activity is purchased at a price of p_a. The individual derives utility from the consumption of a numeraire good X, leisure f, a vector of housing attributes Z, and a location-specific amenity v; v is itself a function of the level of air quality, $v(q)$, $\partial v/\partial q > 0$. The individual also experiences disutility from the days spent sick. In addition to disutility, illness results in lost workdays and reduction in wage income. The individual's choice problem can be represented as

$$\text{Max: } u[X, f, \mathbf{Z}, v, s(q, a)]$$
$$+ \lambda\{I + p_w[T - f - s(q, a)] - X - p_a \cdot a - P_h(q, \mathbf{Z})\}$$

where $P_h(q, \mathbf{Z})$ is the hedonic property value function. The relevant first-order conditions include

$$\left[\frac{\partial u}{\partial v} \cdot \frac{\partial v}{\partial q}\right]/\lambda + \left[\frac{\partial u/\partial s}{\lambda} - p_w\right] \cdot \frac{\partial s}{\partial q} = \frac{\partial P_h}{\partial q} \tag{23}$$

$$\left[\frac{\partial u/\partial s}{\lambda} - p_w\right] = \frac{p_a}{\partial s/\partial a} \tag{24}$$

The left-hand side of equation (23) is the individual's marginal willingness to pay for an improvement in q. This marginal value can be inferred from observations of the marginal implicit price of q, $\partial P_h/\partial q$. The marginal willingness to pay for q has two components, one related to the amenity and the other related to the disutility and income loss associated with illness. Substituting equation (24) into (23) yields another way of looking at marginal value:

$$w_q = \left[\frac{\partial u}{\partial v} \cdot \frac{\partial v}{\partial q}\right]/\lambda + p_a \cdot \frac{\partial s/\partial q}{\partial s/\partial a}$$

$$w_q = \left[\frac{\partial u}{\partial v} \cdot \frac{\partial v}{\partial q}\right]/\lambda + p_a \cdot \frac{\partial a}{\partial q} = \frac{\partial P_h}{\partial q}$$

The marginal willingness to pay for q can be separated into its own amenity- and health-related components. The marginal implicit price of q measures the sum of the components, but it provides no information on the relative importance of the amenity and the health components of marginal value.[19]

Manipulations similar to those described in chapter 10 can be used to derive an alternative expression for w_q:

$$w_q = \left[\frac{\partial u/\partial s}{\lambda} - p_w\right]\frac{ds}{dq} - p_a \cdot \frac{\partial a^*}{\partial q} + \left[\frac{\partial u}{\partial v} \cdot \frac{\partial v}{\partial q}\right]/\lambda$$

---

[19]See Portney (1981) for an earlier analysis of the relationship between property values and the value of reduced mortality risks due to air pollution. He inferred a value of mortality risk reduction by assuming that the amenity component of $\partial P_h/\partial q$ was very small.

As before, the first two terms represent the disutility and the lost wages associated with the observed change in number of sick days and the observed change in averting expenditures after optimal adjustment to the change in q. The marginal implicit price of q captures both the health-related and amenity values associated with the improvement in q. The amenity value could be found by subtraction if the marginal implicit price of q and the health benefit were estimated separately.

This model can be extended to the case where health is affected by air quality both at home and at some other location, say work. Suppose the number of days of illness depended on air quality at home, q_H, air quality at work, q_W, and averting activities. The individual can choose q_H through the selection of a place to live. As long as q_H and q_W vary independently, they can be treated separately for purposes of valuation. The two marginal willingnesses to pay take the by-now familiar form

$$ w_{q_H} = \left[\frac{\partial u \cdot \partial v}{\partial u \cdot \partial q_H} \right] / \lambda + p_a \cdot \frac{\partial s / \partial q_H}{\partial s / \partial a} = \frac{\partial P_h}{\partial q_H} $$

$$ w_{q_W} = p_a \cdot \frac{\partial s / \partial q_W}{\partial s / \partial a} $$

The marginal implicit price of q_H captures the welfare effect of changes in air quality at home, but does not say anything about the welfare effects of changes in q elsewhere. In order to implement either measure, it is necessary to know the separate contributions of pollution at home and at work to inducing illness.

Applying the Model to Agricultural and Commercial Land

Up to this point I have presented the hedonic model in the context of the market for housing. But the model can be applied to the market for any differentiated product. In this section I show how the hedonic price model can be reformulated to deal with the land as an input into production processes. The individual's utility function is replaced by the profit function of the firm. Bid and offer curves for land as a function of its productive characteristics can be derived. And the model can be used for measuring the values of improvements in these characteristics. Possible applications include the markets for commercial land (for example, for shopping and offices) and for agricultural land.[20]

Let Q be a vector of exogenously determined productive characteristics of land and let Z be a vector of characteristics that can be produced by landowners at some cost, $C(Q, Z, P_f)$ where P_f is a vector of factor input prices. Following Rosen's original development (1974) as described in chapter 4, the cost function can be used to derive a family of offer curves for each characteristic.

On the demand side of the property market, there are producers who use land as an input. For ease of exposition, assume that each producer uses one parcel of land. Suppose each of the i producers has a multiproduct production function:

$$X_i = X_i (F_i, Q_i, Z_i)$$

where F_i is the vector of variable factor inputs. Each producer selects output and input levels and characteristics of the parcel of land he rents so as to maximize variable profits, $P \cdot X_i - P_f \cdot F_i$, subject to the production function. The solution to this problem can be expressed as the variable profit function $N(P, F_i, Q_i, Z_i)$. Bid curves for each characteristic can be derived from the variable profit function. An equilibrium for the hedonic rental market is given by the double envelope of the bid and offer curves. The equilibrium is represented by the hedonic rental function, $R(Q, Z)$. This function can be differentiated to give the marginal implicit prices in equilibrium for all of the productive characteristics of land. Inverse demand functions and marginal bid functions (holding profits constant) can be derived for each characteristic. The estimation and identification of these functions are subject to the same kinds of problems discussed in the earlier section on the identification problem; but for present purposes, assume that these problems have been resolved.

Now suppose there has been an increase in one exogenous characteristic, say q_j. If we assume that input and output prices are unaffected by this improvement, the approaches to welfare measurement discussed above can be extended to deal with this problem. To do this, first assume that neither producers nor landowners can adjust the quantities of characteristics taken or produced. Then

---

[20]Palmquist (1989) has developed the analytical model for agricultural land. Miranowski and Hammes (1984) and Palmquist and Danielson (1989) present estimates of hedonic price functions and marginal implicit prices for agricultural land.

the welfare benefit, say W^*, is given by

$$W^* = \sum_{i=1}^{n} \int_{q_j^0}^{q_j^1} b_i\left(q_j, Q^*, Z\right) dq_j$$

where $b_i(\cdot)$ is the marginal bid function of each producer. If the change in q_j reduces landowners' costs of supplying Z, there is a further benefit, say W^{**}.

Now assume that the hedonic price function shifts to its new equilibrium but that producers and landowners are still not permitted to adjust to this change. As discussed earlier, all price changes are transfers, so this stage does not add to the net benefits. If producers and landowners are allowed to adjust their quantities in response to the changed hedonic price function, there will be additional gains, say W^{***}.

The total benefit from the increase in q_j is $W^* + W^{**} + W^{***}$. Since it seems implausible for W^{**} to be negative, the most easily measured W^* provides a lower bound on true benefits.

In the case where the improvement in the environmental characteristic leads to changes in output prices, there will be further changes in producers' surplus and in consumers' surplus. These must be taken into account in deriving a welfare measure. In such a case it may be more feasible to pursue welfare measurement through the output market, as described in chapter 9. The derivative of the variable profit function with respect to the price of each output gives the supply function for that output. Changes in any of the characteristics of land shift the supply functions of each of the outputs. The framework developed in chapter 9 provides a consistent way of deriving welfare measures as areas between shifting supply curves for each of the multiple outputs.

THE REPEAT SALES MODEL

The standard hedonic price model is based on the analysis of a cross section of housing prices at a point in time. The repeat sales model is a variant of the hedonic price model that exploits a time series of sale prices of houses whose structural characteristics have not changed over time. The repeat sales model was developed by Palmquist (1982). It substantially simplifies the task of model specification and estimation, since characteristics that do not change over time can be omitted from the regression equation.

Suppose for simplicity that houses are not bought and sold but are rented for one-year terms; and suppose that rents, R, just clear the market in every year. Suppose for the moment that there is no depreciation of the quality of the housing services provided by each unit of housing and that the overall price level for housing is constant over time. Then, in any particular year, we could estimate a hedonic price function of the form

$$R_i = R(S_i, N_i, Q_i) \tag{25}$$

where i indexes dwellings. Assume that there is one environmental attribute, q_j, which varies across space and is also changing over time at all or most locations. If all of the other characteristics in equation (25) are constant over time and if the hedonic price function itself is unchanged,[21] then for any two years, t and t', we have

$$R_i' - R_i = R(S_i, N_i, Q_i^*, q_{ji}') - R(S_i, N_i, Q_i^*, q_{ji})$$
$$= f(q_{ji}' - q_{ji})$$

where Q_i^* is all elements of Q_i other than q_j and the prime indicates the variable for the second year. All terms involving characteristics other than q_j cancel out. This is obvious for the linear specification. For the semi-log form,

$$\ln R = a + c \cdot q_j + \sum_{k=1}^{n} c_k \cdot Z_k$$

where Z_k is all other characteristics (S_j, N, and Q^*), the change in rents is given by

$$\ln R' - \ln R = \ln(R'/R) = a + c \cdot q_j$$

Even the quadratic Box–Cox specification with all its interaction terms

---

[21]The assumption that the hedonic price function is unchanged is problematic. There is nothing in the theory of hedonic price functions that assures that this assumption will hold in fact. Changes in the level of any characteristic could change the whole hedonic price function. Palmquist (1982, p. 345) presents evidence that supports the assumption of stability of the hedonic price function over time. But Edmonds (1985) found that the hedonic price function for housing in Tokyo changed significantly over time.

substantially simplifies in the repeat sales formulation.[22] This would considerably reduce the data requirements and simplify the task of econometric estimation.

Unfortunately, application of the repeat sales model in the real world is not quite as simple as portrayed here, since other things change over time. Specifically, houses depreciate with age and there may be fluctuations in the general price level for houses. Now let us consider a more realistic situation in which the prices of houses, P_{h}, are observed, houses as assets depreciate over time at a constant percentage rate, d, and the price level for housing in general fluctuates over time.

Let I_t ($t = 0, \ldots t, \ldots, T$) be the price index for housing in this housing market, with $I_0 = 1$. For houses having no changes in characteristics (and still assuming that the basic hedonic price function is stable over time), sales prices for houses at different points in time are governed by

$$P_{h_t} = I_t \cdot e^{-d \cdot t} \cdot P_{h_o}(\cdot)$$

Since price level fluctuations and depreciation cause proportional rather than absolute changes in housing prices, this relationship places restrictions on the way changes in other housing characteristics can be specified, at least if the advantages of the repeat sales model are to be retained. Specifically, the price changes associated with changes in characteristics over time must also be proportionate, that is, the basic hedonic price function must take the semi-log form:

$$P_{ht} = e^{(a + c \cdot q_{jt} + \Sigma c_k \cdot z_k)}$$

Then for a house that sells twice, at t and t', we have

$$\ln(P_{ht'}/P_{ht}) = \ln(I_{t'}/I_t)\, e^{-d(t'-t)} \cdot e^{c(q_{t'}-q_t)}$$

So, although this model does not allow for flexibility in the functional form of the hedonic equation, it does substantially reduce data costs and chances for misspecification of other housing characteristics.

---

[22]With the quadratic Box–Cox functional form of equation (6), the expression for the difference in prices reduces to

$$R'^{[\lambda]} - R^{[\lambda]} = C\,(q_j'^{[\beta]} - q_j^{[\beta]})$$

$$+ \frac{1}{2} b_{jj}\,(q_j'^{[\beta]2} - q_j^{[\beta]2})$$

where $\quad C = a_j + \sum_{k=1}^{n} c_{jk} \cdot Z_k \cdot k^{[\beta]}$

One potential econometric problem must be noted. Suppose a property is sold n times during the period being studied. This results in $n - 1$ independent pairs of sales prices that can be used as observations in the regression equation. If n is greater than 2, the error terms for different pairs of sales of the same house are correlated since they share a common sale price term. Palmquist (1982) shows that estimation of the repeat sales model by ordinary least squares results in biased coefficient estimates, and shows how this problem can be remedied through use of Aitken's estimator.

In conclusion, the repeat sales model has the advantages of reducing the data costs and the possibility of specification error by eliminating from the regression equation those structural and other characteristics that do not change over time. However, in order to use the repeat sales model it is necessary to assume that the coefficients of the hedonic price function are constant over time. And the unavoidable facts of depreciation and fluctuating price levels place restrictions on the way the environmental characteristic of concern can be specified.

DISCRETE CHOICE MODELS

The hedonic price model is based on the assumption that each attribute of the housing bundle is a continuous variable and that an individual can choose any point on the continuous and differentiable hedonic price function in the n-dimensional attribute space. As noted above, this is clearly not a completely realistic assumption, and in some respects it may seriously misrepresent the problem of choosing a bundle of housing attributes. For example, the number of bedrooms in a house is not a continuous variable. There may be no one-bedroom houses on one-acre lots, or four-bedroom houses with swimming pool and attached garage on one-quarter-acre lots. Discrete choice models provide an alternative way of looking at housing choice and inferring values for housing attributes.

Some of the discrete choice models that have been used in the literature focus on the individual's bid function for housing bundles. Such models are based on the probability that an individual will be the highest bidder for a specified bundle of housing attributes. These are known as bid rent models or random bidding models. An alternative approach is to focus on the individual's utility function defined on housing attributes. These models investigate the probability that a specified bundle of housing attributes (including the price of the bundle) will be chosen by the individual, that is, will be revealed to convey the highest level of utility. These are known as random utility models.[23]

Both types of models can be used to derive the marginal bid or marginal willingness-to-pay function for individual attributes from an estimate of the bid function or indirect utility function. Thus, in principle, both types of models allow for the calculation of the benefits of changes in an environmental attribute, at least assuming no relocation and no changes in the hedonic price function. Both types of models start with the assumption of utility maximization subject to the standard budget constraint defined by income, prices of market goods, and the hedonic price function.

In the bid rent model, the utility maximization problem is solved to obtain the individual's bid function, that is, the bid as a function of the housing attributes and income, holding utility constant. Then, for an individual of any given income level, the bid can be written as a function of the observable housing attributes plus a random error term reflecting unobserved attributes of either the individual or the housing bundle:

$$B_i = B_i(Z) + \epsilon_i$$

where, for simplicity, Z represents the vector of all housing attributes and location-specific amenities.[24] For an example of this model, see Ellickson (1981), and Gross (1988). Because this bid function is derived for a given income level, and all households face the same market prices, these terms can be omitted from that function.

The probability that individual i will choose housing bundle k rather than any of the other available bundles, indicated by k', is

$$P(k|Z) = P(B_{ik} > B_{ik'})$$
$$= P[B_i(Z_k) - B_i(Z_{k'}) + \epsilon_{ik} - \epsilon_{ik'}) > 0]$$

If the random error terms are independently and identically distributed with a Type I Extreme Value distribution, this probability can be written in the logit form:

$$P(k|Z) = \exp(B_{ik}) / \Sigma_k \exp(B_{ik}) \tag{26}$$

Finally, if the bid function is specified to be linear in the parameters, maximum likelihood methods can be used to estimate these parame-

---

[23]These two modeling approaches parallel the two approaches to analyzing referendum data described in chapter 6.

[24]See Lerman and Kern (1983) and Gross (1988) for details of this specification of the model.

ters. As Lerman and Kern (1983) and Gross (1988) point out, estimation of equation (26) fixes only the slope of the bid function, not its level. But information on the bids actually paid can be used in the estimation process to fix these values and to make it possible to calculate the marginal bid functions for individual attributes.

In the random utility model, given the multinomial logit specification, the probability that individual i will choose housing bundle k over bundle k' can be written as (following Cropper and her coauthors, 1993)

$$P[V_i(\mathbf{Z}_k) - V_i(\mathbf{Z}_{k'}) + (\epsilon_{ik} - \epsilon_{ik'})]$$
$$= \exp[V_i(\mathbf{Z}_k)]/\Sigma_k \exp[V_i(\mathbf{Z}_{k'})]$$

where $V_i(\cdot)$ is the indirect utility function. If $V_i(\cdot)$ is specified to be linear in the parameters, this, too, can be estimated with maximum likelihood methods. Knowledge of the parameters of the indirect utility function makes it possible to compute welfare measures for changes in any of the housing attributes, including an environmental amenity.

In an interesting simulation study, Cropper and her coauthors (1993) have compared estimates of welfare measures derived from a hedonic price model with those from the random utility model for given known household preferences. After simulating an equilibrium in an urban housing market, they used the resulting hedonic price and individual choice data to estimate both a hedonic price model with its marginal bid functions and a random utility model. They then calculated welfare measures for 25 percent and 100 percent changes in each of ten attributes, including both neighborhood attributes and attributes of individual houses (such as number of bathrooms, lot size, and age). They found that the random utility model provided more accurate estimates of the known welfare measure than the hedonic price model. This was true even when they assumed that the researcher did not know the true form of individuals' utility functions. They suggest that the reason for this is the difficulty in identifying and obtaining accurate estimates of the marginal bid functions with the hedonic price model when data are generated by only a single market.

Other researchers (see for example, Bartik and Smith, 1987, pp. 1224–1225, Palmquist, 1991, p. 119) have suggested that although one the strengths of the discrete choice model is its ability to generate welfare measures for nonmarginal change relatively easily, the model does so because it forces the researcher to make strong assumptions about the functional form of the utility function or bid function. If similar strong assumptions are made about the functional form of the inverse demand functions for attributes in the hedonic model, these functions can be identified too. But the results of Cropper and her coauthors (1993) suggest that even when the functional form of preferences is known, the discrete choice model outperforms the hedonic model as a way of measuring welfare change. However, this conclusion must be considered to be tentative, since at this writing there are no published estimates of welfare values for environmental amenities generated from discrete choice models.

SUMMARY

It is now time to summarize the answers to the two questions raised at the beginning of this chapter. The first question concerned the specification and estimation of a model of housing prices for purposes of valuing environmental amenities. Hedonic price theory provides a coherent basis for explaining the prices of houses in an urban market as a function of the levels of characteristics embedded in each house. And there is a dynamic version of this model which relates changes in the prices of houses to changes in the levels of one or more characteristics, other things being equal. The major limitation of the hedonic model is its assumption that consumers of housing can select their most preferred bundle of characteristics from a complete range of levels of all characteristics. However, the discrete choice model can be adapted to deal with this problem. Both the random bidding model and the random utility model can be developed to explain housing market behavior and choices as functions of, among other things, the characteristics of houses.

The second question concerned the derivation of measures of economic value and welfare change for changes in the levels of environmental amenities. Measures of value for marginal and nonmarginal changes can be derived from a properly specified hedonic price model. Values for marginal changes in amenity levels are found simply by adding up the observed or computed marginal willingnesses to pay for all affected individuals. But for nonmarginal amenity changes, welfare measurement requires knowledge of the inverse demand function or the income-compensated bid function for the amenity. And these, in turn, require a solution to the daunting identification problem.

The discrete choice models provide means for direct estimation of either bid functions or indirect utility functions. Such estimates, in turn, make calculation of welfare changes a straightforward matter.

There are some limitations to the property value models for estimating welfare effects. First, since the property value models are based on the consequences of individuals' choices of residence, they do not capture willingness to pay for improvements in environmental amenities at other points in the urban area—for example, in the work place, shopping areas, or parks and recreational areas. Second, because the property value models are based on observing behavioral responses to differences in amenity levels across houses, they only capture willingness to pay for perceived differences in amenities and their consequences. For example, if there are subtle, long-term health effects associated with reduced environmental quality at some housing sites but people are unaware of the causal link of these effects to the housing site, their willingness to pay to avoid the effects will not be reflected in housing price differences.

On the other hand, property value measures have the capability of capturing the value of all of the possible effects of changes in environmental quality at a housing site in a single number. If air pollution, for example, causes increases in the incidence of respiratory disease, damages ornamental vegetation, and reduces the quality of the view, the property value model summarizes these effects in a single number.

REFERENCES

Anderson, Robert J., and Thomas D. Crocker. 1971. "Air Pollution and Residential Property Values," *Urban Studies* vol. 8, no. 3, pp. 171–180.

Anderson, Robert J., and Thomas D. Crocker. 1972. "Air Pollution and Property Values: A Reply," *Review of Economics and Statistics* vol. 54, no. 4, pp. 470–473.

Atkinson, Scott E., and Thomas D. Crocker. 1987. "A Bayesian Approach to Assessing the Robustness of Hedonic Property Value Studies," *Journal of Applied Econometrics* vol. 2, no. 1, pp. 27–45.

Ball, M. J., and R. N. Kirwan. 1977. "Accessibility and Supply Constraints in the Urban Housing Market," *Urban Studies* vol. 14, no. 1, pp. 10–32.

Bartik, Timothy J. 1987. "The Estimation of Demand Parameters in Hedonic Price Models," *Journal of Political Economy* vol. 95, no. 1, pp. 81–88.

Bartik, Timothy J. 1988. "Measuring the Benefits of Amenity Improvements in Hedonic Price Models," *Land Economics* vol. 64, no. 2, pp. 172–183.

Bartik, Timothy J., and V. Kerry Smith. 1987. "Urban Amenities and Public Policy," in Edwin S. Mills, ed., *Handbook of Regional and Urban Economics*. Amsterdam, The Netherlands: Elsevier.

Brown, James N., and Harvey S. Rosen. 1982. "On the Estimation of Structural Hedonic Price Models," *Econometrica* vol. 50, no. 3, pp. 765–768.

Cassel, Eric, and Robert Mendelsohn. 1985. "The Choice of Functional Forms for Hedonic Price Equations: Comment," *Journal of Urban Economics* vol. 18, no. 2, pp. 135–142.

Court, Lewis M. 1941. "Entrepreneurial and Consumer Demand Theories for Commodities Spectra," *Econometrica* vol. 9, no. 1, pp. 135–162.

Crocker, Thomas D. 1970. "Urban Air Pollution Damage Functions: Theory and Measurement." Riverside, Calif.: University of California, NTIS PB 197–668.

Cropper, Maureen L., Leland B. Deck, and Kenneth E. McConnell. 1988. "On the Choice of Functional Form for Hedonic Price Functions," *Review of Economics and Statistics* vol. 70, no. 4, pp. 668–675.

Cropper, Maureen L., Leland B. Deck, Nalin Kishor, and Kenneth E. McConnell. 1993. "Valuing Product Attributes Using Single Market Data: A Comparison of Hedonic and Discrete Choice Approaches," *Review of Economics and Statistics* vol. 75, no. 2, pp. 225–232.

Edmonds, Radcliffe G., Jr. 1985. "Some Evidence on the Intertemporal Stability of Hedonic Price Functions," *Land Economics* vol. 61, no. 4, pp. 445–451.

Ellickson, Bryan. 1981. "An Alternative Test of the Hedonic Theory of Housing Markets," *Journal of Urban Economics* vol. 9, no. 1, pp. 56–79.

Epple, Dennis. 1987. "Hedonic Prices and Implicit Markets: Estimating Demand and Supply Functions for Differentiated Products," *Journal of Political Economy* vol. 87, no. 1, pp. 59–80.

Freeman, A. Myrick, III. 1971. "Air Pollution and Property Values: A Methodological Comment," *Review of Economics and Statistics* vol. 53, no. 4, pp. 415–416.

Freeman, A. Myrick, III. 1974a. "Air Pollution and Property Values: A Further Comment," *Review of Economics and Statistics* vol. 56, no. 4, pp. 454–456.

Freeman, A. Myrick, III. 1974b. "On Estimating Air Pollution Control Benefits from Land Value Studies," *Journal of Environmental Economics and Management* vol. 1, no. 1, pp. 74–83.

Freeman, A. Myrick, III. 1979. *The Benefits of Environmental Improvement: Theory and Practice*. Baltimore, Md.: Johns Hopkins University Press for Resources for the Future.

Goodman, Allen C. 1978. "Hedonic Prices, Price Indices and Housing Markets," *Journal of Urban Economics* vol. 5, no. 4, pp. 471–484.

Graves, Phil, James C. Murdoch, Mark A. Thayer, and Don Waldman. 1988. "The Robustness of Hedonic Price Estimation: Urban Air Quality," *Land Economics* vol. 64, no. 3, pp. 220–233.

Griliches, Zvi, ed. 1971. *Price Indexes and Quality Change*. Cambridge, Mass.: Harvard University Press.

Gross, David J. 1988. "Estimating Willingness to Pay for Housing Characteristics: An Application of the Ellickson Bid-Rent Model," *Journal of Urban Economics* vol. 24, no. 1, pp. 95–112.

Halvorsen, Robert, and Henry O. Pollakowski. 1981. "Choice of Functional Form for Hedonic Price Equations," *Journal of Urban Economics* vol. 10, no. 1, pp. 37–49.

Harrison, David, Jr., and Daniel L. Rubinfeld. 1978. "Hedonic Housing Prices and the Demand for Clean Air," *Journal of Environmental Economics and Management* vol. 5, no. 2, pp. 81–102.

Horowitz, Joel L. 1984. "Estimating Compensating and Equivalent Income Variations from Hedonic Price Models," *Economics Letters* vol. 14, no. 4, pp. 303–308.

Horowitz, Joel L. 1986. "Bidding Models of Housing Markets," *Journal of Urban Economics* vol. 20, no. 2, pp. 168–190.

Horowitz, Joel L. 1987. "Identification and Stochastic Specification in Rosen's Hedonic Price Model," *Journal of Urban Economics* vol. 22, no. 2, pp. 165–173.

Kanemoto, Yoshitsugu. 1988. "Hedonic Prices and the Benefits of Public Projects," *Econometrica* vol. 56, no. 4, pp. 981–989.

Lerman, Steven R., and Clifford R. Kern. 1983. "Hedonic Theory, Bid Rents, and Willingness-to-pay: Some Extensions of Ellickson's Results," *Journal of Urban Economics* vol. 13, no. 3, pp. 358–363.

Lind, Robert C. 1973. "Spatial Equilibrium, the Theory of Rents, and the Measurement of Benefits from Public Programs," *Quarterly Journal of Economics* vol. 87, no. 2, pp. 188–207.

Mäler, Karl Göran. 1977. "A Note on the Use of Property Values in Estimating Marginal Willingness to Pay for Environmental Quality," *Journal of Environmental Economics and Management* vol. 4, no. 4, pp. 355–369.

McConnell, Kenneth E., and T. T. Phipps. 1987. "Identification of Preference Parameters in Hedonic Models: Consumer Demands with Nonlinear Budgets," *Journal of Urban Economics* vol. 22, no. 1, pp. 35–52.

Mendelsohn, Robert. 1984. "Estimating the Structural Equations of Implicit Markets and Household Production Functions," *Review of Economics and Statistics* vol. 66, no. 4, pp. 673–677.

Mendelsohn, Robert. 1985. "Identifying Structural Equations with Single Market Data," *Review of Economics and Statistics* vol. 67, no. 3, pp. 525–529.

Mendelsohn, Robert. 1987. "A Review of Identification of Hedonic Supply and Demand Functions," *Growth and Change* vol. 18, no. 1, pp. 82–92.

Michaels, R. Gregory, and V. Kerry Smith. 1990. "Market Segmentation and Valuing Amenities with Hedonic Models: The Case of Hazardous Waste Sites," *Journal of Urban Economics* vol. 28, no. 2, pp. 223–242.

Miranowski, John A., and Brian D. Hammes. 1984. "Implicit Prices of Soil Characteristics for Farmland in Iowa," *American Journal of Agricultural Economics* vol. 66, no. 5, pp. 745–749.

Murdoch, James C., and Mark J. Thayer. 1988. "Hedonic Price Estimation of Variable Urban Air Quality," *Journal of Environmental Economics and Management* vol. 15, no. 2, pp. 143–146.

Murray, Michael P. 1983. "Mythical Demands and Mythical Supplies for Proper Estimation of Rosen's Hedonic Price Model," *Journal of Urban Economics* vol. 14, no. 3, pp. 327–337.

Nelson, Jon P. 1978a. *Economic Analysis of Transportation Noise Abatement.* Cambridge, Mass.: Ballinger.

Nelson, Jon P. 1978b. "Residential Choice, Hedonic Prices, and the Demand for Urban Air Quality," *Journal of Urban Economics* vol. 5, no. 3, pp. 357–369.

Niskanen, William A., and Steve H. Hanke. 1977. "Land Prices Substantially Underestimate the Value of Environmental Quality," *Review of Economics and Statistics* vol. 59, no. 3, pp. 375–377.

Ohsfeldt, Robert L., and Barton A. Smith. 1985. "Estimating the Demand for Heterogeneous Goods," *Review of Economics and Statistics* vol. 67, no. 1, pp. 165–171.

Ohsfeldt, Robert L., and Barton A. Smith. 1988. "Assessing the Accuracy of Structural Parameter Estimates in Analyses of Implicit Markets," *Land Economics* vol. 64, no. 2, pp. 135–146.

Palmquist, Raymond B. 1982. "Measuring Environmental Effects on Property Values Without Hedonic Regressions," *Journal of Urban Economics* vol. 11, no. 3, pp. 333–347.

Palmquist, Raymond B. 1984. "Estimating the Demand for Characteristics of Housing," *Review of Economics and Statistics* vol. 64, no. 3, pp. 394–404.

Palmquist, Raymond B. 1989. "Land as a Differentiated Factor of Production: A Hedonic Model and Its Implications for Welfare Measurement," *Land Economics* vol. 65, no. 1, pp. 23–28.

Palmquist, Raymond B. 1991. "Hedonic Methods," in John B. Braden and Charles D. Kolstad, eds., *Measuring the Demand for Environmental Improvement.* Amsterdam, The Netherlands: Elsevier.

Palmquist, Raymond B. 1992. "Valuing Localized Externalities," *Journal of Urban Economics* vol. 31, no. 1, pp. 59–68.

Palmquist, Raymond B., and Leon E. Danielson. 1989. "A Hedonic Study of the Effects of Erosion Control and Drainage on Farmland Values," *American Journal of Agricultural Economics* vol. 71, no. 1, pp. 55–62.

Parsons, George R. 1990. "Hedonic Prices and Public Goods: An Argument for Weighting Locational Attributes in Hedonic Regressions by Lot Size," *Journal of Urban Economics* vol. 27, no. 3, pp. 308–321.

Pines, David, and Yoram Weiss. 1976. "Land Improvement Projects and Land Values," *Journal of Urban Economics* vol. 3, no. 1, pp. 1–13.

Polinsky, A. Mitchell, and Daniel L. Rubinfeld. 1977. "Property Values and the Benefits of Environmental Improvements: Theory and Measurement," in Lowdon Wingo and Alan Evans, eds., *Public Economics and the Quality of Life.* Baltimore, Md.: Johns Hopkins University Press for Resources for the Future.

Polinsky, A. Mitchell, and Steven Shavell. 1976. "Amenities and Property Values in a Model of an Urban Area," *Journal of Public Economics* vol. 5, no. 1–2, pp. 119–129.

Portney, Paul R. 1981. "Housing Prices, Health Effects, and Valuing Reductions in Risk of Death," *Journal of Environmental Economics and Management* vol. 8, no. 1, pp. 72–78.

Quigley, John M. 1982. "Nonlinear Budget Constraints and Consumer Demand: An Application to Public Programs for Residential Housing," *Journal of Urban Economics* vol. 12, no. 2, pp. 177–201.

Ridker, Ronald G. 1967. *Economic Costs of Air Pollution: Studies in Measurement.* New York: Praeger.

Ridker, Ronald G., and John A. Henning. 1967. "The Determinants of Residential Property Values with Special Reference to Air Pollution," *Review of Economics and Statistics* vol. 49, no. 2, pp. 246–257.

Rosen, Sherwin. 1974. "Hedonic Prices and Implicit Markets: Product Differentiation in Perfect Competition," *Journal of Political Economy* vol. 82, no. 1, pp. 34–55.

Schnare, Ann B., and Raymond J. Struyk. 1976. "Segmentation in Urban Housing Markets," *Journal of Urban Economics* vol. 3, no. 2, pp. 146–166.

Smith, Barton A. 1978. "Measuring the Value of Urban Amenities," *Journal of Urban Economics* vol. 5, no. 3, pp. 370–387.

Smith, V. Kerry, and Ju Chin Huang. 1991. "Can Hedonic Models Value Air Quality? A Meta-Analysis," paper presented at the National Bureau of Economic Research University Research Conference, Cambridge, Mass., December 1991.

Sonstelie, Jon C., and Paul R. Portney. 1980. "Gross Rents and Market Values: Testing the Implications of the Tiebout Hypothesis," *Journal of Urban Economics* vol. 7, no. 1, pp. 102–118.

Straszheim, Mahlon. 1974. "Hedonic Estimation of Housing Market Prices: A Further Comment," *Review of Economics and Statistics* vol. 56, no. 3, pp. 404–406.

Strotz, Robert H. 1968. "The Use of Land Value Changes to Measure the Welfare Benefits of Land Improvements," in Joseph E. Haring, ed., *The New Economics of Regulated Industries.* Los Angeles, Calif.: Occidental College.

12

Hedonic Wage Models

The hedonic wage model is a formalization of the concept of compensating wage differentials, which can be traced back to Adam Smith. The basic idea is that, other things being equal, workers will prefer jobs with more pleasant working conditions over those that are less pleasant. The greater supply of workers for pleasant jobs will depress the wage levels of such jobs. And an equilibrium, the difference in wages between two jobs with different working conditions, will reflect the workers' monetary valuations of the differences in working conditions.

The basic hedonic wage model has been refined and applied empirically to two important questions of particular interest to environmental and resource economists and policymakers. One question concerns the value of reducing the risk of death, injury, or illness. The hedonic wage model has been used to estimate the wage–risk trade-off as a revealed preference measure of this value. The other question concerns the values of the environmental and social amenities that vary across regions. Wage differences across regions have been used as indicators of the values of region-specific environmental, cultural, and social amenities.

From a worker's perspective, a job can be viewed as a differentiated product, that is, a good with a bundle of characteristics such as working conditions, prestige, training and enhancement of skills, and levels of risk of accidental injury and exposure to toxic substances. If workers are free to move from one urban area to another, then jobs are also differentiated, in part, by the environmental and other characteristics of the urban areas in which the jobs are located. If workers are free to choose from a menu of differentiated jobs, the hedonic price technique can be applied to the data on wages, job characteristics (including their locations), and worker characteristics to estimate the marginal implicit prices of these job characteristics.

Employers, from their perspective, can be viewed as choosing from among a set of workers of different characteristics. This is a distinguishing feature of labor markets. In the typical application of the hedonic theory to differentiated goods, producers are viewed as selling a good embodying a package of characteristics and as being indifferent to the characteristics of the purchaser of the good. In hedonic wage studies, the employer is viewed as selling a package of job characteristics (including the quality of the work environment); but at the same time the employer is purchasing work effort and cannot be indifferent to the productive characteristics of the firm's employees. Thus the hedonic wage equation must be interpreted as an equilibrium relationship that reflects not only the interaction of supply and demand for job characteristics but also the interaction of supply and demand of worker characteristics (see Lucas, 1977; Rosen, 1979). This means that both worker and job characteristics must be included as arguments in the estimated hedonic wage equation.

As in the case of hedonic property values, the derivative of the hedonic wage function with respect to any job characteristic can be interpreted as the marginal implicit price of that characteristic. And if the worker is maximizing utility, the marginal implicit price can be taken as an estimate of the worker's marginal willingness to pay for the characteristic. It gives the change in income necessary to just compensate for a small change in the characteristic. Since in general the hedonic wage function need not be linear, these marginal values may be different for different workers. Similarly, the derivative of the hedonic wage function with respect to any worker characteristic gives its marginal implicit price and, assuming profit maximization, the marginal value of that characteristic to the employer.

In order to estimate the value of a nonmarginal change in a characteristic, it is necessary to know the compensated inverse demand function for it. As in the case of hedonic property values, the inverse demand function cannot be estimated from data from a single labor market unless additional restrictions are imposed. Some examples of efforts to identify these functions are described in the next section.

The interpretation of the hedonic wage function as revealing marginal implicit prices and marginal values requires that all of the transactions that make up the data be undertaken in the same market. In other words, each buyer (seller) in the market must have had the opportunity to match up with any of the other sellers (buyers) and to choose the most preferred given prices, and so forth. In the terminology of chapter 11, the market must be in equilibrium and must not be segmented into submarkets with incomplete mobility among segments. When hedonic wage equations are estimated using data from several urban areas, it is necessary to assume that these areas are part of a single market. In practice, labor markets can be segmented on the basis of geography, with moving costs and lack of information on job alternatives imposing barriers between labor markets in different parts of the country. Markets can also be segmented on the basis of education and skill requirements, and between blue-collar and professional–managerial workers. Geographic segmentation can lead to different marginal implicit price schedules in different regions. Segmentation on the basis of occupation or education level can lead to different marginal implicit price functions across occupational categories. One approach to the problem of geographic segmentation is to estimate the hedonic wage function only for occupational groups that are believed on a priori grounds to be part of a national labor market. In general, the extent of market segmentation and its significance for empirical estimation of hedonic wage functions are not known.

In the next section of this chapter, I discuss the application of the hedonic wage model to measuring the value of risk reduction and briefly review the results of some of the empirical applications of this model. I then outline and describe some of the models that have been developed to explain and interpret interregional wage differences as reflections of the values people place on regional amenities.

WAGE DIFFERENCES AND THE VALUE OF REDUCING RISKS

In chapter 10, I discussed the concept of the statistical value of life and its role in estimating the value of policies that reduce the risk of death (by reducing, for example, exposures to harmful chemicals); I also presented models that show how it is possible to draw inferences about the value of reductions in risk and the statistical value of life from revealed preferences (through the wage–risk trade-off, for example). Here I discuss the use of the hedonic wage model as one approach to estimating the individuals' willingness to pay for reductions in risk. Most of the applications of the hedonic model to risk valuation have dealt with risks of death due to accidents on the job.

As in the simple static model in chapter 10, suppose that each individual chooses a job so as to maximize expected utility from consumption of the numeraire, X, and from the vector of job characteristics, J. In addition to J, each job is characterized by its risk of accidental death, δ_i, where i indexes jobs. Individuals face a hedonic wage function that is the locus of points at which firms' marginal wage offers (as functions of job characteristics) equal workers' marginal acceptance wages (see chapter 4 for a description of equilibrium in hedonic markets). This function is

$$p_w = p_w(\delta, J)$$

where p_w is the weekly or monthly wage and where hours of work per period could be one of the characteristics in J. The individual chooses job i to maximize expected utility subject to the wage constraint

$$\text{Max } E[u] = \pi \cdot u(X, J)$$
$$+ \lambda[p_w(\delta, J) - X]$$

where π is the probability of surviving the period and being able to consume X. In wage–risk studies it is risk of death, δ, rather than survival probability, π, that is observed. The relationship between the two is given by $\pi = (1-\delta)(1-\phi)$, where ϕ is the probability of dying from non-workrelated causes. Since ϕ is usually small for the working-age population, π is approximately equal to $1 - \delta$.

The first-order conditions governing the choices of X and job risk are

$$\pi_i \cdot \partial u/\partial X = \lambda \qquad (1)$$

$$\frac{u(\cdot)}{\lambda} = \frac{\partial p_w}{\partial \delta_i} \qquad (2)$$

and
$$\frac{\pi_i \cdot \partial u/\partial J_j}{\lambda} = -\partial p_w/\partial J_j \qquad (3)$$

for all job characteristics, J_j.

From equation (1), λ is the expected marginal utility of consumption, which, by assumption, is positive. According to equation (2), the marginal willingness to pay for an increase in the probability of surviving the job risk must equal its marginal implicit price. Equation (2) also implies that wages must be lower for jobs that are safer, that is, the marginal implicit price of an increase in π_i is a decrease in the wage rate. And equation (3) requires that the marginal willingness to pay for each job characteristic equal its marginal implicit price.

If workers know the relationship between market wages and job attributes and risks, then each worker selects the collection of job attributes and risks that equates the marginal benefit of each attribute to its marginal cost. In the case of risk of death, the marginal cost of working in a less risky job is the lower wage received, $\partial p_w/\partial \delta_i$; and this must be equated with the marginal willingness to pay for lower job risk. In other words, the risk premium associated with a higher-risk job must be equal to the individual's marginal willingness to accept compensation for risk.

Estimating Marginal Values for Risk

Data on wages, job attributes, and worker attributes are used to estimate the hedonic wage function—an equilibrium relationship between the wage on the one hand and job characteristics and variables affecting worker productivity on the other. If the hedonic wage function can be estimated satisfactorily, we can calculate the risk premium by evaluating the partial derivative of the function at a given risk level and set of job attributes.

One of the major questions in interpreting estimates of willingness to pay for risk reduction is whether individuals perceive differences in risks across jobs and if so, whether these perceptions are accurate. Compensating wage differentials for risk can exist in the labor market only if workers perceive differences in risks across jobs. The absence of compensating differentials need not mean that workers do not value reducing the risk of accidental death, only that they are unaware of the differences in risks. And if individuals have inaccurate estimates of job risks, then risk premia can exist. But these risk premia will yield biased estimates of individuals' marginal willingness to pay for risk reduction unless the researcher can identify what it was that individuals thought they were buying when they accepted a particular job with its bundle of characteristics. Since most hedonic wage studies use objective measures of job risk, it is important to find out if individuals' perceptions correspond well with these objective measures.

The only evidence pertaining directly to individuals' perceptions of job-related risks is found in research that compares workers' risk perceptions with data on frequency of job-related death and injury. This evidence suggests that workers' perceptions are positively correlated with objective risks, but may overstate them. Viscusi (1979) reported a positive correlation between a dichotomous risk variable (posed as a question, Is your job dangerous?) and accident rate data from the Bureau of Labor Statistics (BLS). In a subsequent study, Viscusi and O'Connor (1984) reported that workers in the chemical industry perceived risk of injury on their job to be about 50 percent higher than BLS estimates. Gerking, de Haan, and Schulze (1988) also found that workers' perceptions of risk of death on the job overstated BLS accidental death rates.

Other studies providing evidence on the accuracy of risk perceptions compare relative frequency of deaths, by cause, with individuals' perceptions of these frequencies. Slovic, Fischhoff, and Lichtenstein (1979) found that, on average, people overestimate the likelihood of infrequent causes of death (deaths due to botulism, floods, tornadoes) but underestimate the probability of deaths with higher frequencies (deaths due to heart disease, cancer). However, Fischhoff and coauthors (1981) note that one must distinguish between an individual's perception of the relative frequency of death in some population and the individual's estimate of his or her *own* risk of death. There is evidence that the latter is often underestimated (see Hamermesh, 1985; Fischhoff and coauthors, 1981). What matters in assessing the evidence provided by the hedonic wage model is the accuracy of people's perceptions of the *differences* in risks among different jobs, not the absolute level of these risks. I am not aware of any evidence on this point, and the lack of evidence makes it hard to reach firm conclusions about the validity of studies that use objective rather than subjective risk estimates. In future research it will be clearly important to measure each respondent's perceptions of his or her own risk of death.

In practice, the biggest difficulties in estimating hedonic wage functions have come from lack of sufficiently accurate and detailed data on job-related risk of death and injury, and from lack of adequate data on other job and worker characteristics, especially those that may be correlated with risk of accidental death. Thus it is useful to categorize hedonic wage studies by data source. This categorization is followed in table 12-1, which summarizes empirical estimates of the value of a statistical life from hedonic wage studies; it has been adapted from Fisher, Violette, and Chestnut (1989).

An early group of studies (Thaler and Rosen, 1976; Brown, 1980; Arnould and Nichols, 1983), which, in general, found low values for a statistical life, used actuarial data on deaths by occupation due to all causes, rather than occupationally related deaths. These studies have generally been considered unreliable because of their use of inappropriate risk data.

A second group of studies has used data on job-related deaths by industry, without, however, controlling for occupation. Most of these studies merge Bureau of Labor Statistics data on fatal accidents, by industry, with micro data on wages and job and worker characteristics (Viscusi, 1978, 1981; Olson, 1981). An exception is the study by Moore and Viscusi (1988), who used National Institute of Occupational Safety and Health (NIOSH) data on job risks by state for broad industry groups. The NIOSH data represent a complete census of occupational deaths; they show generally higher risks than the BLS data and a dif-

Table 12-1. Estimates of the Marginal Willingness to Pay for Reductions in Risks of Accidental Death

Study	Mean risk level for the sample[a]	Range of estimates	Judgemental best estimate
			(millions of 1986 dollars)
Early low-range wage–risk estimates			
Thaler and Rosen (1976)[b]	11.0	.44–84	.64
Arnould and Nichols (1983)[b]	11.0	.72	.72
Early high-range wage–risk estimates (based on BLS accident rates)			
R. Smith (1976)	1.0 & 1.5	3.6–3.9	3.7
V. K. Smith (1983)[c]	3.0	1.9–5.8	3.9
Viscusi (1978)	1.2	4.1–5.2	4.3
Olson (1981)	1.0	8.0	8.0
Viscusi (1981)	1.0 to 1.5	8.5–14.9	8.5
a. without risk-interaction terms	1.04	5.4–7.0	7.0
b. with risk-interaction terms	1.04	4.7–13.4	
New wage–risk studies			
Dillingham (1985)	1.4–8.3	2.1–5.8	2.5
Marin and Psacharopoulos (1982)[d]			
a. manual workers	2.0	2.7–3.1	2.9
b. nonmanual workers	2.0	9.0	
Gegax, Gerking, and Schulze (1991)			
a. all union workers	2.2	1.9	
b. union blue-collar workers	10.1	1.6	1.6
Moore and Viscusi (1988)	0.79	5.0–6.5	5.4

Source: Adapted with permission from Fisher, Violette, and Chestnut (1989).

[a]Approximate annual deaths per 10,000 people.

[b]Based on actuarial risk data.

[c]Assumes that 0.4 percent of all injuries are fatal, as reported by Viscusi (1978) for the BLS injury statistics, and that the risk premium for fatal injuries is 33 to 100 percent of the premium for all risks.

[d]In this study the age-adjusted normalized risk variable is not directly comparable with the risk levels used in other studies. However, the average risk of death for the entire sample was 2 in 10,000.

ferent pattern of relative risks across industries. Most of the studies in this second group (see table 12-1) generally find very high values for a statistical life—in the neighborhood of $4–$8 million (in 1986 dollars), for example. The problem with these risk measures is that, within an industry, there is likely to be significant variation in risk across occupations.

A third group of studies, designed to remedy lack of risk data by occupation, includes those of Dillingham (1985) and Marin and Psacharopoulos (1982). Dillingham combined workman's compensation data on job risk with wage data from the University of Michigan's 1977 Quality of Employment Survey and the 1970 New York State Census. When the same risk-by-occupation variable was used with each of the data sets, and when dummy variables for occupation and industry were included in the equation, the implied value-of-life estimates were $2.1 million and $1.7 million ($1986), respectively. When risk data by occupation and industry were used, the risk variable was not significant using the New York census data but was significant using the Quality of Employment data, and implied a value of life of $2.7 million ($1986). Marin and Psacharopoulos, using a measure of deaths by occupation, estimated wage premia for workers in the United Kingdom. Their value-of-life estimate for manual workers was U.S. $2.5 million ($1986).

A recent hedonic wage study by Gegax, Gerking, and Schulze (1991) used individual worker data on risk perceptions and job characteristics to estimate the hedonic wage function. This approach circumvents the problem of inaccurate worker perceptions and provides a more detailed description of jobs than is available in most labor-market data sets. An additional advantage of this study is that it collected data on workers' announced willingnesses to pay to reduce risk of death (responses to contingent valuation questions), so that these data might be compared with estimates of the value of a statistical life based on hedonic wage functions. In their hedonic wage studies, Gegax, Gerking, and Schulze found a significant value of life of $1.6 million ($1986) for blue-collar, unionized workers and a value of $1.9 million for all workers.

Although the last three studies used more appropriate measures of risk of death, it should be noted that none of the studies controlled for risk of injury. The compensating wage differentials reported by the authors may thus reflect compensation for risk of injury as well as risk of death.

Identifying Marginal Willingness-to-Pay Functions

We know that the partial derivative of the hedonic wage function with respect to each characteristic is its marginal implicit price and that in equilibrium we can take the marginal implicit price to be a point estimate of each worker's marginal willingness to pay for that characteristic. But as explained in chapter 11, a second stage of analysis is required in order to identify the marginal willingness-to-pay function for each characteristic. In contrast to the empirical analysis of property values, there have been very few efforts to identify the willingness-to-pay function for job characteristics.

Two recent studies have dealt with the identification problem in somewhat different ways. In both cases, the authors imposed structure on the problem by making explicit assumptions about the form of the underlying utility functions. In one study (Biddle and Zarkin,

1988), identification of the marginal willingness-to-pay function for reducing risks of an accidental injury on the job was achieved by an instrumental variables approach. The second study (Viscusi and Moore, 1989) is noteworthy for two reasons. The first is that the authors identified the marginal willingness-to-pay function for reduced risks of death by making use of interregional variation in the hedonic wage function. To identify the parameters of the marginal bid function, the marginal price of job risk was estimated for different regions of the United States, thus assuring variation in marginal price that is independent of the variables entering the marginal bid function. The second reason is that the authors explicitly took into account the age of each worker and the number of expected life-years at risk.

Viscusi and Moore specified a lifetime utility maximization problem, which was a simplification of equation (12) in chapter 10 in this book in that the exogenous probability of death was constant across all time periods. This assumption made it possible to derive a simplified expression for the first-order conditions and the marginal willingness to pay for any given explicit utility function. The model also made it possible to derive estimates of the implied discount rate on life-years. Viscusi and Moore used data from the 1982 wave of the University of Michigan Panel Study of Income Dynamics, supplemented by National Traumatic Occupational Fatality data published by NIOSH. Depending upon the specification of the utility function in the estimation technique, the implied discount rate ranged from 11 percent to 17 percent for the sample as a whole.[1]

Conclusions

The finding of a positive wage–risk premium is quite robust across data sets, time periods, theoretical models, and even countries.[2] It appears that (1) workers perceive differences in risks across jobs; (2) these perceptions are correlated with objective measures of on-the-job risks; and (3) workers prefer jobs with lower risks, other things being equal, and are willing to pay for safety in the form of reduced wages. In assessing the accuracy of empirical estimates of the value of risk reduction, there are still questions about the accuracy of workers' perceptions of risks, the accuracy of the available risk data that are aggregated over occupational groupings and/or industry categories,

---

[1]For more recent work by these authors on estimating the implicit discount rate for life-years, see Moore and Viscusi (1990).

[2]For a recent comparison of wage–risk premia across countries, see Kniesner and Leeth (1991).

and the ability to control for other individual and job-related determinants of wage differentials. It has also become clear that there is no one single value for risk reduction; rather, individuals' values depend upon such things as age and income. Understanding these individual determinants of behavior toward risk and the preferences and values that lie behind them is a rich field for further theoretical and empirical research.

All of these studies share the common characteristic of assuming that the location of the job is unimportant; that is, they do not control for differences in urban amenities and the potential effects of amenities on wage levels across cities. However, in a series of papers Smith and Gilbert have included location-specific characteristics such as air pollution levels in their hedonic wage equations; see Smith (1983) and Smith and Gilbert (1984, 1985). The modeling of determinants of interurban wage differentials is discussed in the next section.

INTERURBAN WAGE DIFFERENCES AND
THE VALUE OF AMENITIES

Those cities that are more desirable places to live and work in will attract workers from less desirable cities and regions. The in-migration of labor will exert downward pressure on wage rates in the desirable city. An equilibrium occurs when wages have fallen to the point where the marginal worker is indifferent between moving to this city and staying in his or her next-best alternative location. The difference in wages between this city and the next-best alternative is a compensating wage differential.

The possibility that such compensating differentials could be used as estimates of the monetary value of amenity differences spawned a series of empirical studies during the 1970s.[3] In these studies, measures of average wage rates, earnings, or income were regressed on variables reflecting such things as climate (temperature, humidity, frequency of rain, for example), environmental quality (say, measures of air pollution, water pollution, and access to recreational resources such as beaches), cultural amenities (for example, numbers of museums, newspapers, and universities), the disamenities of urban life (for example, crime rates, population density), and city size itself. Some of these studies controlled for differences in the occupational structure of urban labor forces by estimating sepa-

---

[3]The early studies include Nordhaus and Tobin (1972), Hoch and Drake (1974), Hoch (1977), and Meyer and Leone (1977).

rate equations for individual occupations. Others used more aggregated measures such as average earnings. In none of these early studies was the estimating equation derived from a formal model of individual choice, interurban migration, or supplies and demands of labor in a system of interconnected urban labor markets.

Rosen (1979) was apparently the first to attempt to provide a formal model for deriving the structural equations relating wages to urban amenities and disamenities. Such a formal model is necessary to provide a welfare theoretic interpretation of regressions that explains wage differences across cities. Rosen pointed out that there are really two hedonic markets in which individuals are making choices— one for labor, and one for land or housing. A decision to work in one city is also a decision to purchase housing services in that city. As individuals are drawn toward the more desirable cities and push wages down in those cities, they are also pushing out the demand for land and housing and increasing the prices of land and housing. Not only are there compensating wage differentials, then, but there are also compensating land rent and housing price differentials across cities. The labor market model must also provide a coherent explanation for why firms in some cities are able to pay higher wages and still compete in markets for goods traded among cities. A formal model of both sets of markets is required in order to draw inferences about amenity values and willingnesses to pay from data on wage differentials and housing prices.

A number of such formal models have now been presented in the literature.[4] They share the common feature of explicitly dealing with the interaction between markets for labor and for land across urban areas. But they differ in the way in which they model certain specific features of the determination of the interurban equilibrium. For example, some models treat the size of each city as fixed, while others allow the city boundary to expand to accommodate higher populations. Some models treat the costs of firms as exogenous, while other models allow costs to be affected by environmental amenities and the population of the city itself. And some models treat amenity levels as uniform within each city, while others allow for variation in amenity levels within as well as among urban areas.

---

[4]See, for example, Rosen (1979), Cropper and Arriaga-Salinas (1980), Cropper (1981), Henderson (1982), Hoehn, Berger, and Blomquist (1987), Blomquist, Berger, and Hoehn (1988), and Roback (1982, 1988). Bartik and Smith (1987) review most of these models, discuss similarities and differences in their theoretical structures, and describe some of the major empirical results. See also Palmquist (1991).

Apparently models that allow all of the relevant variables (such as city size, firm location, and cost) to be determined endogenously are analytically intractable. As a consequence, the results of these models depend upon the specific features of the model. None of these models fully captures all of the complexity of the general interurban equilibrium. The models produce what Bartik and Smith have termed ''partial descriptions'' of the more complex reality (1987, p. 1232).

In this section I do not try to explain all of the models, with their different results. Instead, I first employ a simple model of individual choice to show how the interaction between wages and rent or housing price affects our ability to interpret regression coefficients as welfare measures. Then I sketch out one version of a simple model of the equilibrium of the land and labor markets that permits the estimation of a marginal amenity-value measure. I end this discussion by examining in a more qualitative manner the implications of constructing richer models with more endogenous variables and more interactions.

The Welfare Measure with Two Hedonic Markets

Most of the early studies of interurban wage differences were based either implicitly or explicitly on the assumption that wage differences measure the values of amenity differences. For example, Nordhaus and Tobin (1972) and Meyer and Leone (1977) used their results to make adjustments to the national income accounts for nonpriced positive and negative welfare effects of urbanization. And more recently Clark and Kahn (1989) interpreted their coefficient on an interurban wage equation as a marginal implicit price and used it in a second-stage estimation of a marginal bid function for an amenity. However, a very simple model of individual choice of a city to live and work in can be used to show that the assumption that wage differences measure amenity values is not valid in general. This is because the wage differential is also affected by an interaction between the markets for labor and land.

To see this, consider the simplest case of an individual who derives utility from the consumption of a numeraire good, X, the quantity of housing consumed, h, and the level of an urban-specific amenity, q. By selecting a city to live in, the individual determines her annual wage, p_w and the level of the urban amenity. Assume that q_i is the same at all locations within city i. Let X be normalized with a price of one that is constant across cities, since X is a nationally traded good. The individual selects q_i along with a selection of X and h so as to maximize

$$u = u(X, h, q_i) + \lambda[p_w(q_i) - p_h(q_i) \cdot h - X]$$

Note that both wages and the price of housing vary across cities according to the level of the amenity in each city. The intuition for this is as follows. If one city is relatively more desirable as a place to live and work, other things being equal, workers will move to that city; the increase in the supply of labor will push down wages in that city; and, since people must live in the same city in which they work, the increased demand for housing will bid up its price. Of course, a complete model must aggregate across individuals as well as specify the supply sides of these markets in order to solve for the wage and housing price equations. I return to this point below.

The first-order conditions for the individual choice problem are:

$$\partial u/\partial X = \lambda$$

$$\partial u/\partial h = \lambda \cdot p_h$$

$$\partial u/\partial q = \lambda (h \cdot \partial p_h/\partial q - \partial p_w/\partial q)$$

Using the first-order conditions, the individual's optimum choice of X and q can be found. Graphically, the optimum combination is at the tangency between the individual's indifference curve for X and q and the budget line giving the terms at which X and q can be exchanged, both holding the quantity of housing h constant; see figure 12-1. The marginal rate of substitution (MRS) between X and q is

$$MRS_{Xq} = \frac{\partial u/\partial q}{\partial u/\partial X} = -\left.\frac{dX}{dq}\right| u^0$$

The slope of the budget line is found by taking the total differential of the budget constraint and treating h as a constant. The slope is

$$dX/dq = \partial p_w/\partial q - h \cdot \partial p_h/\partial q$$

The marginal rate of substitution can also be interpreted as the marginal willingness to pay for q or w_q. In equilibrium, w_q can be inferred from knowledge of the slope of the budget line:

$$w_q = h \cdot \partial p_h/\partial q - \partial p_w/\partial q \qquad (4)$$

This marginal willingness to pay has two components. The first component is the willingness to spend more on housing in a higher-quality city with a higher price for housing; the second is the willingness to accept a lower wage in the higher-quality city.

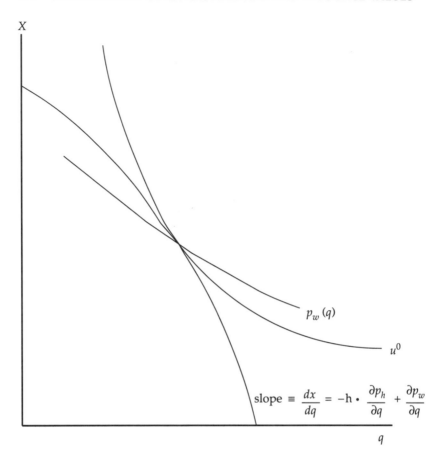

Figure 12-1. The individual's choice of an urban amenity

This analysis makes it clear that knowledge of the hedonic wage gradient across cities is not sufficient for inferring marginal amenity values. The hedonic wage gradient is also shown in figure 12-1. The budget line is more steeply sloped than the wage gradient because it includes the negative term for the effect of q on the price of housing. This means that the marginal implicit price of q in the labor market is an underestimate of the individual's marginal valuation of q when q is in fact purchased simultaneously through two hedonic markets.[5]

---

[5]It is interesting to contrast this analysis with the figures in Henderson (1982) and Clark and Kahn (1989), whose figures implicitly hold the quantity

A General Model of Interurban Equilibrium

In order to estimate a welfare measure such as equation (4), it is necessary to develop a formal model of the interurban equilibrium so that the hedonic wage and housing price equations can be properly specified. The model I present here does not capture all of the complexities of the interurban location equilibrium problem, but it does show the general features of most of the models that have been presented in the literature. This model is patterned most closely after those of Hoehn, Berger, and Blomquist (1987) and Roback (1982). It incorporates the following features:

- It takes account of the variation in housing prices within cities caused by the spatial character of a city. Specifically, since land rents vary inversely with commuting costs, distance from the city center is an argument in the rent function.
- It treats the size of a city as endogenous. As more desirable cities draw workers to them and the price of housing is bid up, it becomes profitable to convert some of the surrounding agricultural land to housing. This helps to dampen the upward pressure on housing prices.
- It treats the number of firms in each city as endogenous. And firms' costs are made to depend on the level of one or more amenities and disamenities and, perhaps, on city size, in order to reflect agglomeration effects.

In this model I first assume that all individuals have identical preferences and identical endowments of wealth, which in turn are assumed to be zero for simplicity. From my initial assumption it follows that an equilibrium must be characterized by equal levels of utility for all individuals.[6] Each individual's preferences are represented by the utility function

$$u = u(X, s, q)$$

where X is the numeraire good traded in a "world" market at a price of one, s is the quantity of land occupied by the individual for hous-

---

of housing constant. But that can be valid only if the price of housing in each city is independent of that city's amenity level.

[6]Roback (1988) has extended this type of model to consider two classes of workers with different labor productivities and different preference functions. A key result is that the equilibrium wage of each type of worker depends in part on the preferences of the other worker class.

ing, and q is the amenity that varies across cities but is uniform within each city. Let there be n cities in this economy. Each city offers a wage of p_{w_i} and an amenity level q_i, $i = 1, \ldots, i, \ldots, n$.

Each city also has the standard circular form in which all jobs are located at the city center. Individuals choose a residential location at distance d, and incur commuting costs of $t \cdot d$ per period, where t is the unit per period transportation cost. In order to compensate for higher commuting costs, the price of land p_s will decline with increasing distance from the city center. The boundary of the city will be at distance d^* where in equilibrium $p_s(d^*)$ is equal to p_{s^*}, the rental price of undeveloped or agricultural land.

Each individual chooses a wage and amenity package by selecting a city in which to live and work, a location for his or her residence, a quantity of residential land, and a quantity of the numeraire. Formally, each individual's choice problem is to maximize:

$$u = u(X, s, q) + \lambda[p_w(q_i) - t \cdot d - p_s(d) \cdot s - X]$$

The solution to this problem gives the indirect utility function:

$$u = v[p_w(q_i) - t \cdot d, p_{si}(d), q_i] \tag{5}$$

An equilibrium is achieved when wages and land prices have adjusted so that all individuals are indifferent as to the choice of a city and their location within the city chosen. From equation (5), this equilibrium must satisfy

$$v[p_w(q_i) - t \cdot d, p_{si}(d), qi] = v^* = v[p_w(q_i) - t \cdot d_i^* p_{s^*}, q_i] \tag{6}$$

The left-hand side of equation (6) can be solved for each city's land price schedule, $p_{si}(d, q_i, t, v^*)$. The right-hand side can be solved for the city's physical size, $d_i^*(t, p_s^*, q_i, v^*)$. The application of Roy's Identity to (6) gives the individual's demand function for land,

$$s = -\frac{\partial v/\partial p_s}{\partial v/\partial p_w} = a(p_w - t \cdot d, p_{si}, q_i)$$

and since the supply of land at any distance is given by $2 \cdot \pi \cdot d$, the total number of people that any city can accommodate is

$$P = \int_0^{d^*} \frac{2 \cdot \pi \cdot d}{d(\cdot)} \cdot dd$$

The model must be closed by specifying the production sector of the economy, which also determines the demand for labor. Assume that the industry producing good X is competitive and subject to constant returns to scale. Then price, which has been normalized at one, equals average cost as well as marginal cost. If firms purchase materials and capital at fixed world prices, then some economic characteristic of each city must be an argument in the firms' cost functions; otherwise, all firms would be forced by competition to pay the same wage. Hoehn, Berger, and Blomquist (1987) assume that q and total population are arguments in the cost function, the latter through an agglomeration effect. Alternatively, Roback (1982) assumes that firms use land in production, so that p_s is an argument in the cost function, along with q.

This system of equations can be solved to determine p_w and P for each city and p_s and d for each individual within the city. These variables depend upon the exogenous q_i, t, and p_{s^*}, which are the variables to be included in hedonic wage and land price equations. Because of all of the interdependencies, the comparative statics of this type of model can be fairly complex. Hoehn, Berger, and Blomquist (1987) analyzed the comparative statics of their model with respect to changes in q. The results depend on how q and P affect the unit cost function.

To derive the marginal willingness to pay for q, we return to equation (5), take its total differential, and set it equal to zero, obtaining

$$du = \frac{\partial v}{\partial p_w} \cdot dp_w + \frac{\partial v}{\partial p_s} dp_s + \frac{\partial v}{\partial q} \cdot dq = 0$$

and

$$w_q \equiv \frac{\partial v/\partial q}{\partial v/\partial p_w} = -\frac{\partial v/\partial p_s}{\partial v/\partial p_w} \cdot \frac{dp_s}{dq} - \frac{dp_w}{dq}$$

Again, employing Roy's Identity, this becomes

$$w_q = s \cdot \frac{dp_s}{dq} - \frac{dp_w}{dq}$$

This expression is similar to the one derived above from the model of individual choice; see equation (4). Here, each individual's marginal willingness to pay for q consists of two components, the change in the expenditure on land associated with an increase in q and the willingness to accept a lower wage rate for an improvement in q.[7] It should

---

[7] The comparative static analysis of Hoehn, Berger, and Blomquist (1987) shows that it is possible for dp_s/dq to be negative or for dp_w/dq to be positive.

be noticed that although the level of q is uniform within each city, individuals can have different marginal willingnesses to pay for q, depending upon the other elements in their consumption bundle—in particular, the quantity of land.

Other models of the interurban equilibrium have been developed to examine other forms of interaction. The additional features that can be captured in these models include the following:

• Intra-urban variation in amenity levels. Some amenities and disamenities that vary on average across cities also vary systematically within each city. For example, crime rates and air pollution tend to be higher in the center of each city than at the city boundary. This variation in amenity levels within cities will affect the spatial pattern of land rents and housing prices within each city; see Cropper (1981).

• The existence of both traded and nontraded goods. With zero transportation costs, traded goods sell at the same prices in all cities. But the prices of nontraded goods can vary across cities, leading to differences in the cost of living. The explicit treatment of nontraded goods is necessary to determine how cost of living differences should be treated in the specification of the hedonic wage function. Cropper (1981) has shown that housing prices should not be included in the index of prices used as a argument in hedonic wage functions.

• Variation in the cost of supplying housing across cities. Although the simpler urban models are formulated in terms of land rent, housing prices are easier to observe for purposes of hedonic price estimation. Since the concept of interurban equilibrium is inherently long run, it is necessary to model the supply side of the housing market. The cost of producing housing will depend on, among other things, the price of land (which varies within and across cities) and the wage rate (which also varies across cities); see for example Roback (1982) and Hoehn, Berger and Blomquist (1987).

SUMMARY

In this chapter we have examined two different types of application of hedonic price theory to labor markets. In one application, the focus is on the intrinsic characteristics of jobs. These applications seek to infer willingnesses to pay for changes in characteristics such as risk of accidental death or injury on the job. In these applications, it is usually assumed implicitly that the location of the job is unimportant. An exception is the work of Smith and Gilbert (1984, 1985) described above. In some wage–risk studies, regional dummy variables may be included to control for differences in wage levels across broad regions of the country, but the location of the job and the amenities or dis-amenities that go with living near the job are not explicitly considered.

In the other application, the focus is on the choice of a city in which to live and work and, in some cases, where in the city to live. Jobs are treated as homogeneous, except for their location. These applications seek to infer willingness to pay for urban amenities. In some of these applications, job characteristics are controlled to some extent by examining interurban differences within broad occupational categories. A key feature of applications of this kind is the need to model explicitly the interaction between the two hedonic markets—those for labor and for land or housing. Introducing heterogeneous jobs into this interurban labor market does not require adding another hedonic market into the analysis. But it does require amore detailed treatment of the determinants of wage differences within a multi-city, multi-characteristic market for differentiated labor. This is an important subject for further theoretical and empirical research.

REFERENCES

Arnould, Richard J., and Len M. Nichols. 1983. "Wage–Risk Premiums and Worker's Compensation: A Refinement of Estimates of Compensating Wage Differential," *Journal of Political Economy* vol. 91, no. 2, pp. 332–340.

Bartik, Timothy J., and V. Kerry Smith. 1987. "Urban Amenities and Public Policy," in Edwin S. Mills, ed., *Handbook of Regional and Urban Economics.* Amsterdam, The Netherlands: Elsevier.

Biddle, Jeff E., and Gary A. Zarkin. 1988. "Worker Preferences and Market Compensation for Job Risk," *Review of Economics and Statistics* vol. 70, no. 4, pp. 660–667.

Blomquist, Glenn C., Mark C. Berger, and John P. Hoehn. 1988. "New Estimates of Quality of Life in Urban Areas," *American Economic Review* vol. 78, no. 1, pp. 89–107.

Brown, Charles 1980. "Equalizing Differences in the Labor Market," *Quarterly Journal of Economics* vol. 94, no. 1, pp. 113–134.

Clark, David E., and James R. Kahn. 1989. "The Two-Stage Hedonic Wage Approach: A Methodology for the Evaluation of Environmental Amenities," *Journal of Environmental Economics and Management* vol. 16, no. 2, pp. 106–120.

Cropper, Maureen L. 1981. "The Value of Urban Amenities," *Journal of Regional Science* vol. 21, no. 3, pp. 359–373.

Cropper, Maureen L., and A. S. Arriaga-Salinas. 1980. "Inter-City Wage Differentials and the Value of Air Quality," *Journal of Urban Economics* vol. 8, no. 3, pp. 236–254.

Dillingham, Alan E. 1985. "The Influence of Risk Variable Definition on Value-of-Life Estimates," *Economic Inquiry* vol. 23, no. 2, pp. 277–294.

Fischhoff, Baruch, Sarah Lichtenstein, Paul Slovic, Stephen L. Derby, and Ralph L. Keeney. 1981. *Acceptable Risk*. New York: Cambridge University Press.

Fisher, Ann, Dan Violette, and Lauraine Chestnut. 1989. "The Value of Reducing Risks of Death: A Note on New Evidence," *Journal of Policy Analysis and Management* vol. 8, no. 1, pp. 88–100.

Gegax, Douglas, Shelby Gerking, and William Schulze. 1991. "Perceived Risk and the Marginal Value of Safety," *Review of Economics and Statistics* vol. 73, no. 4, pp. 589–596.

Gerking, Shelby, Menno de Haan, and William Schulze. 1988. "The Marginal Value of Job Safety: A Contingent Valuation Study," *Journal of Risk and Uncertainty* vol. 1, no. 2, pp. 185–199.

Hamermesh, Daniel S. 1985. "Expectations, Life Expectancy, and Economic Behavior," *Quarterly Journal of Economics* vol. 100, no. 2, pp. 389–408.

Henderson, J. Vernon. 1982. "Evaluating Consumer Welfare Amenities and Interregional Welfare Differences," *Journal of Urban Economics* vol. 11, no. 1, pp. 32–59.

Hoch, Irving. 1977. "Variations in the Quality of Urban Life Among Cities and Regions," in Lowdon Wingo and Alan Evans, eds., *Public Economics and the Quality of Life*. Baltimore, Md.: The Johns Hopkins University Press for Resources for the Future and the Centre for Environmental Studies.

Hoch, Irving, and Judith Drake. 1974. "Wages, Climate, and the Quality of Life," *Journal of Environmental Economics and Management* vol. 1, no. 3, pp. 268–295.

Hoehn, John P., Mark C. Berger, and Glenn C. Blomquist. 1987. "A Hedonic Model of Interregional Wages, Rents, and Amenity Values," *Journal of Regional Science* vol. 27, no. 4, pp. 605–620.

Kniesner, Thomas J., and John D. Leeth. 1991. "Compensating Wage Differentials for Fatal Injury Risk in Australia, Japan, and the United States," *Journal of Risk and Uncertainty* vol. 4, no. 1, pp. 75–90.

Lucas, Robert E. B. 1977. "Hedonic Wage Equations and Psychic Wages in the Returns to Schooling," *American Economic Review* vol. 67, no. 4, pp. 549–558.

Marin, Alan, and George Psacharopoulos. 1982. "The Reward for Risk in the Labor Market: Evidence from the United Kingdom and a Reconciliation with Other Studies," *Journal of Political Economy* vol. 90, no. 4, pp. 827–853.

Meyer, John R., and Robert A. Leone. 1977. "The Urban Disamenity Revisited," in Lowdon Wingo and Alan Evans, eds., *Public Economics and the Quality of Life*. Baltimore, Md.: The Johns Hopkins University Press for Resources for the Future and the Centre for Environmental Studies.

Moore, Michael J., and W. Kip Viscusi. 1988. "Doubling the Estimated Value of Life: Results Using New Occupational Fatality Data," *Journal of Policy Analysis and Management* vol. 7, no. 3, pp. 476–490.

Moore, Michael J., and W. Kip Viscusi. 1990. "Models for Estimating Discount Rates for Long-Term Health Risks Using Labor Market Data," *Journal of Risk and Uncertainty* vol. 3, no. 4, pp. 381–401.

Nordhaus, William D., and James Tobin. 1972. "Is Growth Obsolete?" in *Economic Research: Retrospect and Prospect, Economic Growth, Fiftieth Anniversary Colloquium*, vol. 5. New York: National Bureau of Economic Research.

Olson, Craig A., 1981. "An Analysis of Wage Differentials Received by Workers on Dangerous Jobs," *Journal of Human Resources* vol. 16, no. 2, pp. 167–185.

Palmquist, Raymond B. 1991. "Hedonic Methods," in John B. Braden and Charles D. Kolstad, eds., *Measuring the Demand for Environmental Quality*. Amsterdam, The Netherlands: North-Holland.

Roback, Jennifer. 1982. "Wages, Rents, and the Quality of Life," *Journal of Political Economy* vol. 90, no. 6, pp. 1257–1278.

Roback, Jennifer. 1988. "Wages, Rents, and Amenities: Differences Among Workers and Regions," *Economic Inquiry* vol. 26, no. 1, pp. 23–41.

Rosen, Sherwin. 1979. "Wage-Based Indices of Urban Quality of Life," in Peter Mieszkowski and Mahlon Straszheim, eds., *Current Issues in Urban Economics*. Baltimore, Md.: The Johns Hopkins University Press.

Slovic, Paul, Baruch Fischhoff, and Sarah Lichtenstein. 1979. "Rating the Risks," *Environment* vol. 21, no. 3, pp. 14–20, 36–39.

Smith, Robert S. 1976. *The Occupational Safety and Health Act*. Washington, D.C.: American Enterprise Institute for Public Policy Research.

Smith, V. Kerry. 1983. "The Role of Site and Job Characteristics in Hedonic Wage Models," *Journal of Urban Economics* vol. 13, no. 4, pp. 296–321.

Smith, V. Kerry, and Carol C. S. Gilbert. 1984. "The Implicit Valuation of Risks to Life: A Comparative Analysis," *Economics Letters* vol. 16, no. 3–4, pp. 393–399.

Smith, V. Kerry, and Carol C. S. Gilbert. 1985. "The Valuation of Environmental Risks Using Hedonic Wage Models," in Martin David and Timothy Smeeding, eds., *Horizontal Equity, Uncertainty, and Economic Well-Being*. Chicago, Ill.: University of Chicago Press.

Thaler, Richard, and Sherwin Rosen. 1976. "The Value of Life Savings," in Nester Terleckyj, ed., *Household Production and Consumption*. New York: Columbia University Press.

Viscusi, W. Kip. 1978. "Labor Market Valuations of Life and Limb: Empirical Evidence and Policy Implications," *Public Policy* vol. 26, no. 3, pp. 359–386.

Viscusi, W. Kip. 1979. Employment Hazards: *An Investigation of Market Performance*. Cambridge, Mass.: Harvard University Press.

Viscusi, W. Kip. 1981. "Occupational Safety and Health Regulation: Its Impact and Policy Alternatives," in J. Crecine, ed., *Research in Public Policy Analysis and Management*, vol. 2. Greenwich, Conn.: JAI Press.

Viscusi, W. Kip, and Charles J. O'Connor. 1984. "Adaptive Responses to Chemical Labeling: Are Workers Bayesian Decision Makers?" *American Economic Review* vol. 74, no. 5, pp. 942–956.

Viscusi, W. Kip, and Michael J. Moore. 1989. "Rates of Time Preference and Valuation of the Duration of Life," *Journal of Public Economics* vol. 38, no. 3, pp. 297–317.

13

Recreational Uses of Natural Resource Systems

Many natural resource systems such as lakes, rivers and streams, estuaries, and forests are used extensively by people for various kinds of recreation activities, including fishing, hunting, boating, hiking, and camping. As places to conduct such activities, natural resource systems provide valuable services to people. From an economic perspective, these services have two important features. The first is that the economic value of these services depends upon the characteristics of the natural resource system. The characteristics determining value can be affected by air and water pollution and by resource management decisions about such things as the rates of harvest of timber and fish, the extraction of minerals and petroleum, and the allocation of water flows between diversionary uses and various instream uses. Knowledge of the values of these services may be important for a variety of resource management decisions.

The second important feature is that access to the resource for recreation is typically not allocated through markets. Rather, access is typically open to all comers at a zero price or a nominal entrance fee that bears no relationship to the cost of providing access. And there is little or no variation in these access prices over time or across sites to provide data for econometric estimation of demand functions.

In this chapter I address three questions. First, how is the value of the flow of recreation services from an existing natural resource system to be defined and measured? Second, how is the value of introducing a new recreation site to be estimated ex ante? And third, how can we estimate the value of changes in the quality of a recreation site or changes in the quality of the flow of recreation services from the system?

All of the models for answering these questions take advantage of the fact that each individual visit to a recreation site involves an implicit transaction in which the cost of traveling to the site is incurred in return for access to the site. Different individuals will face different travel costs to any site, and one individual will face different travel costs for the different sites he or she might visit. The responses of people to this variation in the implicit prices of visits are the basis for estimating the values of recreation sites and changes in site quality.[1]

The choices that people make in response to differences in travel costs have usually been modeled from one of two perspectives regarding time. In one perspective, there is a substantial period of time such as a season or a year during which people can make several trips, perhaps to more than one site. What people choose is the *number* of trips to make to each site. In the other perspective, there is a point in time at which people decide *whether* to visit any site, and if so, *which* site to visit.

In models using the first perspective, variation in the cost of traveling to the site (the implicit price of access) can be exploited to estimate a travel cost demand function. The value of the flow of services from the site is the area under the compensated demand curve for these services or for access to the site, aggregated over all who visit the site. The value of a change in the site's services can be found by invoking the concept of weak complementarity and estimating the area between two compensated demand curves that are conditioned on different levels or qualities of services at the site.

Models using the second perspective are a form of discrete choice or random utility model. Welfare values for individuals are calculated from the parameters of the indirect utility function, which is estimated from observations of individual choices. Since the characteristics of sites are arguments in the indirect utility function, it is a straightforward matter to calculate welfare values for changes in site characteristics or qualities.

In the first section of this chapter I outline the basic travel cost model of the demand for a single site. In the second section I discuss some issues that must be addressed in implementing and estimating the model, such as the measurement of travel cost (including the cost of time), functional form, and the role of substitute sites. In the third and fourth sections I explain how the standard travel cost model can be used to value recreation sites and changes in the characteristics of sites. In the fifth section I discuss the random utility model of recreation choice, and in the final section, I examine the hybrid hedonic

---

[1]Responsibility for the original insight belongs to Harold Hotelling; this insight was formalized by Clawson (1959).

travel cost model of recreation demand and other approaches to estimating recreation values.

THE SIMPLE TRAVEL COST MODEL OF DEMAND FOR A SITE

Although the demand for a site can be modeled as an aggregate or market demand, the usual practice is to estimate demand functions at the level of the individual and to calculate site values by adding up individuals' values for the site. As noted, the value of a recreation site to an individual is defined as the area under the individual's compensated demand curve for that site. However, the compensated demand may not be observable, so the usual practice is to begin with the ordinary demand curve for the site. The ordinary demand curve gives the number of visits an individual of given socioeconomic characteristics makes to the site as a function of the admission fee at the site. At any point in time, all individuals will face the same admission fee or price of access. This fee typically does not change over time. Thus estimation of individuals' demand functions may not be feasible from the available data on entry price and number of visits.

The travel cost model is based on the recognition that the cost of traveling to a site is one important component of the full cost of a visit and that, for any given site, there will usually be wide variation in travel cost across any sample of visitors to that site. To formalize this insight, let us look at a model of an individual's choice of the number of visits to make to a recreation site.[2] Assume, for now, that there is only one site available and that all visits have the same duration. (Modeling of the determination of duration of a visit to a site, and of the choice of a site or sites to visit when there are alternatives, will be taken up in later sections of this chapter.)

We assume that the individual's utility depends on the total time spent at the site, the quality of the site, and the quantity of a numeraire. With the duration of a visit fixed for simplicity, the time on site can be represented by the number of visits. The individual solves the following utility maximization problem:

$$\text{Max: } u\,(X, r, q) \tag{1}$$

subject to the twin constraints of monetary and time budgets:

$$M + p_w \cdot t_w = X + c \cdot r \tag{2}$$

---

[2]This model is adapted from McConnell (1985).

and
$$t^* = t_w + (t_1 + t_2)\, r \tag{3}$$

where X = the quantity of the numeraire whose price is one,
  r = number of visits to the recreation site,
  q = environmental quality at the site,
  M = exogenous income,
  p_w = wage rate,
  c = monetary cost of a trip,
  t^* = total discretionary time,
  t_w = hours worked,
  t_1 = round-trip travel time, and
  t_2 = time spent on site.

Assume that r and q are complements in the utility function. As discussed in chapter 4, this means that the number of visits will be an increasing function of the site's environmental quality. The time constraint reflects the fact that both travel to the site and time spent on the site take time away from other activities. Thus there is an opportunity cost to the time spent in the recreation activity. Assume also that the individual is free to choose the amount of time spent at work and that work does not convey utility (or disutility) directly. Thus the opportunity cost of time is the wage rate. (I return to the nature of the time constraint and the opportunity cost of time in the next section.) Finally, assume that the monetary cost of a trip to the site has two components: the admission fee f, which could be zero, and the monetary cost of travel. This cost is $p_d \cdot d$, where p_d is the per-mile cost of travel and d is the distance to the site and return from it.

Substituting the time constraint (3) into the monetary budget constraint (2) yields:

$$M + p_w \cdot t^* = X + p_r \cdot r \tag{4}$$

where p_r is the full price of a visit given by

$$p_r = c + p_w (t_1 + t_2)$$
$$= f + p_d \cdot d + p_w (t_1 + t_2) \tag{5}$$

As equation (5) makes clear, the full price of a visit consists of four components: the admission fee, the monetary cost of travel to the site, the time cost of travel to the site, and the cost of time spent at the site. On the assumption that individuals are free to choose the number of hours worked at a given wage rate, the two time costs are valued at the wage rate. In a more realistic model with income and payroll taxes, time would be valued at the after-tax wage rate.

Maximizing equation (1) subject to the constraint of equation (4) will yield the individual's demand function for visits:

$$r = r(p_r, M, q) \tag{6}$$

If all individuals spend the same amount of time at the site, and have the same wage, then this component of the price of a visit is the same for all individuals. Given these assumptions, the data on rates of visitation, travel costs, and variation in entry fees (if any) can be used to estimate the coefficient on p_r in a travel cost–visitation function. Because of the linearity of equation (5), the coefficient on p_r can be used to derive the individual's demand for visits to a site as a function of the admission fee.

Let us review the assumptions that make this basic version of the travel cost model "work."

- First, it is assumed that individuals perceive and respond to changes in the travel-related component of the cost of a visit in the same way they would respond to a change in the admission price. The assumption that travel cost and entry fee have equal effects on behavior makes it very important that the monetary value of travel cost be measured accurately.
- Second, it is assumed that each trip to the site is for the sole purpose of visiting the site. If the purpose of the trip is to visit two or more sites or to visit a relative en route, then at least part of the travel cost would be a joint cost that cannot be properly allocated among different purposes.
- Third, it is assumed that all visits entail the same amount of time spent on the site. This assumption plays two important roles in the simple form of the model: it makes it possible to measure site usage by the scalar v, the number of visits; and it makes the full price of a visit, p_r, a parameter to the individual. If the individual chooses the amount of time of each visit, then p_r is an endogenous variable.
- Fourth, it is assumed that there is no utility or disutility derived from the time spent traveling to the site. If part of the trip involves the pleasures of driving through a scenic countryside, then travel cost is overestimated by equation (5).
- Fifth, it is assumed that the wage rate is the relevant opportunity cost of time.
- Sixth, it is assumed that there are no alternative recreation sites available to these individuals.

I return to the issues raised by many of these assumptions in subsequent sections of this chapter.

ISSUES IN MODEL SPECIFICATION

The travel cost model of recreation demand described above was kept quite simple in order to focus attention on the underlying intuition. In this section I discuss several issues of model specification that must be addressed in actually carrying out a travel cost study. These issues are the role of time as a cost and as an element of choice, and how to account for substitute sites and the opportunity to choose from among a set of alternative sites.

Time

Both travel to a site and the recreation activity itself take time; and time is scarce. Also, the time spent in any activity could be a separate argument in individuals' utility functions. For example, time at work could yield either positive or negative utility; and almost certainly time spent at a recreation site has a positive marginal utility, at least up to some point. As we shall see, the opportunity cost of time must be included in a properly specified model of the demand for recreation visits; and the appropriate shadow price of time will depend on the alternative uses to which the time could be put and on the nature of the constraints on individual choice.

In this subsection, I first show why it is important to include the correct measure of time cost in the travel cost demand equation and what bias results from omitting the time cost of travel. I also discuss alternative approaches to estimating the shadow price of time. Then I take up the problem of modeling the choice of the length of the visit.

THE TIME COST OF TRAVEL AND THE DEMAND FOR VISITS

It has long been recognized that the time spent in travel to a site should be included as a component of travel cost for purposes of estimating the demand for visits,[3] but it has not been clear what price of time should be used to convert the time cost of travel to its monetary equivalent. A simple example will show that the price used to value travel time as a cost can have a substantial effect on the estimates of the coefficient on the price of a visit, the elasticity of demand, and, therefore, on the estimated value of visits to the site. Consider two individuals whose distances, travel costs, and so forth

---

[3]Early studies include Knetsch (1963), Scott (1965), and Cesario and Knetsch (1970).

are as follows:

Individual	Distance	Travel cost at $.10/mile	Time for trip
A	10 miles	$1	15 minutes
B	20 miles	$2	30 minutes

Suppose that the individual's demand for visits has been estimated as a function of only the monetary cost of travel. One step in the computation of the demand function is to assume the imposition of an entry fee of, say, $1, and to examine its effects on visitation as a function of distance. The hypothetical entry fee raises the money cost of a visit for A to $2. The conventional assumption is that individual A would have the same visitation rate as was originally observed for B (when her money travel cost was $2). The full cost of the visit for A is $2 plus 15 minutes; but for B the original full cost of travel was $2 plus 30 minutes. If the difference in time cost between 10 miles with the fee and 20 miles without the fee is ignored, the reduction in use associated with the hypothetical admission fee will be overestimated. As a consequence, the estimated demand curve will be more elastic than the true demand curve, and the benefits of the recreation site (the area under the demand curve) will be underestimated.

Since time is a cost that must be measured in monetary units that are commensurable with the entry fee, the observed time cost must be converted to a monetary value by the use of an appropriate shadow price of time. Let us assume that time spent in travel is valued at $2 per hour by recreationists. Then the full cost of travel for a visit by A is $1.50, while for B it is $3. A hypothetical $1 entry fee raises the money cost of a visit for A to $2.50. Since the new money cost for this distance is less than the original money cost for B, the number of visits predicted from the shorter distance with the fee will be higher than the number predicted when time is ignored as a separate element in cost.

The choice of the shadow price of time is critical to the estimation of the elasticity of demand for the site and the calculation of the value of the site. The choice of a high shadow price of time raises the importance of time cost in explaining visits as a function of distance. With a higher shadow price of time, the predicted reduction in the number of visits because of an increase in the entry fee is smaller, and the estimated demand curve is less elastic. Ignoring time cost has the effect of biasing the elasticity estimate upward and the benefit estimate downward. But clearly, using a shadow price of time that is too high can have the opposite effect on estimates of the elasticity of demand and benefits.

One of the key assumptions in the simple travel cost model described above is that individuals are free to choose the hours they work at a given wage and that this wage governs the trade-off between work and leisure. But what if there are institutional constraints on the choice of hours of work, such as the standard forty-hour work week? Models in which there is no trade-off between work and recreation time have been analyzed by Smith, Desvousges, and McGivney (1983b) and by Bockstael, Strand, and Hanemann (1987). In the Smith, Desvousges, and McGivney model, individuals could allocate a fixed nonrecreational time budget between work and other activities. However, the recreation time budget was fixed and exogenous. Individuals could only choose to allocate time between two or more sites. Smith, Desvousges, and McGivney found that the relevant price of a visit to a site was the sum of the money travel cost and a proportion of the total time per visit ($t_1 + t_2$ in our notation). This proportionality factor depended upon the wage rate and the shadow prices attached to the constraints on money and time. Thus the proportionality factor was not observable, but it could be estimated from the data.

The Bockstael, Strand, and Hanemann model was flexible in that it could accommodate either of two different cases regarding the individual's labor-market equilibrium. These authors assumed that the duration of a visit was fixed. If the individual could alter hours of work at the margin, the relevant price was the sum of the money travel cost and the total time of the visit ($t_1 + t_2$) valued at the marginal wage rate. If the individual could not make marginal adjustments in the number of hours worked, the money travel cost and time cost per visit entered the demand function for visits separately. The parameter for the time cost component was a function of parameters of the preference function. Bockstael, Strand, and Hanemann assumed a specific functional form for preferences, derived the alternative demand functions, and estimated the parameters from a data set that distinguished between those who could and those who could not alter their work time at the margin.

One practical difficulty in using the wage rate in recreation demand studies is that surveys often provide data on family income rather than the hourly wage rate. If the wage is inferred by dividing family income by some estimate of hours worked (typically, 2,000 hours per year), measurement error is introduced. One way to avoid this problem is to use a separately estimated hedonic wage equation to predict the wage rate for each individual in the sample (see Smith, Desvousges, and McGivney, 1983b).

A more important question is whether the wage rate, however measured, is the appropriate shadow price of time. Another key assumption in the standard model, in which the wage rate is made the appropriate shadow price, is that there are no time variables appearing directly or implicitly as arguments in the utility function. Suppose that the time spent at work has either a positive or a negative utility at the margin. Then there will be a divergence between the shadow price or scarcity value of time in other activities and the wage rate as a measure of the opportunity cost of time at work.[4] This point can be illustrated by modifying the simple travel cost model outlined above to include time at work as an argument in the utility function.

The constrained maximization problem becomes:

$$\text{max: } u\,(X, r, t_w) + \lambda\,(M + p_w \cdot t_w - X - c \cdot r)$$
$$+ \mu\,[t^* - t_w - (t_1 + t_2)\,r]$$

where site quality q has been omitted for simplicity. Since μ is the marginal utility of time, μ/λ is the scarcity value or marginal willingness to pay for time. The relevant first-order conditions include

$$\frac{\partial u/\partial r}{\lambda} = c + \frac{\mu}{\lambda}(t_1 + t_2) \tag{7}$$

and
$$\frac{\partial u/\partial t_w}{\lambda} + p_w = \frac{\mu}{\lambda} \tag{8}$$

Equation (7) says that the marginal willingness to pay for a visit must equal the full cost of a visit. The full cost is the sum of the monetary cost c and the time cost of the visit, where the time cost $t_1 + t_2$ should be valued at the scarcity value of time, μ/λ. But equation (8) shows that the wage rate may not be a good measure of the scarcity value of time. If the marginal utility of time spent working is negative, then the wage rate is an overestimate of the scarcity value of time. In other words, the opportunity cost of diverting an hour of time away from work is less than the wage rate. After reviewing evidence from transportation and urban commuting studies, Cesario (1976) concluded that the scarcity value of time was approximately one-third of the average wage rate.

In contrast to this suggestion, McConnell and Strand (1981) derived an alternative model that allowed for the estimation of the rela-

---

[4]This basic idea can be traced back to Johnson (1966). Cesario (1976) first analyzed its implications in the context of the travel cost model of recreation demand.

tionship between the scarcity value of time and the wage rate from the sample data. Suppose the demand for visits can be specified as a linear function of travel cost:

$$r = a + b_1 [c + (\mu/\lambda) (t_1 + t_2)] \tag{9}$$

First, let s be the ratio of the shadow price of time to the wage rate. Substitute $s \cdot p_w$ for μ/λ in equation (9). Then estimate

$$r = a + b_1 \cdot c + b_2 \cdot p_w (t_1 + t_2)$$

Since $b_2 \cdot p_w$ must equal $b_1 \cdot s \cdot p_w$, then $s = b_2/b_1$. For McConnell and Strand's sample, s was about .6.

In the McConnell and Strand model, although the scarcity value of time depends in part on the wage rate, it depends also on the alternative use of time, specifically on the marginal utility of work. A richer model of time and choice would allow for other utility-yielding uses of time. For example, there could be other leisure activities; or time could be an input into the household production of utility-yielding goods.

TIME ON SITE

As the models described above have shown, time spent on the recreation site is part of the cost of the recreation visit. Therefore, in principle, time on site should be included in the estimated demand function for recreation visits. However, there are some assumptions under which this is not necessary. As McConnell (1985) has shown, if all individuals in a sample choose visits of the same duration and if they all have the same opportunity cost of time, time cost on site becomes part of the constant term in the estimated equation. The assumption of identical duration of visits may be reasonable in some cases, for example, where day-trip activities predominate, or if separate equations are estimated for visits of different duration. But the assumption of identical opportunity costs of time is probably not reasonable, especially given that most travel cost studies now are based on observations of individuals rather than aggregates of people grouped by distance zone.

A related issue is how to model the choice of the duration of a visit. McConnell (1992) has investigated the implications of incorporating the choice of time on site for the proper specification of the recreation demand function. He specifies the following utility function,

$$u = u (X, r, t_2) \tag{10}$$

and assumes that it displays what he calls joint weak complementarity between r and t_2. By this he means that if $r = 0$, the marginal utility of on-site time is zero, and if on-site time is zero, the marginal utility of a trip is zero. The individual must choose r, t_2, and X to maximize equation (10), subject to the constraint of equation (4),

$$M + p_w \cdot t^* = X + p_r \cdot r$$

where $p_r = f + p_d \cdot d + p_w (t_1 + t_2)$. Since p_r is not exogenous to the choice problem, it cannot be an argument in a Marshallian demand function. McConnell shows that it is still possible to define a Marshallian demand function based on the exogenous components of trip cost. First, define a new term:

$$p_r^* = f + p_d \cdot d + p_w \cdot t_1$$

After maximizing equation (10) and solving for the demand functions, the indirect utility function can be obtained:

$$u = v(p_r^*, p_w, M)$$

McConnell then shows that Roy's Identity can be employed to derive a Marshallian demand function that contains only exogenous variables:

$$r(p_r^*, p_w, M)$$

This expression can be estimated.[5]

Substitute Sites

In the simple travel cost model described at the beginning of this chapter, it was assumed that there was only one site to visit. But suppose there are several sites available to the population whose recreation behavior is being studied. Then it is plausible that the number of visits that any individual makes to the site in question will depend not only on its implicit price but also on the implicit prices of any substitute sites in the region. If the travel costs of the relevant substitute sites are omitted from the estimated equation, its parameters will be biased.

---

[5]For other efforts at modeling the determinants of time on site, see Smith, Desvousges, and McGivney (1983b), Kealy and Bishop (1986), and Wilman (1987).

Some authors have suggested that the effect of omitting the price of a substitute site would bias the estimate of the coefficient on the price of the site toward zero. As a consequence, the estimate of the price elasticity of demand would be biased toward zero and consumer surplus or site value measures would be biased upward. Caulkins, Bishop, and Bouwes (1985) pointed out that the sign of the bias on the own price coefficient depends on the correlation between it and the omitted price variable. If the correlation between the two travel cost variables is positive, then omitting the substitute price biases the own price elasticity toward zero. But if the two travel costs are inversely correlated, the estimated own price coefficient is subject to a negative bias and the price elasticity of demand for visits is biased upward. The sign of the correlation between the two travel costs depends upon the spatial distribution of the population relative to the two sites. If the population is arrayed along a line between the two sites, travel costs to the two sites will be negatively correlated; but if the population is arrayed along a line that is perpendicular to the line connecting the two sites, the travel cost variables will be positively correlated in the population.[6]

When the recreation site being investigated is part of a regional recreation resource system, there is the difficult question of deciding which other sites are substitutes whose prices should be included in the site demand equation. The comprehensive approach of including all alternative sites is cumbersome and may require more data than are available. An alternative is to ask each individual what other site that person visits most frequently, and include only that site's price as the relevant substitute price. But different respondents could name different sites. And this approach may miss important substitution effects. For example, if an individual alternates between fishing at the site in question and swimming at another site, the swimming site would be selected as the alternative. But the more relevant alternative would be the next-best fishing site. There is no single right answer to the question of how to select substitute sites. This is a problem that requires good judgment on the part of the researcher.

Multi-Site Models

Analysis of the demand for a single site would be appropriate if the researcher is interested in valuing the availability of that single site or changes in the characteristics of only that site. But many policy-

---

[6]See Kling (1989) and McKean and Revier (1990) for further discussions of omitted variable bias in the travel cost model.

relevant questions involve changes in the value of a set of sites due to changes in the number and availability of sites, or to changes in the qualities of these sites. In such cases, the interactions and the substitution effects among sites must be modeled explicitly. This calls for some form of multi-site model.

Multi-site models are estimated as systems of demand equations. For each site j ($j = 1, ..., j, ..., m$), a demand equation of the following form is specified:

$$r_{ji} = r_j (p_{r_{ji}}, p_{r_{ki}}, M_i, q_j)$$
$$(i = (i = 1, ..., i, ..., n), (k = 1, ..., k, ..., m), \text{ and } k \neq j$$

where r_{ji} is the number of visits individual i makes to the site, $p_{r_{ji}}$ is the full price of a visit by i to j, and $p_{r_{ki}}$ is the set of substitute prices for visits to other sites. This type of model can be estimated from data on individual observations (see, for example, Burt and Brewer, 1971) or from data aggregated into zones of origin (see, for example, Cicchetti, Fisher, and Smith, 1976).

As the number of sites and demand equations increases, multi-site models can become cumbersome and difficult to estimate. And, as will be discussed in more detail in the section on Valuing Changes in Site Quality, it is not possible to take into account differences in site qualities and characteristics in this form of model. These difficulties have led researchers to experiment with several forms of alternative multi-site models.

One modeling simplification is to aggregate all of the visits made by an individual to all of the available sites into a number of visits to a "typical site," where the typical site is the site most frequently visited by that individual. Specifically, a single demand equation is estimated from individual data, where that equation takes the form

$$r_i = r (p_{r_{ti}}, q_t, M_i)$$

where $p_{r_{ti}}$ and q_t are the price and quality of the site most frequently visited by the individual. Such typical-site models closely resemble the recreation participation models described below in a later section.

In one application of this approach (Caulkins, Bishop, and Bouwes, 1986), an effort was made to control for substitutes by including proxy variables for substitute prices and qualities. The proxies were weighted averages of the prices and qualities of other sites actually visited. The weights were the relative frequencies of visits to the alternative sites.

The typical-site model does not fully capture the complexity of recreation behavior in a multi-site setting. It provides no basis for explaining the allocation of visits across sites in the recreation system. And if people visit different sites to engage in different types of activities, the proxy variables for substitute sites are not likely to capture the true substitution behavior of individuals faced with changes in the availability or quality of their most frequently visited site. The example of the individual who visits one site for fishing and another site for swimming applies here as well.

Another type of simplification of the multi-site model is to treat all observations of r_{ji} as belonging to a single travel cost demand equation. In this "pooled" model, a single equation of the following form would be estimated:

$$r_{ji} = r(p_{r_{ji}}, q_j, M_i)$$

For one example of this type class of model, see Smith, Desvousges, and Fisher (1986).

In another approach, Morey, in a series of papers (1981, 1984, 1985), has developed a model of how individuals allocate a *given* recreation time and money budget among alternative sites on the basis of site prices and qualities. But this ability to model site substitution behavior formally comes at the expense of losing the ability to explain the total level of recreation activity for each individual. In Morey's model, utility is a function of the number of visits to each site and the attributes of these sites. Utility is maximized subject to a fixed, exogenous time budget, where time is allocated to the on-site activity, travel to the site, and working to earn the money costs of travel and admission to the sites visited. A specific functional form for the utility function is chosen so that share equations can be derived as solutions to the constrained maximization problem. The share of visits made to site j ($s_j \equiv r_j / \Sigma r_j$) is an estimable function of site prices and characteristics. Since the share equations are derived from an explicit utility function, it is possible to calculate exact measures of welfare for changes in the availability or qualities of sites (Morey, 1985). In its focus on the allocation of activities among sites, the Morey share model has some similarity to the random utility models described later in this chapter.

VALUING RECREATION SITES

The value of a recreation resource is the discounted present value of the flow of recreation services to those who visit the site, as measured by the area under the compensated demand curve for visits to the site. Since the travel cost model yields a Marshallian demand curve, the usual practice is to rely on Willig's (1976) analysis to argue that the ordinary consumer surplus measure is a close approximation to the true value. However, Willig's bounds may not be appropriate for valuing recreation sites. His analysis is based on price changes for which positive quantities would be demanded at both the initial and second prices. But the value of a site is found by postulating a price change sufficient to drive the quantity demanded to zero. For a relatively small price increase, such as from f^0 to f^1 as shown in figure 13-1, the error in using consumer surplus is the area a. But the error in employing the ordinary consumer surplus as an approximation to the compensating variation measure of the site value is the area $a + b$. Since this error could be substantial, some investigators have begun to apply the methods of exact welfare measurement for determining the values of recreation sites. The functional form for the travel cost demand equation can be derived from an explicit specification of the utility function. Then the methods of Hausman (1981) or Vartia (1983) can be used to obtain the indirect utility function.[7]

The statement that the area under the demand curve is the measure of the value of the flow of services is based implicitly on the assumption that the demand curve is known with certainty. This implies that the welfare measure is deterministic. But since the demand curve is estimated from a statistical model, there are random errors in the estimates of the parameters. This fact makes the welfare measure itself a random variable; and it has implications for the way in which welfare measures are calculated. Bockstael and Strand (1987) have shown how different assumptions about the sources of error in the regression estimates lead to different methods for calculating the consumer surplus measure.[8] More recently, Smith (1990) has argued that when the purpose of the statistical estimation of demand functions is the derivation of welfare measures, estimators should be selected on the basis of the statistical properties of the welfare estimates rather than the statistical properties of the demand function parameters.

Sometimes policymakers wish to know the value of adding a new site to an existing recreation resource system. Conceptually, the value is the area under the compensated demand curve for the new site. But there are no data for estimating this demand curve directly

---

[7]This approach to modeling recreation site demand is discussed in Bockstael, Hanemann, and Strand (1986). For an application, see Bockstael, Strand, and Hanemann (1987).

[8]See also Graham-Tomasi, Adamowicz, and Fletcher (1990).

Price ($)

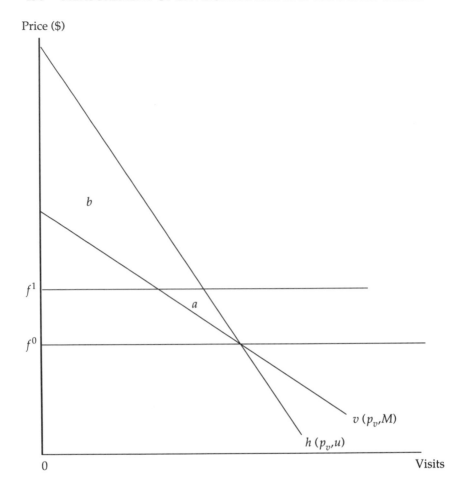

Figure 13-1. The difference between the Marshallian and compensating measures of the value of a recreation site

before the site is built. However, knowledge of the pattern of demands for the existing sites can be exploited to obtain an estimate of the value of adding a site to the system, at least if the new site is qualitatively similar to one of the existing sites.

The first step is to estimate a system of demand functions for the set of existing sites using a multi-site model of the type described above.[9] The second step is to identify one of the existing sites as the

---

[9]For applications of this approach, see Burt and Brewer (1971) and Cicchetti, Fisher, and Smith (1976).

closest substitute to the new site on the basis of nearby location or similar qualitative characteristics, or both.[10] Then the travel costs to the new site are used to generate a set of price reductions for those for whom the new site is closer than the existing site. These price changes are inserted in the demand function for the existing site to calculate a welfare measure. The principal limitation of this approach arises from the inability of the multi-site travel cost model to explicitly incorporate quality differences across sites. Also, the approach is implicitly based on the assumption that the only source of value associated with the new site is the reduction in travel costs for some visitors.

VALUING CHANGES IN SITE QUALITY

An improvement in an attribute of environmental quality at a site shifts the demand curve for visits to the site out and to the right. Given the assumptions of weak complementarity (that there is a choke price and that at the choke price, the marginal utility of quality is zero), the area between the two compensated demand curves for the site is an exact measure of the welfare change. However, the travel cost model provides only Marshallian demand curves. As pointed out in chapter 4, the Willig error bounds do not apply to areas between ordinary demand curves. So value measures based on the travel cost model must be used with caution. An alternative approach, as discussed above, is to utilize one of the methods of exact welfare measurement.

In utilizing an exact welfare measure of value, there is no need to take into account changes in recreation use at other sites or savings in travel cost as people substitute visits to the improved site for visits to other sites. These components of benefits are fully captured in the area between the demand curves for the improved site. For example, if recreation is switched from an alternative site, the demand curve for that site shifts to the left. But it would be incorrect to measure the area between those demand curves to adjust the measure of benefits. There has been no change in any quality measure at the alternative site. So the integral of the demand function for that site over the range of quality change at that site is zero.

---

[10]As Hof and King (1982) point out, since all of the information required for this approach is contained in the demand function for the substitute site, a properly specified single site model including substitute site prices is sufficient.

If there are simultaneous changes in quality at several recreation sites, the theory of welfare change under multiple-price or quantity changes applies. Consider, for example, two sites where there is a simultaneous improvement in quality. In principle, benefits are measured by evaluating the welfare change for site A, holding (by assumption) the quality at site B at its original level, then evaluating the welfare change for site B given the improved level of quality at site A. The welfare measure is independent of the order in which the quality changes are evaluated.

Two questions must still be resolved in implementing this measure of the value of site-quality changes. First, what attributes of the site matter to people, and how are they to be measured empirically? And second, how are the relevant quality measures to be incorporated into models of site demand so that shifts in demand curves can be predicted? In this section, I take up each of these questions in turn. I then consider the special problems raised by on-site congestion as one relevant measure of the quality of an individual's experience at a site.

Measuring Site Quality

The word "quality" is a proxy for a variety of characteristics of a site, some of which are affected negatively by pollution or by competing economic activities such as logging and mineral extraction, or positively by investments in enhancing the flow of recreation services from a resource. Quality measures can be based on scientific data, such as the dissolved oxygen concentration or pH of a water body. Or quality can be represented by some measure of perceived quality derived from questions asked of recreation visitors. Also to be considered is whether a single quality measure should be employed, or whether multiple measures are appropriate, either entered separately or aggregated into some kind of quality index.

Objective measures of quality are reproducible in the sense that two observers should obtain the same measure, within the bounds of measurement error. But people might make choices about recreation on the basis of their perceptions of quality rather than the objective measure. If individuals' perceptions are functions of objective measures and personal characteristics, then it may be possible to estimate a "perception function" and to use this function to model choices and measure welfare values. For some kinds of recreation activity, individuals' choices may depend on an interaction between an objective measure of quality and an individual characteristic such as skill or ownership of specialized equipment. An increase in the abundance of a particular species of fish, for example, will have a bigger impact on the behavior of an angler who has acquired skill in fishing for that species than on anglers who have not.

A wide variety of measures have been employed in empirical estimates of the value of site quality, among them:

- angler success rates, as measured by fish caught per visit (Stevens, 1966; Vaughan and Russell, 1982b; Samples and Bishop, 1985);
- water quality, as measured by such things as oil and bacteria at beach sites (Feenberg and Mills, 1980);
- bacteria and turbidity at beaches on the Chesapeake Bay (Kling, 1988a, 1988b);
- number of developed access sites to water, area of water body relative to total site area, and amount of dissolved oxygen present in water (Smith, Desvousges, and McGivney, 1983a);
- Lake Condition Index based on dissolved oxygen, turbidity, winter fish-kill, and algae growth (Bouwes and Schneider, 1979; Caulkins, Bishop, and Bouwes, 1986);
- annual snowfall at a ski area, vertical transport of lift, and acres of terrain of varying difficulty for skiing (Morey, 1981).

Incorporating Quality in the Recreation Demand Model

In order to estimate the value of a quality change at a site, we need to know how the demand curve for that site shifts with the change in the relevant quality characteristic. Consider first the case in which recreation demand is being modeled for only a single site. If the demand function of equation (6) were estimated from data collected during one season, there would be no variation in site quality attributes across the sample. Thus there would be no way of identifying the influence of objective measures of site quality on recreation demand. If the data were collected over a longer period of time, during which site quality changed, it might still be difficult to separate the effects of quality change from other factors that were also changing over time. However, if behavior were based on perceptions of quality, and if perceptions varied independently of travel cost, then the influence of perceived quality on behavior could be measured.

One way to obtain an estimate of the impact of quality changes at a single site on the demand for the site is to ask users contingent activity questions. For example, McConnell (1986) asked residents of New Bedford, Massachusetts, how many times they actually visited beaches around New Bedford Harbor and how many times they would expect to visit those same beaches if the pollution due to poly-chlorinated biphenyls (PCBs) were eliminated. Responses to these questions, along with estimates of travel costs for respondents, provided the data to estimate a demand function for visits to these beaches that incorporated the effect of PCB pollution on demand.

When a number of different sites are being studied using a multi-site recreation demand model, there may be substantial variation in qualities across sites. However, it may be difficult to disentangle the effects of differences in site qualities from the effects of differences in travel costs. All of the empirical attempts to incorporate site quality in multi-site demand models have involved some kind of simplification; thus they also have various limitations in their ability to characterize recreation demand accurately.

One approach has been to pool all of the observed visitation data to estimate a single visitation demand function. The number of times individual i visits site j is a function only of the price and quality of site j. Specifically,

$$r_{ji} = r(p_{r_{ji}}, q_j, M_i) \text{ for all } i, j.$$

The linear and semi-log specification would be

$$r_{ji} = a + b \cdot p_{r_{ji}} + c \cdot q_j$$
$$= (a + c \cdot q_j) + b \cdot p_{r_{ji}}$$

and $$\log r_{ji} = a + b \cdot p_{r_{ji}} + c \cdot q_j$$
$$= (a + c \cdot q_j) + b \cdot p_{r_{ji}}$$

As these equations show, the coefficients on own price and site quality are constrained to be the same across all sites. Furthermore, the effect of a quality change is limited to shifting the intercept of the site demand function. For an example of this specification, see Smith, Desvousges, and Fisher (1986).

An even more restrictive simplification is the "typical trip" (or "average travel cost") model. In this model, it is assumed that all of the trips made by individual i are made to the same site, that site being the one most frequently visited by the individual. Formally,

$$r_i^* = r(p_{r_i}^*, q_i^*, M_i)$$

where $$r_i^* = \sum_{j=1}^{m} r_{ji}$$

and $p_{r_i}^*$ and q_i^* are the price and quality of i's most frequently visited site.

This specification has restrictions similar to those of the pooled model. The price and quantity coefficients are constrained to be the same across sites; quality effects are limited to the intercept term of the demand functions; and there are no site substitution effects. For an example of this specification, see Caulkins, Bishop, and Bouwes (1986).

The procedures for calculating the value of a change in site qualities are essentially the same for both models. For a given vector of price and qualities, consumer surpluses are estimated as areas under the visit demand functions. Then these consumer surpluses are recalculated for the alternative vector of site qualities. If quality changes at only one site, there are no conceptual problems with this procedure. In the typical-trip model, only those individuals for whom the typical site is the improved site experience a welfare change. In the pooled model, all individuals who visit the improved site experience a welfare change. But when qualities change at several sites, the procedures for the pooled model are inconsistent with welfare theory. This is because theory requires that a path be specified for calculating the welfare effects of a sequence of changes in individual site qualities. For example, if qualities change at sites A, B, and C, one possible path is to calculate the value of the change at site A, conditioned on B and C being at their original qualities, then calculating the value of the change at B conditioned on A being at its new value and C being unchanged, and so forth. But since the pooled model does not include qualities at substitute sites, no such sequence can be specified in that model. Thus, if quality changes at one site do affect the number of visits at other sites, these models may yield incorrect estimates of welfare changes.

Kling (1988a, 1988b) has used a Monte Carlo simulation strategy to investigate the properties of welfare measures derived from the pooled and typical-trip models. She specified three alternative forms and parameters for utility functions, solved the constrained maximization problems to generate site visitation data, and estimated these two forms of recreation demand model with the simulated data. Using the two estimated models, she then estimated welfare values for specified changes in site qualities and compared these estimates with the known "true" welfare values calculated from the specified utility functions. The pooled model generally performed poorly. But when the pooled model was reestimated in a Tobit form, it gave good results for one of the specified utility models. The typical-trip model also gave good results for one utility function, but poor results for the other two utility functions.

A third approach to incorporating site qualities in a multi-site model has been termed the "generalized travel cost" or "varying parameters model." In this approach, the number of visits to each site is estimated separately as a function of its own price, without controlling for site quality and prices of substitute sites. Then the coefficients on price terms for each site are regressed on the site quality variables. Specifically,

$$r_j = \sum_{i=1}^{n} r_{ji} = a + b_j \cdot p_{r_{ji}} \qquad \text{for all } j$$

and $b_j = c + d \cdot q_j$ \hfill (11)

By substitution, this is equivalent to writing the site demand as a function of own price and a price-quality interaction term:

$$r_j = a + c \cdot p_{r_{ji}} + d \cdot q_j \cdot p_{r_{ji}}$$

Welfare measures for a set of quality changes can be calculated using the same procedures as those described above. These welfare measures are also subject to the same problems as those described above, at least if site substitution effects due to quality differences are important. For applications of this model, see Vaughan and Russell (1982a), Smith, Desvousges, and McGivney (1983a), Smith and Desvousges (1985), and Smith, Desvousges, and Fisher (1986).

The last model to be considered here has some similarities to the varying parameter model just described. In this model, values for each site are estimated from a multi-site demand model that does not take account of quality differences explicitly. But it does take account of substitute site prices. The first step is to estimate a system of site demand functions where visits to each site depend on all site prices. Then the value of each site is calculated as the area under its site demand curve at the given set of prices. Finally, the estimated site values are regressed on site quality in an equation analogous to (10) above. Specifically, if W_j is the welfare value of site j, then the equation explaining value is

$$W_j = c + d \cdot q_j \hfill (12)$$

for the linear specification. Equation (11) can be used to calculate new site values for different site qualities for any single site. However, because substitution effects among sites due to quality changes are not accounted for, this model is inappropriate for valuing quality changes at multiple sites. For an application of this model, see Samples and Bishop (1985).

Congestion as a Quality Attribute

Congestion of a recreation site occurs when the number of users is so large that it diminishes the utility and therefore the willingness to pay of those users. The degree of crowding at a site can be considered as one of the site attributes that influence the quality of the recreation services each individual experiences and the utility each individual obtains from visiting the site. The presence of congestion at a recreation site has implications for the estimation of recreation demand and the measurement of the value of environmental quality change.

Congestion, like pollution, is a negative externality. It affects all users of a site. Users may have different marginal willingnesses to pay to avoid congestion. And also like pollution, the congestion case satisfies the conditions of weak complementarity. Thus the benefits of avoiding congestion can be evaluated by measuring areas between demand curves. In this subsection, a simple model of congestion is developed to show the effects of congestion on demand for a site and on the prediction of recreation use.[11] The model is then used to show how the value of improving quality at a site is reduced if congestion is present, and how improving quality at one site can lead to additional benefits by reducing congestion at other sites.

Suppose we have recorded the level of recreation use at a particular site that we know to be congested; the case can be graphically illustrated as in figure 13-2. Let this level of use be V^1. Suppose also that the admission fee at the site is zero. Assume that by application of the travel cost model, we have estimated individuals' demand curves for the site and aggregated them to obtain D^1–V^1. This will be called a constant-congestion demand curve because it reflects the behavior of a set of users, all of whom are experiencing the same level of congestion when they visit this site. The triangular area, 0–D^1–V^1 is the consumer surplus associated with having this recreation site available, given its present level of use, congestion, and site quality.

Suppose now that it is known how much each individual would be willing to pay to use this site if it were uncongested. Aggregating the individuals' willingnesses to pay for an uncongested site gives an

---

[11]This model of congestion is developed more fully in Freeman and Haveman (1977). The main purpose of that paper was to develop rules for optimal pricing and show how these are affected by assumptions about the incidence of congestion costs. The earliest analyses of the problem of congested recreation facilities include Fisher and Krutilla (1972) and Cicchetti and Smith (1973). Other theoretical analyses can be found in McConnell and Duff (1976), McConnell (1980), and Smith (1981). Dorfman (1984) provides an elegant treatment of the optimal congestion and optimal pricing problems.

Price ($)

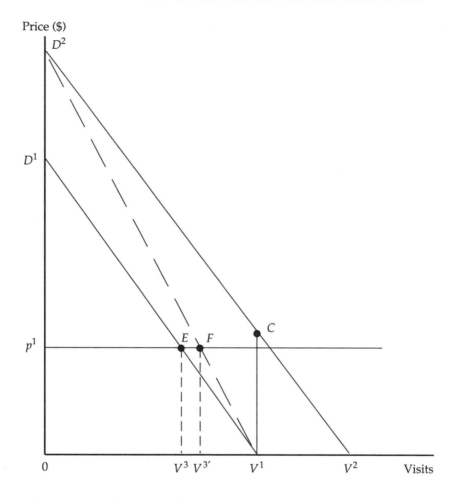

Figure 13-2.   The effect of congestion on the demand for a recreation site

uncongested demand curve, D^2–V^2. According to figure 13-2, congestion deters some users (V^2–V^1) from using the site; their loss is the area C–V^1–V^2. We can also identify the costs of congestion to existing users. At use level 0–V^1, those users incur congestion costs of D^1–D^2–C–V^1, the area between their uncongested demand curve and the observed constant-congestion demand curve out to V^1. This is their total willingness to pay to avoid the present level of congestion at the site.

Now suppose a gate fee of p^1 is imposed. Can the new level of use be predicted from the information at hand? The answer is no. The constant-congestion demand curve D^1–V^1 would predict a reduction in recreation use to $0-V^3$. But with lower use, congestion is reduced. With less congestion, willingness to pay is higher, and more people are willing to use the site at a given price. Thus we would expect use to be greater than that predicted by the constant-congestion demand curve. Suppose that actual use is $0-V^{3\prime}$. There is another constant-congestion demand curve through point F. This curve is conditioned on the level of congestion associated with the observed level of use, $0-V^3$. This curve is not shown in figure 13-2. In fact, there is a whole family of constant-congestion demand curves. Each one represents the aggregation of individual willingnesses to pay conditional on a level of total use and its associated congestion. Only one point on a particular constant-congestion demand curve can ever be observed; that is the point that corresponds to the level of use on which the curve is conditioned. However, any one of these constant-congestion demand curves could be estimated by the travel cost technique if that level of use were actually observed, since the travel cost technique predicts individuals' responses to price changes conditioned on a given observed level of total use.

The locus of these observable points traces out another kind of demand curve, a congestion-adjusted demand curve. This is represented as D^2-V^1 in figure 13-2. The congestion-adjusted demand curve can be used for predicting the effect of changes in entry fees on use levels. But the area under this curve has no significance for benefit evaluation. The curve could not be derived from a travel cost study, since for the latter all recreationists experience the same congestion level.

Now let us examine the benefits of improving some quality characteristic at a site when congestion exists; the case is illustrated in figure 13-3. The constant-congestion demand curve D^1-V^1 in figure 13-3 has been computed from travel cost data based on a level of use of $0-V^1$. Now let there be a quality improvement at the site. Suppose the constant-congestion demand curve were shifted out to the right to D^1-V^1. This would indicate the potential increase in use and potential benefits if increases in congestion could be avoided; it is the area $D^1-D^{1\prime}-V^{1\prime}-V^1$. But of course, the additional users impose congestion costs on one another, which partially offset the effects of improved quality on use and benefits.

Suppose now that the actual increase in use is to $0-V^{1\prime\prime}$. There is then a new constant-congestion demand curve through this point, $D^{1\prime\prime}-V^{1\prime\prime}$. The net benefit of improving site quality, taking into account the increased congestion, is the area $D^1-D^{1\prime\prime}-V^{1\prime\prime}-V^1$. The increase in congestion has dissipated some of the potential benefits from improving site quality.

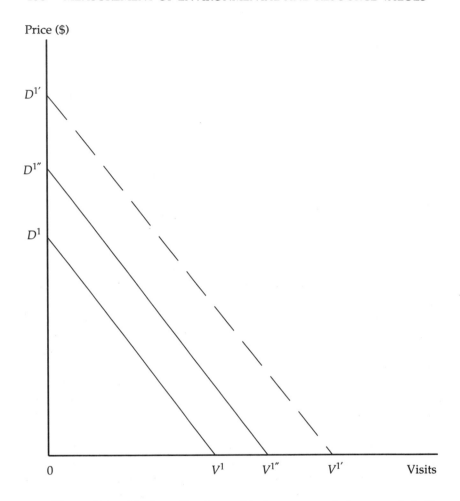

Figure 13-3. The benefits of a quality improvement at a congested site

In order to empirically take into account the separate effect of congestion, it is necessary to include terms for congestion as a site characteristic in the system of visitation equations for a set of sites in a given region. The estimation of the coefficients on the congestion term poses the same set of problems as those found in the estimation of the effects of site quality on visitation, discussed above.

Consider two sites, A and B, where A presently is uncongested, perhaps because of poor site quality, and B is congested. Suppose that there is an improvement in quality at site A. We expect visits to A to increase and visits to B to decrease. The reduction in visits would be reflected in a leftward shift of the constant-congestion demand curve for site B. If B were uncongested, the shift in the demand curve for B would be ignored in the calculation of the increase in value for site A. But given that B is congested, the shift in recreation use from B toward A reduces that congestion. Since this reduction in congestion is a consequence of the site quality improvement at A, the value of this reduction should be added to the value of the quality improvement at site A in computing the total welfare value of the change in quality at site A.

RANDOM UTILITY MODELS

As we have seen, it is difficult to adapt the standard travel cost model of recreation site demand to fully capture the values of site characteristics or the effects of variations in site quality on the demand for a site. The discrete choice or random utility model (RUM), with its emphasis on explaining the choice among sites as a function of the characteristics of the available sites, seems to be well suited to this task. But the gain in the ability to explain choices among sites comes at the expense of an inability to explain the total demand for a recreation activity, that is, the number of activity days in a season. In this section I briefly sketch the logic of the simple random utility model and its use in estimating the value of a site and the benefits of changes in characteristics. I then describe alternative approaches to modeling overall recreation activity within the RUM framework.[12]

Consider a set of n individuals ($i = 1, ..., i, ..., n$). For each recreation occasion, these individuals can choose one site from a set of m alternatives ($j = 1, ..., j, ..., m$). Each individual has an income M_i and vector of socioeconomic characteristics S_i. Each site has a vector of environmental quality attributes Q_j. The travel cost for individual i to site j is C_{ji}. For individual i choosing site j, utility is given by the conditional indirect utility function:

$$u_{ji} = v_i (M_i - C_{ji}, Q_j, S_i) + \epsilon_{ji}$$

where ϵ_{ji} is the unobservable component of utility and is assumed to

---

[12]For more detailed treatments of the theory and estimation of RUMs, readers should consult Bockstael, Hanemann, and Strand (1986) and Bockstael, McConnell, and Strand (1991). Interesting examples of applications include studies by Feenberg and Mills (1980), Caulkins, Bishop, and Bouwes (1986), Bockstael, Hanemann, and Kling (1987), and Morey, Shaw, and Rowe (1991).

be random. The prices of market goods are omitted for notational simplicity. The probability of individual i choosing to visit site j is the probability that u_{ji} is greater than u_{ki} for all $k \neq j$. If all of the ϵ_{ji} are independently and identically distributed with a Type I Extreme Value distribution with zero mean, the probability of individual i visiting site j^* is

$$Pr_{j^*}(i) = \frac{e^{v_i(M_i - C_{ji}, Q_j, S_i)}}{\sum_{j=1}^{m} e^{v_i(M_i - C_{ji}, Q_j, S_i)}}$$

Once a functional form for $v_i(\cdot)$ has been specified, its parameters can be estimated using maximum likelihood methods.

Care must be taken in specifying the choice set to avoid violation of the property of Independence of Irrelevant Alternatives (IIA). Such violations are likely to occur if two or more different types of recreation resource or recreation activity are included in the choice set. For example, suppose there are three alternatives to choose from: Lake A, Beach B, and Beach C. An improvement in the quality of Beach C is likely to reduce the probability of visiting Beach B relative to the probability of visiting Lake A, a violation of IIA. One way to avoid this problem is to specify a nested model of choice in which the individual is assumed, first, to choose a type of recreation resource or activity and then, second, to choose a site within that category. For an example, see Bockstael, Hanemann, and Strand (1986).

Once the parameters of $v_i(\cdot)$ have been estimated, the monetary value of a change in Q_j from Q_j^0 to Q_j^1 can be calculated. For individual i, this value is defined implicitly by

$$v_i(M_i - C_{ji} - CS_{ji}, Q_j^1, S_i) + \epsilon_{ji} = v_i(M_i - C_{ji}, Q_j^0, S_i) + \epsilon_{ji} \quad (13)$$

where CS_{ji} is the compensating surplus measure of the welfare change associated with the change in Q_j. Before CS_{ji} can be calculated from equation (12), two hurdles must be overcome. The first hurdle arises because for each individual there is no variation in income across alternatives in the choice set, so there is no independent estimate of the parameter on income that gives the marginal utility of income. This hurdle can be gotten over by recognizing that the relevant income measure is total income less the cost of the activity, C_{ji}. Thus the coefficient on income is the negative of the estimated coefficient on the cost of the activity.

The second hurdle arises because of the unobserved component of utility. Because of that, it is not possible to know whether any one individual will visit the site in question. If individual i does not visit the site, CS_{ji} is zero. As Bockstael, McConnell, and Strand (1991, pp. 261–262) show, the researcher's uncertainty can be dealt with by defining the welfare measure as the payment that equates the researcher's expected value of realized utility with and without the change in Q_j. Specifically, CS_{ji} is defined implicitly by

$$E\left[v^*\left(M_i - C_{ji} - CS_{ji}^*, Q_j^1, S_i\right)\right] = E\left[v^*\left(M_i - C_{ji}, Q_j^0, S_i\right)\right]$$

where $v^*(\cdot)$ is the maximum of

$$v_i\left(M_i - C_{ji}, Q_j, S_i\right) \qquad \text{all } j.$$

Given the assumptions about the distribution of ϵ_{ji} and assuming that the marginal utility of income is constant,

$$CS_{ji}^* = \frac{\ln\left[\sum_{j=1}^{m} e^{v(Q_j^1)}\right] - \ln\left[\sum_{j=1}^{m} e^{v(Q_j^0)}\right]}{b}$$

where b is the coefficient on income and the other arguments in $v_i(\cdot)$ are omitted for simplicity.

Similar calculations can be used to obtain the value of adding or deleting a site with a specified set of characteristics from the individual's choice set. For the addition of site $m + 1$, the expression is

$$CS_{m+1,i}^* = \frac{\ln\left[\sum_{j=1}^{m+1} e^{v(Q_j^1)}\right] - \ln\left[\sum_{j=1}^{m} e^{v(Q_j^0)}\right]}{b}$$

and for deleting site m, the expression is

$$CS_{n-1,i}^* = \frac{\ln\left[\sum_{j=1}^{m} e^{v(Q^j)}\right] - \ln\left[\sum_{j=1}^{m-1} e^{v(Q_j^1)}\right]}{b}$$

which is equal to or less than zero.[13]

Since these expressions give the welfare measure associated with one choice occasion, it is necessary to find some way of predicting

[13]Again, see Bockstael, McConnell, and Strand (1991), especially pp. 262–263.

each individual's total number of visits per year or per recreation season. One approach is to model the determination of the number of choice occasions or activity days as a separate problem;[14] this approach is discussed in a later section on the estimation of participation rates. A second approach is an extension of the discrete choice framework to include undertaking no recreation activity as an option for each choice occasion. There are at least several ways of doing this. One is to specify that one of the m alternatives is no activity and to estimate a single logit equation (see, for example, Feenberg and Mills, 1980). Another way is to specify a nested form of logit model in which the first equation predicts whether the individual will undertake any activity on a given choice occasion, while the second equation gives the probability that the individual will choose a particular site or activity (see, for example, Caulkins, Bishop, and Bouwes, 1986). The first equation would include as explanatory variables some measures of the average availability and/or quality of recreation sites accessible to the individual. The second equation would include measures of the costs and qualities of each of the available alternatives. This latter model specification is more consistent with the property of independence of irrelevant alternatives. The expected number of visits to a site is given by the product of the number of choice occasions and the joint probability of undertaking the activity and choosing site i, given that some activity is chosen.

OTHER APPROACHES TO ESTIMATING RECREATION VALUES

The Hedonic Travel Cost Model

Up to this point, the modeling strategies we have discussed have focused on the demand for visits to a site. Where site demand depends in part on the characteristics of the site, information on site demand can be used to value site characteristics. An alternative strategy for estimating the value of a site characteristic is to examine the implicit prices of characteristics themselves. When the issue is formulated in this manner, hedonic methods immediately come to mind. Brown and Mendelsohn (1984) developed a formal hedonic model of the relationship between the travel cost to a site as a form of price and the characteristics of the site. They implemented the model using

---

[14]See, for example, Bockstael, Hanemann, and Strand (1986).

data on visits to sites of different characteristics by steelhead fishermen.[15]

The key feature of the hedonic travel cost model is the assumption that the cost of visiting any site j from origin i is functionally related to the characteristics of site j. Specifically,

$$C_{ji} = C_i(Q_j) \qquad (14)$$

where C_{ji} is the per-trip travel cost from origin i to site j and Q_j is a vector of site characteristics q_k, $k = 1, ..., k, ..., p$. Equation (14) is estimated separately for each origin. So, in effect, each origin and the sites visited by people from that origin are treated as belonging to a single hedonic market. If an individual visits more than one site, each site is treated as a separate observation for purposes of estimating equation (14). It should be noted that estimating (14) from data on all visitors from the origin requires assuming that they all have the same opportunity cost of time used for travel.[16]

For the representative individual at origin i, utility depends upon the characteristics of the sites visited, the number of visits to each site, and the quantity of the numeraire good. The individual's choice problem can be represented by

$$\text{Max } u(r_j, Q_j, X)$$

subject to $\qquad M = X + \sum_{j=1}^{m} r_j \cdot C(Q_j)$

where r_j is the number of visits to site j and the subscript indexing origins has been omitted for simplicity. The first-order conditions for a maximum include:

$$\frac{\partial u/\partial q_k}{\lambda} = r_j \cdot (\partial C/\partial q_k) \qquad \text{for all } k \qquad (15)$$

---

[15]Other empirical applications of the model include studies by Mendelsohn (1983), Bockstael, Hanemann, and Kling (1987), Smith and Kaoru (1987), and Englin and Mendelsohn (1991).

[16]Smith and Kaoru (1987) are critical of the hedonic travel cost model for requiring this assumption, among other reasons. However, in principle, it should be possible to estimate separate hedonic price functions for subsets of the origin population that are homogeneous on the basis of the opportunity cost of time. This approach would be equivalent to recognizing that the hedonic market is segmented or stratified on the basis of the cost of time. In practice, this approach to the problem would require far more observations and detailed individual data than are likely to be available.

and
$$\frac{\partial u/\partial r_j}{\lambda} = C(Q_j) \qquad \text{for all } j \qquad (16)$$

Given equation (15), the marginal implicit price of q_k can be used in a second-stage regression to estimate marginal willingness-to-pay functions for characteristics. Implementation of the second stage must deal with the problems of identification discussed in chapter 11 in the context of the hedonic property value model.

The hedonic travel cost model as described here has come under significant criticism, especially by Bockstael, McConnell, and Strand (1991, pp. 233–234) and Smith and Kaoru (1987).[17] Much of this criticism has focused on the following problem. Unlike the hedonic property value model, in the hedonic travel cost model there is no marketlike process that brings the marginal implicit travel costs of each characteristic into a market-clearing relationship with the quantities of those characteristics. While travel cost is an implicit price affecting individuals' behavior, there is no feedback from this behavior to adjust the implicit price of the characteristic. In the property value model, an increase in the quantity of the characteristic at one property makes that property relatively more attractive. The market bids its price up until a new equilibrium is reached. In the hedonic travel cost model, there is no similar mechanism operating to clear the market should there be an increase in the level of some characteristic at a recreation site.

All of the hedonic travel cost studies cited above report negative implicit prices for at least some characteristics. Negative implicit prices for good characteristics are economically implausible; they are not consistent with the model of choice laid out in equations (14) through (16). It is the absence of any market adjustment process for implicit prices that explains the observations of negative prices. But the corollary is that in the absence of appropriate adjustment mechanisms, observations of marginal implicit prices do not necessarily reveal individuals' marginal values.

Estimating Participation Rates

So far, our discussion has been about site-specific analyses that have as their objective the measurement of economic demand and value.

---

[17]But see Smith, Palmquist, and Jakus (1991) for a more optimistic assessment of the usefulness of the hedonic travel cost model. They calculate marginal implicit prices by estimating a "best-practice" or least-cost frontier of the availability of characteristics. This approach constrains the marginal implicit prices of desirable characteristics to be non-negative.

Another approach to the analysis of recreation demand and value involves less stringent data requirements, assumptions, and estimation techniques. Such less stringent requirements can be considered advantages; but they entail loss of the ability to infer values from the empirical analysis. This approach is to estimate equations relating the participation in specified recreation activities by a given population to the socioeconomic characteristics of that population and to the supply and quality of recreation opportunities available to that population. If population-specific participation equations can be estimated, it would be possible to predict the increase in participation to be expected with an increase in the supply of recreation opportunities or with an improvement in site amenities.

Examples of this approach include Miller and Hay's analysis of the determinants of participation in duck hunting (Miller and Hay, 1981) and Russell and Vaughan's analysis of the effects of improved water quality on recreational fishing activity (Russell and Vaughan, 1982; Vaughan and Russell, 1982b). In these applications, levels of recreation activity are usually predicted with a two-equation model. The first equation predicts an individual's probability of undertaking the recreation activity in question. This equation is usually specified as a logit equation. The second equation predicts the number of days of activity conditional on being a participant in the activity. Explanatory variables would include one or more measures of the quantity and/or quality of the recreation resource available, along with socioeconomic variables. For example, Miller and Hay used acres of public recreation land and acres of waterfowl habitat in the individual's state of residence. Russell and Vaughan used surface area of water bodies capable of supporting different types of fishing activity in each individual's state of residence. Any of these supply and/or quality measures can be criticized because of their ad hoc nature. Theory gives no guidance on how to define the areas within which supply or quality are measured, or on how different measures might be aggregated into an index of availability.[18]

---

[18]One alternative to the two-stage model described here is the Tobit specification, in which an individual's number of activity days is given by

$$r_i = v_i(s, p_w, p_d, M_i) + \epsilon_i \quad \text{if} \quad r_i(s, p_w, p_d, M_i) > 0$$

and $\quad r_i = 0 \quad \text{if} \quad r_i(s, p_w, p_d, M_i) \leq 0$

where s is a measure of the availability of the resource (for example, acres of hunting habitat or fishable water within a given distance of the individual's residence) and p_w and p_d are the unit prices of time and distance. This specification allows for the large number of individuals with zero activity levels that is typical of recreation data for the general population.

Once the participation equations have been estimated, changes in participation can be calculated by plugging in alternative values for the relevant quantity and/or quality variables explaining the probability of participation and the number of days of the activity. Values for the predicted change in participation could be obtained by multiplying the change in participation by some unit value per recreation day obtained from some other source. Where unit values might be obtained, and some problems in their use, are discussed in the next section.

Valuing Increases in Activities

In valuing changes in the number of recreation days, the typical practice is to use a unit-day value derived independently and to multiply the predicted change in activity days by the (assumed) constant unit value; see, for example, Miller and Hay (1981) and Vaughan and Russell (1982a). This practice is authorized by the U.S. Water Resources Council guidelines for the economic evaluation of water resource development projects (U.S. Water Resources Council, 1983). That document includes suggested ranges of unit values to be applied to different qualities of resources and types of activities. These unit values can be derived from the results of either travel cost studies or contingent valuation studies of recreation values.[19] When travel cost studies are the source of unit values, the total value or consumer surplus of the site is divided by the number of activity days or visits to obtain an average consumer surplus or consumer surplus per visit or per day. This approach to valuing changes in activity levels can be criticized on the grounds that it is based upon average values when marginal values are relevant. Moreover, if the average quality of the recreation resources influences activity rates, applying a unit value to the change in activity levels does not capture the value of the change in quality to the existing users.

Vaughan and Russell (1982a) were the first to examine rigorously the implications of using average consumer surplus to value marginal changes in the availability or quality of a recreation resource. They used the household production function framework to examine the implications of changes in the nonmarket input—the recreation

---

[19]See Walsh, Johnson, and McKean (1990) for a survey and review of unit values derived from these two types of study. Smith and Kaoru (1990) describe the results of a meta-analysis of travel cost models. They use consumer surplus per trip as a dependent variable to characterize the results of the studies they look at.

resource—on the cost and value of the home-produced good—recreation activity days. If activity days, v, are produced by combining time on site, travel to a site, and access to the site, the cost function for home production is

$$c(r, s, p_w, p_d)$$

where, again, s is a measure of availability or supply of recreation sites or surface area within a given distance of the individual's residence. The individual's demand for activity days is

$$r = r(P, p_r, M) \tag{17}$$

where p_r is the implicit price of an activity day. In equilibrium, $p_r = \partial c/\partial v$. The solution to the individual's utility maximization problem yields a reduced form equation explaining the number of days chosen,

$$r = f(s, p_r, M)$$

where other prices are omitted for simplicity. This is the participation rate equation estimated in the second stage of the participation model described in the preceding section.

An increase in the availability of the resource shifts the marginal cost of home production downward, resulting in an increase in the number of activity days chosen. The welfare value of the increase in the resource is the area between the two marginal cost curves bounded by the compensated demand curve for activity days. The question analyzed by Vaughan and Russell is whether this welfare measure can be approximated by the product of the change in activity days and the average consumer surplus.

Let us assume that the marginal cost curve of home production is horizontal. This is a necessary assumption for Vaughan and Russell's analysis. Given this assumption, the true welfare measure is given by

$$W_r^* = -\int_{p_r^1}^{p_r^2} r(p_r, M)dp_r$$

Let the welfare measure obtained from the unit value approach be A^*, where

$$A^* = \int_{p_r^1}^{\infty} r(p_r, M)\, dp_r/(r^2 - r^1)$$

The question is whether A^* is a good approximation of W_r^*, or to put it differently, whether A^*/W_r^* is approximately equal to 1. As Vaughan and Russell (1982a) show, the answer depends on the functional form of the demand function for activity days, equation (17). In the case of a linear demand, they show that

$$\frac{A^*}{W^*_r} = \frac{1}{1 + r^2/r^1} < \frac{1}{2}$$

In other words, even for very small increases in activity levels, the unit value approximation is no more than half of the desired welfare measure. And for nonmarginal changes in activity levels, the approximation gets worse. Vaughan and Russell also show that in the case of a constant elasticity demand function for v, the ratio A^*/W_r^* is virtually independent of the magnitude of the change in activities, but varies inversely with the magnitude of the elasticity parameter. For elasticities below about 1.5, the unit value approach overestimates the true welfare change. And for elasticities in the range of 0.6–0.8, it overestimates the true welfare measure by a factor of 2 to 3.

These results do not support the use of average consumer surplus unit values for valuing changes in activity levels. As the results show, the magnitude of the error can be quite large. And even the sign of the error depends upon the functional form of the demand function over plausible values for the elasticity of the implicit demand function. Furthermore, Vaughan and Russell compared the true welfare measure with a unit value derived from the implicit demand for activities, although the usual practice is to use average consumer surpluses derived from studies of specific recreation sites. Finally, the Vaughan and Russell analysis was developed for the case of an increase in the number of sites that lowered the cost of visits. It does not apply to the valuation of an increase in activities resulting from an improvement in the quality of existing sites.

SUMMARY

Both of the behavioral models presented in this chapter, the travel cost model and the random utility model, draw inferences about individuals' willingness to pay for the recreation services of a site by treating the cost of travel to the site as an implicit price. But the two models are based on quite different views of how individuals make recreation choices. The standard travel cost model is a direct offshoot of conventional consumer demand theory. It views individuals as selecting quantities of visits to different sites as a function of the vector of prices and quantities. It models behavior over a time period long enough so that individuals can choose many visits, perhaps to several different sites. The strength of the standard model is that it captures individuals' choices of total visits over a recreation season. But the weakness of the standard model is its difficulty in capturing the role of site- specific characteristics or qualities in influencing the distribution of total visits across different sites. And one might question whether the implicit assumption that people choose a total quantity of visits and their distribution across sites at the beginning of the recreation season reflects a realistic picture of recreation behavior.

In contrast, the random utility model focuses specifically on each separate choice occasion when individuals are deciding which site to visit. The selection of a site is explained by the price and characteristics of that site relative to the alternatives in the choice set. So site characteristics play a central role in explaining recreation decisions. This is the strength of the random utility model. But the weakness of the model is its difficulty in explaining total visits over a season.

The travel cost model is better suited to answering the first of the three questions posed at the beginning of this chapter, that is, how is the value of an existing site to be measured. This is because the travel cost model is better at predicting the total number of trips to a site. The random utility model does a better job of answering the third question, how is the value of a change in site quality to be measured. This is because the random utility model does a better job of capturing the role of site characteristics in influencing the decisions about whether to participate in recreation and the choice of a site to visit. The random utility model also probably does a better job of answering the second question, how is the value of adding a site to a system to be estimated. This is because the random utility model provides an estimate of the indirect utility function itself. And this function can be used to calculate the welfare values of changes in the vectors of available characteristics and their costs or prices.

REFERENCES

Bockstael, Nancy E., W. Michael Hanemann, and Catherine L. Kling. 1987. "Estimating the Value of Water Quality Improvements in a Recreational Demand Framework," *Water Resources Research* vol. 23, no. 5, pp. 951–960.

Bockstael, Nancy E., W. Michael Hanemann, and Ivar E. Strand, Jr. 1986. *Measuring the Benefits of Water Quality Improvements Using Recreation De-*

mand Models. Report to the U.S. Environmental Protection Agency. College Park, Md.: University of Maryland.

Bockstael, Nancy E., Kenneth A. McConnell, and Ivar [E.] Strand. 1991. "Recreation," in John B. Braden and Charles D. Kolstead, eds., *Measuring the Demand for Environmental Quality.* Amsterdam, The Netherlands: North-Holland.

Bockstael, Nancy E., and Ivar E. Strand. 1987. "The Effect of Common Sources of Regression Error on Benefit Estimates," *Land Economics* vol. 63, no. 1, pp. 11–20.

Bockstael, Nancy E., Ivar E. Strand, and W. Michael Hanemann. 1987. "Time and the Recreational Demand Model," *American Journal of Agricultural Economics* vol. 69, no. 2, pp. 293–302.

Bouwes, Nicolaas W., Sr., and Robert Schneider. 1979. "Procedures in Estimating Benefits of Water Quality Change," *American Journal of Agricultural Economics* vol. 61, no. 3, pp. 535–539.

Brown, Gardner, Jr. and Robert Mendelsohn. 1984. "The Hedonic Travel Cost Method," *Review of Economics and Statistics* vol. 66, no. 3, pp. 427–433.

Burt, Oscar R., and Durward Brewer. 1971. "Estimation of Net Social Benefits from Outdoor Recreation," *Econometrica* vol. 39, no. 5, pp. 813–827.

Caulkins, Peter P., Richard C. Bishop, and Nicolaas W. Bouwes [Sr.]. 1985. "Omitted Cross-Price Variables in the Linear Travel Cost Model: Correcting Common Misperceptions," *Land Economics* vol. 61, no. 2, pp. 182–187.

Caulkins, Peter P., Richard C. Bishop, and Nicolaas W. Bouwes, Sr. 1986. "The Travel Cost Model for Lake Recreation: A Comparison of Two Methods for Incorporating Site Quality and Substitution Effects," *American Journal of Agricultural Economics* vol. 68, no. 2, pp. 291–297.

Cesario, Frank J. 1976. "Value of Time and Recreation Benefit Studies," *Land Economics* vol. 52, no. 1, pp. 32–41.

Cesario, Frank J., and Jack L. Knetsch. 1970. "Time Bias in Recreation Benefits Estimates," *Water Resources Research* vol. 6, no. 3, pp. 700–704.

Cicchetti, Charles J., Anthony C. Fisher, and V. Kerry Smith. 1976. "An Econometric Valuation of a Generalized Consumer Surplus Measure: The Mineral King Controversy," *Econometrica* vol. 44, no. 6, pp. 1259–1276.

Cicchetti, Charles J., and V. Kerry Smith. 1973. "Congestion, Quality Deterioration and Optimal Use: Wilderness Recreation in the Spanish Peaks Primitive Area," *Social Science Research* vol. 2, no. 1, pp. 15–30.

Clawson, Marion. 1959. *Methods of Measuring the Demand for and Value of Outdoor Recreation.* RFF Reprint No. 10. Washington, D.C.: Resources for the Future.

Dorfman, Robert. 1984. "On Optimal Congestion," *Journal of Environmental Economics and Management* vol. 11, no. 2, pp. 91–106.

Englin, Jeffrey, and Robert Mendelsohn. 1991. "A Hedonic Travel Cost Analysis for Valuation of Multiple Components of Site Quality: The Recreation Value of Forest Management," *Journal of Environmental Economics and Management* vol. 21, no. 3, pp. 275–290.

Feenberg, Daniel, and Edwin S. Mills. 1980. *Measuring the Benefits of Water Pollution Abatement*. New York: Academic Press.

Fisher, Anthony C., and John V. Krutilla. 1972. "Determination of Optimal Capacity of Resource-based Recreation Facilities," *Natural Resources Journal* vol. 12, no. 3, pp. 417–444.

Freeman, A. Myrick, III, and Robert H. Haveman. 1977. "Congestion, Quality Deterioration, and Heterogeneous Tastes," *Journal of Public Economics* vol. 8, no. 2, pp. 225–232.

Graham-Tomasi, Theodore, Wiktor L. Adamowicz, and Jerald J. Fletcher. 1990. "Errors of Truncation in Approximations to Expected Consumer Surplus," *Land Economics* vol. 66, no. 1, pp. 50–55.

Hausman, Jerry A. 1981. "Exact Consumer's Surplus and Dead Weight Loss," *American Economic Review* vol. 71, no. 4, pp. 662–676.

Hof, John G., and David A. King. 1982. "On the Necessity of Simultaneous Recreation Demand Equation Estimation," *Land Economics* vol. 58, no. 4, pp. 547–552.

Johnson, M. Bruce. 1966. "Travel Time and the Price of Leisure," *Western Economics Journal* vol. 4, no. 1, pp. 135–145.

Kealy, Mary Jo, and Richard C. Bishop. 1986. "Theoretical and Empirical Specifications Issues and Travel Cost Demand Studies," *American Journal of Agricultural Economics* vol. 68, no. 3, pp. 660–667.

Kling, Catherine L. 1988a. "Comparing Welfare Estimates of Environmental Quality Changes from Recreation Demand Models," *Journal of Environmental Economics and Management* vol. 15, no. 3, pp. 331–340.

Kling, Catherine L. 1988b. "The Reliability of Estimates of Environmental Benefits from Recreation Demand Models," *American Journal of Agricultural Economics* vol. 70, no. 4, pp. 892–901.

Kling, Catherine L. 1989. "A Note on the Welfare Effects of Omitting Substitute Prices and Qualities from Travel Cost Models," *Land Economics* vol. 65, no. 3, pp. 290–296.

Knetsch, Jack L. 1963. "Outdoor Recreation Demands and Benefits," *Land Economics* vol. 39, no. 4, pp. 387–396.

McConnell, Kenneth E. 1980. "Valuing Congested Recreation Sites," *Journal of Environmental Economics and Management* vol. 7, no. 4, pp. 289–294.

McConnell, Kenneth E. 1985. "The Economics of Outdoor Recreation," in Allen V. Kneese and James L. Sweeney, eds., *Handbook of Natural Resource and Energy Economics*, vol. 1. Amsterdam, The Netherlands: North-Holland.

McConnell, Kenneth E. 1986. *The Damages to Recreational Activities from PCBs in New Bedford Harbor*. Cambridge, Mass.: Industrial Economics.

McConnell, Kenneth E. 1992. "On-Site Time in the Demand for Recreation," *American Journal of Agricultural Economics* vol. 74, no. 4, pp. 918–925.

McConnell, Kenneth E., and Virginia A. Duff. 1976. "Estimating Net Benefits of Outdoor Recreation Under Conditions of Excess Demand," *Journal of Environmental Economics and Management* vol. 2, no. 1, pp. 24–30.

McConnell, Kenneth E., and Ivar [E.] Strand. 1981. "Measuring the Cost of Time in Recreation Demand Analysis: An Application to Sport Fishing," *American Journal of Agricultural Economics* vol. 63, no. 1, pp. 153–156.

McKean, John R., and Charles F. Revier. 1990. "An Extension of: 'Omitted Cross-Price Variable Biases in the Linear Travel Cost Model: Correcting Common Misperceptions,'" *Land Economics* vol. 66, no. 4, pp. 431–436.

Mendelsohn, Robert. 1983. "An Application of the Hedonic Travel Cost Framework for Recreation Modeling to the Valuation of Deer," in V. Kerry Smith and Anne D. Witte, eds., *Advances in Applied Microeconomics*. Greenwich, Conn: JAI Press.

Miller, Jon R., and Michael J. Hay. 1981. "Determinants of Hunter Participation: Duck Hunting in the Mississippi Flyway," *American Journal of Agricultural Economics* vol. 63, no. 4, pp. 677–684.

Morey, Edward R. 1981. "The Demand for Site-Specific Recreational Activities: A Characteristics Approach," *Journal of Environmental Economics and Management* vol. 8, no. 4, pp. 345–371.

Morey, Edward R. 1984. "The Choice of Ski Areas: Estimation of a Generalized CES Preference Ordering with Characteristics," *Review of Economics and Statistics* vol. 66, no. 4, pp. 584–590.

Morey, Edward R. 1985. "Characteristics, Consumer Surplus, and New Activities: A Proposed Ski Area," *Journal of Public Economics* vol. 26, no. 2, pp. 221–236.

Morey, Edward R., W. Douglass Shaw, and Robert D. Rowe. 1991. "A Discrete-Choice Model of Recreational Participation, Site Choice, and Activity Valuation When Complete Trip Data Are Not Available," *Journal of Environmental Economics and Management* vol. 20, no. 2, pp. 181–201.

Russell, Clifford S., and William J. Vaughan. 1982. "The National Recreational Fishing Benefits of Water Pollution Control," *Journal of Environmental Economics and Management* vol. 9, no. 4, pp. 328–354.

Samples, Karl C., and Richard C. Bishop. 1985. "Estimating the Value of Variations in Anglers' Success Rates; An Application of the Multiple-Site Travel Cost Method," *Marine Resource Economics* vol. 2, no. 1, pp. 55–74.

Scott, Anthony D. 1965. "The Valuation of Game Resources: Some Theoretical Aspects," *Canadian Fisheries Reports*, No. 4, Department of Fisheries, Ottawa, pp. 27–47.

Smith, V. Kerry. 1981. "Congestion, Travel Cost Recreational Demand Models, and Benefit Evaluation," *Journal of Environmental Economics and Management* vol. 8, no. 1, pp. 92–96.

Smith, V. Kerry. 1990. "Estimating Recreation Demand Using the Properties of the Implied Consumer Surplus," *Land Economics* vol. 66, no. 2, pp. 111–120.

Smith, V. Kerry, and William H. Desvousges. 1985. "The Generalized Travel Cost Model and Water Quality Benefits: A Reconsideration," *Southern Economics Journal* vol. 52, no. 2, pp. 371–381.

Smith, V. Kerry, William H. Desvousges, and Ann Fisher. 1986. "A Comparison of Direct and Indirect Methods for Estimating Environmental Benefits," *American Journal of Agricultural Economics* vol. 68, no. 2, pp. 280–290.

Smith, V. Kerry, William H. Desvousges, and Matthew P. McGivney. 1983a. "Estimating Water Quality Benefits: An Econometric Analysis," *Southern Economic Journal* vol. 50, no. 2, pp. 422–437.

Smith, V. Kerry, William H. Desvousges, and Matthew P. McGivney. 1983b. "The Opportunity Cost of Travel Time in Recreation Demand Models," *Land Economics* vol. 59, no. 3, pp. 259–278.

Smith, V. Kerry, and Yoshiaki Kaoru. 1987. "The Hedonic Travel Cost Model: A View from the Trenches," *Land Economics* vol. 63, no. 2, pp. 179–192.

Smith, V. Kerry, and Yoshiaki Kaoru. 1990. "Signals or Noise? Explaining the Variation in Recreation Benefit Estimates," *American Journal of Agricultural Economics* vol. 72, no. 2, pp. 419–433.

Smith, V. Kerry, Raymond B. Palmquist, and Paul Jakus. 1991. "Combining Farrell Frontier and Hedonic Travel Cost Models for Valuing Estuarine Quality," *Review of Economics and Statistics* vol. 73, no. 4, pp. 694–699.

Stevens, Joe B. 1966. "Recreation Benefits from Water Pollution Control," *Water Resources Research* vol. 2, no. 2, pp. 167–182.

U.S. Water Resources Council. 1983. *Economic and Environmental Principals and Guidelines for Water and Related Agencies*. Washington, D.C.: U.S. Government Printing Office.

Vartia, Yrjö O. 1983. "Efficient Methods of Measuring Welfare Change and Compensated Income in Terms of Ordinary Demand Functions," *Econometrica* vol. 51, no. 1, pp. 79–98.

Vaughan, William J., and Clifford S. Russell. 1982a. *Freshwater Recreational Fishing: The National Benefits of Water Pollution Control*. Washington, D.C.: Resources for the Future.

Vaughan, William J., and Clifford S. Russell. 1982b. "Valuing a Fishing Day: An Application of a Systematic Varying Parameter Model," *Land Economics* vol. 58, no. 4, pp. 451–463.

Walsh, Richard G., Donn M. Johnson, and John R. McKean. 1990. "Nonmarket Values from Two Decades of Research on Recreation Demand," in Arthur N. Link and V. Kerry Smith, eds., *Advances in Applied Micro-Economics*, vol. 5. Greenwich, Conn.: JAI Press.

Willig, Robert D. 1976. "Consumers' Surplus Without Apology," *American Economic Review* vol. 66, no. 4, pp. 589–597.

Wilman, Elizabeth A. 1987. "A Simple Repackaging Model of Recreational Choices," *American Journal of Agricultural Economics* vol. 69, no. 3, pp. 604–612.

14

Conclusions

I n the concluding chapter of *The Benefits of Environmental Improve-ment: Theory and Practice* (Freeman, 1979), I wrote:

> Suppose the administrator of EPA wished to know the mag-nitude of the benefits accruing from a given pollution control policy. If asked, I believe an economist could specify the economic theory and models he would use, the data he would like to have, and the empirical techniques he would apply to the data to obtain measures of benefits. (p. 248)

I added five qualifications to this optimistic assessment of the state of the art. Briefly, they were:

1. Where indirect methods were not available (for example, for measuring nonuse values), the economist would have to resort to hypothetical direct measures. And I was concerned about "the accu-racy of responses of individuals in the necessarily hypothetical situa-tions they pose" (Freeman, 1979, p. 249).

2. In the valuation of reductions in mortality risks, there did not appear to be broad acceptance of the idea that individuals' behavior could or should be the basis of welfare measures.

3. Since all welfare measures are conditioned upon the existing distribution of income, to accept these welfare measures for making public policy choices involved accepting the existing distribution as satisfactory.

4. Economic measures of welfare change "must be built upon noneconomic data such as dose-response functions" (Freeman, 1979, p. 250), which often were not available.

5. The data required to implement many of the models and methods would often be difficult and costly to obtain.

In looking back over this list of qualifications today, I see less reason to be concerned about the first two of them. As regards the first, I think that substantial progress has been made in the development of hypothetical methods. I am now much more optimistic about the ability of properly framed questions to generate meaningful data on preferences and values. This does not mean that all the investigator has to do is go out and ask people some questions. But well-designed hypothetical studies do have a place in economists' tool boxes. As to the second, although there is still some debate about it, I think that there is now fairly wide acceptance of the use of revealed preference measures of the value of risk reduction for evaluating public policies that affect health and safety. Such measures are sanctioned, for example, by the U.S. Environmental Protection Agency for use in performing regulatory impact analyses (U.S. Environmental Protection Agency, 1983). And we have a substantial body of evidence on the magnitude of risk premia revealed in various "markets" (see Fisher, Violette, and Chestnut, 1989).

But having said this, I want also to add two new qualifications to the list. First, the economic framework, with its focus on the welfare of humans, is inadequate to the task of valuing such things as biodiversity, the reduction of ecological risks, and the protection of basic ecosystem functions. When policies to protect biodiversity or ecosystems are proposed, economists may be able to say something sensible about the costs of the policies; but except where nonuse values are involved and where people use ecosystems (for example, for commercial harvesting of fish or for recreation), economists will not be able to contribute comparable welfare measures on the benefit side of the equation.

The second new qualification has to do with the process of fitting data to a model to generate a welfare measure, and the effects of the choice of a model and functional form on the welfare measure. There has been a virtual explosion in the number of models available to be exploited. And a number of rich data sets have been created. Several studies show that welfare measures can be sensitive to the choice of a model and/or the choice of a functional form for a specific model. Some of these studies involve simulations in which the "true" welfare measure is known and can be compared with estimates derived by fitting the simulated data to alternative models; for examples, see Cropper, Deck, and McConnell (1988) and Cropper, Deck, Kishor, and McConnell (1993). Other studies involve examining the sensitivity of welfare estimates to changes in some aspect of the specification of the model being applied to real data; for examples, see Kaoru and Smith (1990) and Cooper and Loomis (1992). Since the true model cannot be known, we must add model uncertainty to the list of sources of uncertainty in welfare measures. An important area for future research is to obtain a better understanding of the sources and properties of model uncertainty.

THE STATE OF THE ART

Methods and Models

In this subsection, following the classifications of environmental service flows presented in chapter 1, I briefly outline the types of models that can be applied in each case. This information is summarized in table 14-1.

The first thing to note is that contingent valuation models apply to the estimation of all types of service flows except those that work through changes in market prices (producers' and consumers' surpluses). It should also be noted that contingent valuation is the only method available for estimating nonuse values and the values of ecological stability and biodiversity. By the latter, I refer to ecological functions that do not affect human welfare directly and for which people do not have a willingness to pay.

For valuing mortality risks, some version of the hedonic wage model is probably the method of choice, although averting behavior studies and contingent valuation studies have been conducted in this area with some success. Some form of averting behavior model would be most appropriate for valuing the nonfatal health effects as well. Even the health production function model can be considered a special case of the averting behavior models in that it is based on substitutability among market and nonmarket inputs into the production of health.

Hedonic property value models capture the value of differences in amenity and disamenity levels across residential locations. But they might also be measuring other environmental effects caused by the same environmental agent. For example, air pollution may reduce visibility and cause adverse health effects and damages to household materials, all of which would be reflected in property prices. Hedonic wage models can be used to value differences in the average levels of amenities or disamenities among cities or regions.

Human exploitation of natural and managed ecosystems in agriculture, commercial fisheries, and forestry is largely managed through market mechanisms. Thus the values of changes in the ecological services will be reflected in changes in prices and incomes. The cost function and production function models described in chapter 9 provide the basis for estimating the resulting changes in producers' and consumers' surpluses. However, these models are also applicable where changes in other environmental services affect the productivity and costs of firms. Examples include the effects of air pollution on manufacturing costs and the costs of repair and maintenance of structures, and the effects of water quality on processing costs.

Table 14-1. Models for Estimating the Values of Environmental Service Flows

Type of environmental service flow	Type of model	Described in:
Human health		
Mortality risk	Hedonic wages	Chapters 10, 12
	Averting behavior	Chapter 10
	Contingent valuation	Chapter 6
	Contingent ranking	′ ′
	Contingent activity	′ ′
Chronic morbidity	Averting behavior	Chapter 10
	Contingent valuation	Chapter 6
	Contingent ranking	′ ′
	Contingent activity	′ ′
Acute morbidity	Averting behavior	Chapter 10
	Health production function	′ ′
	Contingent valuation	Chapter 6
	Contingent ranking	′ ′
	Contingent activity	′ ′
Other direct impacts on humans: amenities, visibility, noise	Hedonic property values	Chapter 11
	Hedonic wages, interurban	Chapter 12
	Averting behavior	Chapter 4
	Contingent valuation	Chapter 6
	Contingent ranking	′ ′
	Contingent activity	′ ′
Economic productivity of ecological systems	Changes in producers' and consumers' surpluses	Chapter 9
Other ecological services		
Recreation	Travel cost model with weak complementarity	Chapter 13
	Random utility model	Chapter 13
	Contingent valuation	Chapter 6
	Contingent ranking	′ ′
	Contingent activity	′ ′
Ecological stability, biodiversity	Contingent valuation	Chapter 6
	Contingent ranking	′ ′
	Contingent activity	′ ′
Effects on nonliving systems, such as materials	Changes in producers' and consumers' surpluses	Chapter 9
	Hedonic property values	Chapter 11
	Averting behavior	Chapter 4
	Contingent valuation	Chapter 6
	Contingent ranking	′ ′
	Contingent activity	′ ′
Nonuse values	Contingent valuation	Chapter 6
	Contingent ranking	′ ′
	Contingent activity	′ ′

Finally, humans use both natural and managed ecosystems for various forms of recreation. The standard travel cost and random utility models are available to exploit information on the implicit price of access and on the cost of traveling to a site in order to estimate the values of sites and changes in their characteristics.

The Costs of Environmental Improvements

It is a commonly held view both within and outside the economics profession that the costs of environmental regulations are relatively easy to measure, at least in comparison with the task of measuring environmental benefits. This optimistic view is consistent with what I would call a naive theory of cost, which takes the following form. Firms respond to pollution control regulations by purchasing and installing waste treatment equipment and control systems that are, in effect, bolted on to the existing factory. The purchase and installation costs of this equipment, plus the added operating and maintenance costs it would entail, are readily identified in the firm's accounts. These expenditures can be taken, at least as a first approximation, to be the social costs of complying with the regulation.

The naive theory fails to recognize the fundamental symmetry between benefits and costs as changes in the utilities of individuals. And it also neglects several important realities concerning the ways in which government regulations can affect people's welfares. The symmetry of benefits and costs stems from the fact that ultimately all costs take the form of utility losses to individuals in their dual roles as receivers of income and consumers of market and nonmarket goods and services. These losses have their monetary counterparts in compensating variations and compensating surpluses. Because of this fundamental symmetry, proper measurement of costs involves the same kinds of problems as, and is likely to be as difficult as, the measurement of the benefits of environmental improvement. Once the symmetry is acknowledged, the whole economist's tool kit of direct and indirect and observed and hypothetical methods of measuring welfare changes becomes available for the cost analyst.

The naive view about cost estimation is also implicitly based on the presumption that only firms cause pollution, and that therefore only firms must incur costs in the process of moving to meet environmental quality objectives. But this is not the case. Regulations, for example, may be placed on household activities, and these regulations might not require any identifiable expenditure of money. Suppose that commuters are required to form car pools as part of a program to meet air quality standards for ozone. Household expenditures on commuting may, in fact, be lower as a consequence of the regulation. The cost of such regulation takes the form of increased commuting time and loss of convenience. These may be difficult to quantify and value. Consider also a regulation that decreases the quality or stream of services from a consumer good. In the 1970s, for example, new cars complying with tail-pipe emissions standards had lower performance and fuel efficiency than earlier models. The true cost of complying with these standards was the sum of the cost of the pollution control equipment and the decreased value of the services from the car.

When an environmental regulation affects household activities or the availability of nonmarket goods and services, there may be opportunities for employing indirect methods to draw inferences about the negative values or costs of these changes. For example, if the quality of a good is reduced as a consequence of a regulation, hedonic price models might be used to estimate the marginal implicit price of the relevant attribute. In the example of imposed car-pooling, an estimate of the shadow price of time could be used to derive the costs of the increased commuting time. Also, hypothetical methods such as referenda or contingent ranking might be used where a comprehensive set of regulations affects a wide range of activities.

Another factor neglected in the naive theory of costs is that market mechanisms are likely both to shift the burden of firms' expenditures and to change the magnitude of the burden so that the true costs are not accurately measured by summing the expenditures by firms. At least for nonmarginal changes, firms will raise prices and will experience decreases in quantity demanded. Price increases cause losses of consumer surplus, which are part of the social costs. And there may be losses in producers' surpluses and factor incomes as well. Thus it is not correct simply to equate pollution control expenditures and social costs (Portney, 1981).

As an extreme example, consider the case of a tax on the carbon content of fuels as a means of reducing the emissions of carbon dioxide (CO_2). Since there do not appear to be any economically feasible technologies for controlling CO_2 emissions from combustion, the tax would work entirely through raising the prices of carbon-based fuels and reducing demands for them. Consumers would bear at least part of the costs of the revenues raised by the tax; the rest would come from decreases in resource rents. But since these tax revenues represent a transfer to the government, they are not part of the social cost of reducing CO_2 emissions. The social cost of controlling CO_2 by taxation comes entirely in the form of the equivalent of the deadweight losses or welfare triangles associated with the tax.

Where the direct impact of a regulation is on firms, and where its effects are transmitted to individuals through changes in prices and incomes, the process of cost estimation must call on two types of models. The first is a model of the firm's production technology and costs. The second is a market model that can be used to calculate the changes in prices and incomes of the affected individuals. This model could in some circumstances be a partial equilibrium model. For broad social regulations, a general equilibrium model of the economy may be required (Hazilla and Kopp, 1990).

In modeling the behavior of the firm, a number of questions have to be considered. For example, can the technology be modeled as additively separable so that total cost is the sum of the costs of producing the marketed outputs and the cost of treating the waste? Or must the technology and costs be modeled as fully joint? The assumption of additive separability may be appropriate if, in fact, firms would respond to a regulation by purchasing add-on control or treatment equipment. But other kinds of technological responses, such as input substitution and recycling, may have to be modeled in a joint production framework. Another question is, can the model be built up from engineering data, as in the case of linear programming simulation models of plants (for an example, see Russell and Vaughan, 1976)? Or, alternatively, can an econometric estimate of the cost or production function be obtained from available data?

Once a model of firms' costs has been obtained, the next step is to embed that model in a model of the market economy so that the changes in all relevant quantities, prices, and incomes can be predicted. In the simplest case, where the regulation involves a marginal change in costs and where the economy is perfectly competitive, costs can be measured by the predicted change in expenditures on factor inputs. This is because the invariant factor prices are equal to the values of the marginal products of those inputs in other uses; and, in turn, those values of marginal products measure the opportunity losses to consumers associated with the marginal reallocation of inputs.

With nonmarginal changes, it can be expected that there will also be changes in product prices and perhaps factor prices. If a regulation affects only one industry, a partial equilibrium model of the form described in chapter 9 may be appropriate. Rather than have costs be a function of q, among other things, a parameter reflecting the stringency of the regulation or the degree of pollution control required could be included as a shifter of the cost or production function.[1]

---

[1]In some cases regulations might better be modeled as imposing inequality constraints on firms' inputs or outputs.

Implicit in this approach are the assumptions that the regulation does not cause any shifts in either the output demand functions or the factor supply functions, either immediately or over time. But if either of these assumptions is not valid, then a general equilibrium framework would be required. For example, if the regulation affected both the X and Y industries, the regulation-induced change in the price of Y could shift the demand function for X. Also, if Y were an intermediate product and an input in the X industry, the increase in its price or the upward shift in its supply function would have a secondary impact on the cost of producing X. Hazilla and Kopp (1990) have shown that estimating true social costs in a general equilibrium framework can lead to quite different results in comparison with the pollution control expenditure approach utilized by the Environmental Protection Agency (EPA).

Uncertainty

The state of the art in measurement is not sufficiently far advanced to produce exact measures of value for many kinds of environmental and resource changes. This leads to the question, must policymakers wait for further research to produce exact measures before they can use value and benefit information to guide decisionmaking? If not, how should they interpret the ranges of values that current research has produced?

To counsel waiting for exact measures is equivalent to saying that in many cases value measures should never be used. The state of the art cannot be expected to advance to the point of producing exact values for all kinds of environmental change. This is because of the inherent uncertainty and imprecision in measurement techniques based on statistical inference, and because of the fact that the true values held by individuals will vary with their circumstances (age, income, and so forth) and with the description of the specific changes being valued. So how are policymakers to proceed in the face of continued and inherent uncertainty about values?

A simple approach with a basic intuitive appeal is to perform sets of calculations with the upper and lower bounds of the range, and perhaps with the midpoint of the range as well. This is the essence of the approach for valuing reductions in mortality risks outlined in the EPA guidelines for performing regulatory impact analyses:

> If mortality is to be valued directly, a range of values can be used to determine the sensitivity of the results to alternative

values. Recent studies that measure how much people need to be compensated to incur small risks provide a means for selecting such values. ... They have found that annual wages are between about $4 and $70 higher for jobs with 1 in 100,000 greater risk. This translates into a value for a statistical life of roughly $400,000 to $7,000,000 (in 1982 dollars). (U.S. Environmental Protection Agency, 1983, p. 11)

Clearly, if benefits of a policy calculated with the upper end of the range are less than the lower end of its range of estimated costs, the policy is unlikely to be justifiable on economic grounds; and if the benefits calculated with the lower end of the range exceed the upper end of the range of costs, the economic case for the policy is quite strong.

This simple-minded approach is a step in the right direction, but it can be criticized because it does not make use of all of the relevant information contained in the estimates making up the range of values. Formally, the range reflects only the information contained in the two estimates yielding the highest and lowest values. It ignores information on the quality of these two estimates; and it ignores all of the information contained in the other estimates that yielded values within the range. There is a way to make use of the results of all of the available estimates and to incorporate judgments about the quality of each of these estimates. This formal approach is based on viewing probabilities as statements about the degree of confidence held about the occurrence of some possible event. The approach involves assigning probabilities to all of the values produced by the available estimates, where a higher probability reflects a greater degree of confidence in that estimate.[2] For example, the assignment of a probability of one to a particular estimate means we can be certain that this study has produced the correct value.

Once the probabilities have been assigned, various statistical manipulations can be performed. For example, the expected value of the parameter in question (the mean of its probability distribution) can be calculated and used for benefit-cost calculations. The variance of the distribution can be used to determine confidence intervals on the value to be used, thus preserving for policymakers information on the uncertainty about values.

---

[2]For a discussion of this approach, see von Winterfeldt and Edwards (1986), particularly chapter 4.

REFERENCES

Cooper, Joseph, and John Loomis. 1992. "Sensitivity of Willingness-to-Pay Estimates to Bid Design in Dichotomous Choice Contingent Valuation Models," *Land Economics* vol. 68, no. 2, pp. 211–224.

Cropper, Maureen L., Leland B. Deck, and Kenneth E. McConnell. 1988. "On the Choice of Functional Form for Hedonic Price Functions," *Review of Economics and Statistics* vol. 70, no. 4, pp. 668–675.

Cropper, Maureen L., Leland B. Deck, Nalin Kishor, and Kenneth E. McConnell. 1993. "Valuing Product Attributes Using Single Market Data: A Comparison of Hedonic and Discrete Choice Approaches," *Review of Economics and Statistics* vol. 75, no. 2, pp. 225–232.

Fisher, Ann, Dan Violette, and Lauraine Chestnut. 1989. "The Value of Reducing Risks to Death: A Note on New Evidence," *Journal of Policy Analysis and Management* vol. 8, no. 1, pp. 88–100.

Freeman, A. Myrick, III. 1979. *The Benefits of Environmental Improvement: Theory and Practice*. Baltimore, Md.: The Johns Hopkins University Press for Resources for the Future.

Hazilla, Michael, and Raymond J. Kopp. 1990. "The Social Cost of Environmental Quality Regulations: A General Equilibrium Analysis," *Journal of Political Economy* vol. 98, no. 4, pp. 853–873.

Kaoru, Yoshiaki, and V. Kerry Smith. 1990. *"Black Mayonnaise" and Marine Recreation: Methodological Issues in Valuing an Cleanup*, RFF Discussion Paper QE91-02. Washington, D.C.: Resources for the Future.

Portney, Paul R. 1981. "The Macroeconomic Impacts of Federal Environmental Regulation," in Henry M. Peskin, Paul R. Portney, and Allen V. Kneese, eds., *Environmental Regulation and the U. S. Economy*. Washington, D.C.: Resources for the Future.

Russell, Clifford S., and William J. Vaughan. 1976. *Steel Production: Processes, Products, and Residuals*. Baltimore, Md.: The Johns Hopkins University Press for Resources for the Future.

Smith, V. Kerry. 1990. "Estimating Recreation Demand Using the Properties of the Implied Consumer Surplus," *Land Economics* vol. 66, no. 2, pp. 111–120.

U.S. Environmental Protection Agency. 1983. *Guidelines for Performing Regulatory Impact Analyses*. Washington, D.C.: U.S. Environmental Protection Agency.

von Winterfeldt, Detlof, and Ward Edwards. 1986. *Decision Analysis and Behavioral Research*. Cambridge: Cambridge University Press.

Name Index

Adamowicz, Viktor L., 184, 457n8
Adams, Richard M., 27n6, 282, 290, 291, 297n8
Ainslie, George, 210
Anderson, James E., 257n24, 275n3
Anderson, Robert J., 262, 369, 375
Andreoni, James, 72n28
Arnould, Richard J., 426
Arriga-Salinas, A. S., 431n4
Arrow, Kenneth, 167, 220n2, 251, 264, 267
Arthur, W. B., 335, 341
Atkinson, Scott E., 331, 379
Ayres, Robert U., 5

Bailey, Martin J., 11n8, 202, 319n3
Ball, M. J., 387n8
Barro, Robert J., 209
Bartik, Timothy, 117, 370n3, 388n9, 389, 390n11, 391n12, 394, 395n13, 399, 400n15, 414, 431n4, 432
Baumol, William J., 20n1, 282n6
Becker, Gary S., 223
Beggs, S., 175
Bell, Frederick W., 305, 309n17
Belsley, David A., 187
Ben-Akiva, Moshe, 134n21
Berger, Mark C., 249n18, 431n4, 435, 437, 438
Bergland, Olvar, 68n23, 81
Bergson, Abram, 40
Bergstrom, Theodore C., 338
Berkowitz, Monroe, 316n2
Biddle, Jeff E., 428
Bishop, Richard C., 24n4, 146, 185, 190n20, 224n4, 234n13, 235, 262, 263, 453n5, 454, 455, 461, 461, 463, 464, 469n12, 472
Blackorby, Charles, 204, 206n6
Blomquist, Glenn, 330, 339, 431n4, 435, 437, 438

Boadway, Robin, 39n1, 42n3, 42–43n5, 85n33
Bockstael, Nancy E., 66n21, 108, 109n8, 110, 133, 149, 450, 457, 469n12, 471n13, 472n14, 473n15, 474
Bohm, Peter, 185n17, 262
Bouwes, Nicolaas W., Sr., 454, 455, 461, 463, 469n12, 472
Bowes, Michael D., 5n3, 302n9, 304n12
Boyd, J. Hayden, 4n2
Boyle, Kevin J., 146, 185
Bradford, David F., 105n5, 214n11
Brewer, Durward, 455, 458n9
Brookshire, David S., 167, 168, 170n4, 172n8, 177n11, 179, 183, 185, 250n20
Broome, John, 322
Brown, Charles, 426
Brown, Gardner, Jr., 472
Brown, James N., 131, 388
Brown, Thomas C., 6n5
Bruce, Neil, 39n1, 42n3, 42–43n5, 85n33
Buchanan, James M., 322n6
Burt, Oscar R., 455, 458n9

Cameron, Trudy A., 136n26, 188, 191n21, 192
Cardell, S., 175
Carson, Richard T., 12, 23, 167, 168, 169, 170, 171, 172n7, 174, 175n10, 176, 178, 179, 182, 183, 185, 187, 190n20, 191n21
Cassel, Eric, 380
Caulkins, Peter P., 454, 455, 461, 463, 469n12, 472
Cesario, Frank J., 448n3, 451
Chameides, W. L., 5
Charles River Associates, 25

495

Subject Index